Canadian Mathematical Society
Société mathématique du Canada

Jonathan M. Borwein   Qiji J. Zhu

Techniques of
Variational Analysis

With 12 Figures

 Springer

Jonathan M. Borwein
Faculty of Computer Science
Dalhousie University
Halifax, NS B3H 1W5
Canada

Qiji J. Zhu
Department of Mathematics and Statistics
Western Michigan University
Kalamazoo, MI 49008
USA

Editors-in-Chief
Rédacteurs-en-chef
Jonathan Borwein
Karl Dilcher
Department of Mathematics and Statistics
Dalhousie University
Halifax, Nova Scotia B3H 3J5
Canada
cbs-editors@cms.math.ca

Mathematics Subject Classification (2000): 49-02

Library of Congress Cataloging-in-Publication Data
On file.

ISBN 978-1-4419-2026-3   e-ISBN 978-0-387-28271-8

springeronline.com

J. M. Borwein
and
Q. J. Zhu

Techniques of Variational Analysis

An Introduction

March 10, 2005

Springer

Berlin Heidelberg New York
Hong Kong London
Milan Paris Tokyo

To Tova, Naomi, Rachel and Judith.

To Charles and Lilly.

And in fond and respectful memory of
Simon Fitzpatrick (1953–2004).

Preface

Variational arguments are classical techniques whose use can be traced back to the early development of the calculus of variations and further. Rooted in the physical principle of least action they have wide applications in diverse fields. The discovery of modern variational principles and nonsmooth analysis further expand the range of applications of these techniques. The motivation to write this book came from a desire to share our pleasure in applying such variational techniques and promoting these powerful tools. Potential readers of this book will be researchers and graduate students who might benefit from using variational methods.

The only broad prerequisite we anticipate is a working knowledge of undergraduate analysis and of the basic principles of functional analysis (e.g., those encountered in a typical introductory functional analysis course). We hope to attract researchers from diverse areas – who may fruitfully use variational techniques – by providing them with a relatively systematical account of the principles of variational analysis. We also hope to give further insight to graduate students whose research already concentrates on variational analysis. Keeping these two different reader groups in mind we arrange the material into relatively independent blocks. We discuss various forms of variational principles early in Chapter 2. We then discuss applications of variational techniques in different areas in Chapters 3–7. These applications can be read relatively independently. We also try to put general principles and their applications together.

The recent monograph "Variational Analysis" by Rockafellar and Wets [237] has already provided an authoritative and systematical account of variational analysis in finite dimensional spaces. We hope to supplement this with a concise account of the essential tools of infinite-dimensional first-order variational analysis; these tools are presently scattered in the literature. We also aim to illustrate applications in many different parts of analysis, optimization and approximation, dynamical systems, mathematical economics and elsewhere. Much of the material we present grows out of talks and short lecture series we have given in the past several years. Thus, chapters in this book can easily be arranged to form material for a graduate level topics course. A fair collection of suitable exercises is provided for this purpose. For many reasons, we avoid pursuing maximum generality in the main corpus. We do, however, aim at selecting proofs of results that best represent the general technique.

In addition, in order to make this book a useful reference for researchers who use variational techniques, or think they might, we have included many more extended guided exercises (with corresponding references) that either give useful generalizations of the main text or illustrate significant relationships with other results. Harder problems are marked by a ∗. The forthcoming book "Variational Analysis and Generalized Differentiation" by Boris Mordukhovich [204], to our great pleasure, is a comprehensive complement to the present work.

We are indebted to many of our colleagues and students who read various versions of our manuscript and provided us with valuable suggestions. Particularly, we thank Heinz Bauschke, Kirsty Eisenhart, Ovidiu Furdui, Warren Hare, Marc Lassonde, Yuri Ledyaev, Boris Mordukhovich, Jean Paul Penot, Jay Treiman, Xianfu Wang, Jack Warga, and Herre Wiersma. We also thank Jiongmin Yong for organizing a short lecture series in 2002 at Fudan university which provided an excellent environment for the second author to test preliminary materials for this book.

We hope our readers get as much pleasure from reading this material as we have had during its writing. The website `www.cs.dal.ca/~borwein/ToVA` will record additional information and addenda for the book, and we invite feedback.

Halifax, Nova Scotia   *Jonathan Borwein*
Kalamazoo, Michigan   *Qiji Zhu*
December 31, 2004

Contents

1 **Introduction** 1
   1.1 Introduction 1
   1.2 Notation 2
   1.3 Exercises 4

2 **Variational Principles** 5
   2.1 Ekeland Variational Principles 5
   2.2 Geometric Forms of the Variational Principle 10
   2.3 Applications to Fixed Point Theorems 15
   2.4 Finite Dimensional Variational Principles 19
   2.5 Borwein–Preiss Variational Principles 30

3 **Variational Techniques in Subdifferential Theory** 37
   3.1 The Fréchet Subdifferential and Normal Cone 39
   3.2 Nonlocal Sum Rule and Viscosity Solutions 47
   3.3 Local Sum Rules and Constrained Minimization 54
   3.4 Mean Value Theorems and Applications 78
   3.5 Chain Rules and Lyapunov Functions 87
   3.6 Multidirectional MVI and Solvability 95
   3.7 Extremal Principles 103

4 **Variational Techniques in Convex Analysis** 111
   4.1 Convex Functions and Sets 111
   4.2 Subdifferential 117
   4.3 Sandwich Theorems and Calculus 127
   4.4 Fenchel Conjugate 134
   4.5 Convex Feasibility Problems 140
   4.6 Duality Inequalities for Sandwiched Functions.............150
   4.7 Entropy Maximization 157

5   Variational Techniques and Multifunctions 165
   5.1   Multifunctions 165
   5.2   Subdifferentials as Multifunctions 188
   5.3   Distance Functions 214
   5.4   Coderivatives of Multifunctions 220
   5.5   Implicit Multifunction Theorems 229

6   Variational Principles in Nonlinear Functional Analysis 243
   6.1   Subdifferential and Asplund Spaces 243
   6.2   Nonconvex Separation Theorems 259
   6.3   Stegall Variational Principles 266
   6.4   Mountain Pass Theorem 274
   6.5   One-Perturbation Variational Principles 280

7   Variational Techniques in the Presence of Symmetry 291
   7.1   Nonsmooth Functions on Smooth Manifolds 291
   7.2   Manifolds of Matrices and Spectral Functions 299
   7.3   Convex Spectral Functions 316

References 339

Index 353

1

Introduction and Notation

1.1 Introduction

In this book, *variational techniques* refer to proofs by way of establishing that an appropriate auxiliary function attains a minimum. This can be viewed as a mathematical form of the principle of least action in physics. Since so many important results in mathematics, in particular, in analysis have their origins in the physical sciences, it is entirely natural that they can be related in one way or another to variational techniques. The purpose of this book is to provide an introduction to this powerful method, and its applications, to researchers who are interested in using this method. The use of variational arguments in mathematical proofs has a long history. This can be traced back to Johann Bernoulli's problem of the Brachistochrone and its solutions leading to the development of the calculus of variations. Since then the method has found numerous applications in various branches of mathematics. A simple illustration of the variational argument is the following example.

Example 1.1.1 (Surjectivity of Derivatives) Suppose that $f\colon \mathbb{R} \to \mathbb{R}$ is differentiable everywhere and suppose that

$$\lim_{|x|\to\infty} f(x)/|x| = +\infty.$$

Then $\{f'(x) \mid x \in \mathbb{R}\} = \mathbb{R}$.

Proof. Let r be an arbitrary real number. Define $g(x) := f(x) - rx$. We easily check that g is coercive, i.e., $g(x) \to +\infty$ as $|x| \to \infty$ and therefore attains a (global) minimum at, say, $\bar{x}$. Then $0 = g'(\bar{x}) = f'(\bar{x}) - r$. ●

Two conditions are essential in this variational argument. The first is *compactness* (to ensure the existence of the minimum) and the second is *differentiability* of the auxiliary function (so that the differential characterization of the results is possible). Two important discoveries in the 1970's led to significant useful relaxation on both conditions. First, the discovery of general variational principles led to the relaxation of the compactness assumptions. Such principles typically assert that any lower semicontinuous (lsc) function, bounded from below, may be perturbed slightly to ensure the existence of the minimum. Second, the development of the nonsmooth analysis made possible the use of nonsmooth auxiliary functions.

The emphasis in this book is on the new developments and applications of variational techniques in the past several decades. Besides the use of variational principles and concepts that generalize that of a derivative for smooth functions, one often needs to combine a variational principle with other suitable tools. For example, a decoupling method that mimics in nonconvex settings the role of Fenchel duality or the Hahn–Banach theorem is an essential element in deriving many calculus rules for subdifferentials; minimax theorems play a crucial role alongside the variational principle in several important results in nonlinear functional analysis; and the analysis of spectral functions is a combination of the variational principles with the symmetric property of these functions with respect to certain groups. This is reflected in our arrangement of the chapters. An important feature of the new variational techniques is that they can handle nonsmooth functions, sets and multifunctions equally well. In this book we emphasize the role of nonsmooth, most of the time extended valued lower semicontinuous functions and their subdifferential. We illustrate that sets and multifunctions can be handled by using related nonsmooth functions. Other approaches are possible. For example Mordukhovich [204] starts with variational geometry on closed sets and deals with functions and multifunctions by examining their epigraphs and graphs.

Our intention in this book is to provide a concise introduction to the essential tools of infinite-dimensional first-order variational analysis, tools that are presently scattered in the literature. We also aim to illustrate applications in many different parts of analysis, optimization and approximation, dynamic systems and mathematical economics. To make the book more appealing to readers who are not experts in the area of variational analysis we arrange the applications right after general principles wherever possible. Materials here can be used flexibly for a short lecture series or a topics course for graduate students. They can also serve as a reference for researchers who are interested in the theory or applications of the variational analysis methods.

1.2 Notation

We introduce some common notations in this section.

Let (X, d) be a metric space. We denote the closed ball centered at x with radius r by $B_r(x)$. We will often work in a real Banach space. When X is a Banach space we use X^* and $\langle \cdot, \cdot \rangle$ to denote its (topological) dual and the duality pairing, respectively. The closed unit ball of a Banach space X is often denoted by B_X or B when the space is clear from the context.

Let $\mathbb{R}$ be the real numbers. Consider an extended-real-valued function $f \colon X \to \mathbb{R} \cup \{+\infty\}$. The *domain* of f is the set where it is finite and is denoted by $\operatorname{dom} f := \{x \mid f(x) < +\infty\}$. The *range* of f is the set of all the values of f and is denoted by $\operatorname{range} f := \{f(x) \mid x \in \operatorname{dom} f\}$. We call an extended-valued function f *proper* provided that its domain is nonempty. We say $f \colon X \to \mathbb{R} \cup \{+\infty\}$ is *lower semicontinuous* (lsc) at x provided that $\liminf_{y \to x} f(y) \geq f(x)$. We say that f is lsc if it is lsc everywhere in its domain.

A subset S of a metric space (X, d) can often be better studied by using related functions. The extended-valued *indicator function* of S,

$$\iota_S(x) = \iota(S; x) := \begin{cases} 0 & x \in S, \\ +\infty & \text{otherwise,} \end{cases}$$

characterizes S. We also use the *distance function*

$$d_S(x) = d(S; x) := \inf\{d(x, y) \mid y \in S\}.$$

The distance function determines closed sets as shown in Exercises 1.3.1 and 1.3.2. On the other hand, to study a function $f \colon X \to \mathbb{R} \cup \{+\infty\}$ it is often equally helpful to examine its *epigraph* and *graph*, related sets in $X \times \mathbb{R}$, defined by

$$\operatorname{epi} f := \{(x, r) \in X \times \mathbb{R} \mid f(x) \leq r\}$$

and

$$\operatorname{graph} f := \{(x, f(x)) \in X \times \mathbb{R} \mid x \in \operatorname{dom} f\}.$$

We denote the *preimage* of $f \colon X \to \mathbb{R} \cup \{+\infty\}$ of a subset S in $\mathbb{R}$ by

$$f^{-1}(S) := \{x \in X \mid f(x) \in S\}.$$

Two special cases which will be used often are $f^{-1}((-\infty, a])$, the *sublevel set*, and $f^{-1}(a)$, the *level set*, of f at $a \in \mathbb{R}$. For a set S in a Banach space X, we denote by $\operatorname{int} S$, $\overline{S}$, $\operatorname{bd} S$, $\operatorname{conv} S$, $\overline{\operatorname{conv}} S$ its *interior, closure, boundary, convex hull, closed convex hull*, respectively, and we denote by $\operatorname{diam}(S) := \sup\{\|x - y\| \mid x, y \in S\}$ its *diameter* and by $B_r(S) := \{x \in X \mid d(S; x) \leq r\}$ its *r-enlargement*. Closed sets and lsc functions are closely related as illustrated in Exercises 1.3.3, 1.3.4 and 1.3.5.

Another valuable tool in studying lsc functions is the *inf-convolution* of two functions f and g on a Banach space X defined by $(f \square g)(x) := \inf_{y \in X}[f(y) + g(x - y)]$. Exercise 1.3.7 shows how this operation generates nice functions.

Multifunctions (set-valued functions) are equally interesting and useful. Denote by 2^Y the collection of all subsets of Y. A multifunction $F \colon X \to 2^Y$ maps each $x \in X$ to a subset $F(x)$ of Y. It is completely determined by its *graph*,

$$\operatorname{graph} F := \{(x, y) \in X \times Y \mid y \in F(x)\},$$

a subset of the product space $X \times Y$ and, hence, by the indicator function $\iota_{\text{graph } F}$. The domain of a multifunction F is defined by $\text{dom} F := \{x \in X \mid F(x) \neq \emptyset\}$. The *inverse of a multifunction* $F \colon X \to 2^Y$ is defined by

$$F^{-1}(y) = \{x \in X \mid y \in F(x)\}.$$

Note that F^{-1} is a multifunction from Y to X. We say a multifunction F is *closed-valued* provided that for every $x \in \text{dom} F$, $F(x)$ is a closed set. We say the multifunction is *closed* if indeed the graph is a closed set in the product space. These two concepts are different (Exercise 1.3.8).

The ability to use extended-valued functions to relate sets, functions and multifunctions is one of the great advantages of the variational technique which is designed to deal fluently with such functions. In this book, for the most part, we shall focus on the theory for extended-valued functions. Corresponding results for sets and multifunctions are most often derivable by reducing them to appropriate function formulations.

1.3 Exercises

Exercise 1.3.1 Show that $x \in \overline{S}$ if and only if $d_S(x) = 0$.

Exercise 1.3.2 Suppose that S_1 and S_2 are two subsets of X. Show that $d_{S_1} = d_{S_2}$ if and only if $\overline{S_1} = \overline{S_2}$.

Exercise 1.3.3 Prove that S is a closed set if and only if ι_S is lsc.

Exercise 1.3.4 Prove that f is lsc if and only if epi f is closed.

Exercise 1.3.5 Prove that f is lsc if and only if its sublevel set at a, $f^{-1}((-\infty, a])$, is closed for all $a \in \mathbb{R}$.

These results can be used to show the supremum of lsc functions is lsc.

Exercise 1.3.6 Let $\{f_a\}_{a \in A}$ be a family of lsc functions. Prove that $f := \sup\{f_a, a \in A\}$ is lsc. Hint: epi $f = \bigcap_{a \in A} \text{epi } f_a$.

Exercise 1.3.7 Let f be a lsc function bounded from below. Prove that if g is Lipschitz with rank L, then so is $f \Box g$.

Exercise 1.3.8 Let $F \colon X \to 2^Y$ be a multifunction. Show that if F has a closed graph then F is closed-valued, but the converse is not true.

2

Variational Principles

A lsc function on a noncompact set may well not attain its minimum. Roughly speaking, a variational principle asserts that, for any extended-valued lsc function which is bounded below, one can add a small perturbation to make it attain a minimum. Variational principles allow us to apply the variational technique to extended-valued lsc functions systematically, and therefore significantly extend the power of the variational technique. Usually, in a variational principle the better the geometric (smoothness) property of the underlying space the nicer the perturbation function. There are many possible settings. In this chapter, we focus on two of them: the Ekeland variational principle which holds in any complete metric space and the Borwein–Preiss smooth variational principle which ensures a smooth perturbation suffices in any Banach space with a smooth norm. We will also present a variant of the Borwein–Preiss variational principle derived by Deville, Godefroy and Zizler with an elegant category proof.

These variational principles provide powerful tools in modern variational analysis. Their applications cover numerous areas in both theory and applications of analysis including optimization, Banach space geometry, nonsmooth analysis, economics, control theory and game theory, to name a few. As a first taste we discuss some of their applications; these require minimum prerequisites in Banach space geometry, fixed point theory, an analytic proof of the Gordan theorem of the alternative, a characterization of the level sets associated with majorization and a variational proof of Birkhoff's theorem on the doubly stochastic matrices. Many other applications will be discussed in subsequent chapters.

2.1 Ekeland Variational Principles

2.1.1 The Geometric Picture

Consider a lsc function f bounded below on a Banach space $(X, \|\cdot\|)$. Clearly f may not attain its minimum or, to put it geometrically, f may not have

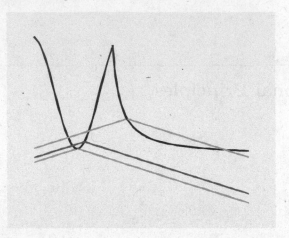

Fig. 2.1. Ekeland variational principle. Top cone: $f(x_0) - \varepsilon|x - x_0|$; Middle cone: $f(x_1) - \varepsilon|x - x_1|$; Lower cone: $f(y) - \varepsilon|x - y|$.

a supporting hyperplane. Ekeland's variational principle provides a kind of approximate substitute for the attainment of a minimum by asserting that, for any $\varepsilon > 0$, f must have a supporting cone of the form $f(y) - \varepsilon\|x - y\|$. One way to see how this happens geometrically is illustrated by Figure 2.1. We start with a point z_0 with $f(z_0) < \inf_X f + \varepsilon$ and consider the cone $f(z_0) - \varepsilon\|x - z_0\|$. If this cone does not support f then one can always find a point $z_1 \in S_0 := \{x \in X \mid f(x) \le f(z) - \varepsilon\|x - z\|)\}$ such that

$$f(z_1) < \inf_{S_0} f + \frac{1}{2}[f(z_0) - \inf_{S_0} f].$$

If $f(z_1) - \varepsilon\|x - z_1\|$ still does not support f then we repeat the above process. Such a procedure either finds the desired supporting cone or generates a sequence of nested closed sets (S_i) whose diameters shrink to 0. In the latter case, $f(y) - \varepsilon\|x - y\|$ is a supporting cone of f, where $\{y\} = \bigcap_{i=1}^{\infty} S_i$. This line of reasoning works similarly in a complete metric space. Moreover, it also provides a useful estimate on the distance between y and the initial ε-minimum z_0.

2.1.2 The Basic Form

We now turn to the analytic form of the geometric picture described above – the Ekeland variational principle and its proof.

Theorem 2.1.1 (Ekeland Variational Principle) *Let (X, d) be a complete metric space and let $f \colon X \to \mathbb{R} \cup \{+\infty\}$ be a lsc function bounded from below. Suppose that $\varepsilon > 0$ and $z \in X$ satisfy*

$$f(z) < \inf_X f + \varepsilon.$$

Then there exists $y \in X$ such that

(i) $d(z,y) \le 1$,
(ii) $f(y) + \varepsilon d(z,y) \le f(z)$, *and*
(iii) $f(x) + \varepsilon d(x,y) \ge f(y)$, *for all $x \in X$.*

Proof. Define a sequence (z_i) by induction starting with $z_0 := z$. Suppose that we have defined z_i. Set

$$S_i := \{x \in X \mid f(x) + \varepsilon d(x, z_i) \le f(z_i)\}$$

and consider two possible cases: (a) $\inf_{S_i} f = f(z_i)$. Then we define $z_{i+1} := z_i$. (b) $\inf_{S_i} f < f(z_i)$. We choose $z_{i+1} \in S_i$ such that

$$f(z_{i+1}) < \inf_{S_i} f + \frac{1}{2}[f(z_i) - \inf_{S_i} f] = \frac{1}{2}[f(z_i) + \inf_{S_i} f] < f(z_i). \quad (2.1.1)$$

We show that (z_i) is a Cauchy sequence. In fact, if (a) ever happens then z_i is stationary for i large. Otherwise,

$$\varepsilon d(z_i, z_{i+1}) \le f(z_i) - f(z_{i+1}). \quad (2.1.2)$$

Adding (2.1.2) up from i to $j - 1 > i$ we have

$$\varepsilon d(z_i, z_j) \le f(z_i) - f(z_j). \quad (2.1.3)$$

Observe that the sequence $(f(z_i))$ is decreasing and bounded from below by $\inf_X f$, and therefore convergent. We conclude from (2.1.3) that (z_i) is Cauchy. Let $y := \lim_{i \to \infty} z_i$. We show that y satisfies the conclusions of the theorem. Setting $i = 0$ in (2.1.3) we have

$$\varepsilon d(z, z_j) + f(z_j) \le f(z). \quad (2.1.4)$$

Taking limits as $j \to \infty$ yields (ii). Since $f(z) - f(y) \le f(z) - \inf_X f < \varepsilon$, (i) follows from (ii). It remains to show that y satisfies (iii). Fixing i in (2.1.3) and taking limits as $j \to \infty$ yields $y \in S_i$. That is to say

$$y \in \bigcap_{i=1}^{\infty} S_i.$$

On the other hand, if $x \in \bigcap_{i=1}^{\infty} S_i$ then, for all $i = 1, 2, \ldots$,

$$\varepsilon d(x, z_{i+1}) \le f(z_{i+1}) - f(x) \le f(z_{i+1}) - \inf_{S_i} f. \quad (2.1.5)$$

It follows from (2.1.1) that $f(z_{i+1}) - \inf_{S_i} f \le f(z_i) - f(z_{i+1})$, and therefore $\lim_i[f(z_{i+1}) - \inf_{S_i} f] = 0$. Taking limits in (2.1.5) as $i \to \infty$ we have $\varepsilon d(x,y) = 0$. It follows that

$$\bigcap_{i=1}^{\infty} S_i = \{y\}. \tag{2.1.6}$$

Notice that the sequence of sets (S_i) is nested, i.e., for any i, $S_{i+1} \subset S_i$. In fact, for any $x \in S_{i+1}$, $f(x) + \varepsilon d(x, z_{i+1}) \le f(z_{i+1})$ and $z_{i+1} \in S_i$ yields

$$\begin{aligned} f(x) + \varepsilon d(x, z_i) &\le f(x) + \varepsilon d(x, z_{i+1}) + \varepsilon d(z_i, z_{i+1}) \\ &\le f(z_{i+1}) + \varepsilon d(z_i, z_{i+1}) \le f(z_i), \end{aligned} \tag{2.1.7}$$

which implies that $x \in S_i$. Now, for any $x \ne y$, it follows from (2.1.6) that when i sufficiently large $x \notin S_i$. Thus, $f(x) + \varepsilon d(x, z_i) \ge f(z_i)$. Taking limits as $i \to \infty$ we arrive at (iii). $\qquad\qquad\bullet$

2.1.3 Other Forms

Since $\varepsilon > 0$ is arbitrary the supporting cone in the Ekeland's variational principle can be made as "flat" as one wishes. It turns out that in many applications such a flat supporting cone is enough to replace the possibly non-existent support plane. Another useful geometric observation is that one can trade between a flatter supporting cone and a smaller distance between the supporting point y and the initial ε-minimum z. The following form of this tradeoff can easily be derived from Theorem 2.1.1 by an analytic argument.

Theorem 2.1.2 *Let (X, d) be a complete metric space and let $f\colon X \to \mathbb{R} \cup \{+\infty\}$ be a lsc function bounded from below. Suppose that $\varepsilon > 0$ and $z \in X$ satisfy*

$$f(z) < \inf_X f + \varepsilon.$$

Then, for any $\lambda > 0$ there exists y such that

(i) $d(z, y) \le \lambda$,
(ii) $f(y) + (\varepsilon/\lambda)d(z, y) \le f(z)$, *and*
(iii) $f(x) + (\varepsilon/\lambda)d(x, y) > f(y)$, *for all $x \in X \setminus \{y\}$.*

Proof.   Exercise 2.1.1. $\qquad\qquad\bullet$

The constant λ in Theorem 2.1.2 makes it very flexible. A frequent choice is to take $\lambda = \sqrt{\varepsilon}$ and so to balance the perturbations in (ii) and (iii).

Theorem 2.1.3 *Let (X, d) be a complete metric space and let $f\colon X \to \mathbb{R} \cup \{+\infty\}$ be a lsc function bounded from below. Suppose that $\varepsilon > 0$ and $z \in X$ satisfy*

$$f(z) < \inf_X f + \varepsilon.$$

Then, there exists y such that

(i) $d(z, y) \le \sqrt{\varepsilon}$,

(ii) $f(y) + \sqrt{\varepsilon}d(z,y) \leq f(z)$, and

(iii) $f(x) + \sqrt{\varepsilon}d(x,y) > f(y)$,   for all $x \in X \setminus \{y\}$.

Proof. Set $\lambda = \sqrt{\varepsilon}$ in Theorem 2.1.2.   ●

When the approximate minimization point z in Theorem 2.1.2 is not explicitly known or is not important the following weak form of the Ekeland variational principle is useful.

Theorem 2.1.4 *Let (X,d) be a complete metric space and let $f\colon X \to \mathbb{R} \cup \{+\infty\}$ be a lsc function bounded from below. Then, for any $\varepsilon > 0$, there exists y such that*

$$f(x) + \sqrt{\varepsilon}d(x,y) > f(y).$$

Proof. Exercise 2.1.6.   ●

2.1.4 Commentary and Exercises

Ekeland's variational principle, appeared in [106], is inspired by the Bishop–Phelps Theorem [24, 25] (see the next section). The original proof of the Ekeland variational principle in [106] is similar to that of the Bishop–Phelps Theorem using Zorn's lemma. J. Lasry pointed out transfinite induction is not needed and the proof given here is taken from the survey paper [107] and was credited to M. Crandall. As an immediate application we can derive a version of the results in Example 1.1.1 in infinite dimensional spaces (Exercises 2.1.2).

The lsc condition on f in the Ekeland variational principle can be relaxed somewhat. We leave the details in Exercises 2.1.4 and 2.1.5.

Exercise 2.1.1 Prove Theorem 2.1.2. Hint: Apply Theorem 2.1.1 with the metric $d(\cdot,\cdot)/\lambda$.

Exercise 2.1.2 Let X be a Banach space and let $f\colon X \to \mathbb{R}$ be a Fréchet differentiable function (see Section 3.1.1). Suppose that f is bounded from below on any bounded set and satisfies

$$\lim_{\|x\|\to\infty} \frac{f(x)}{\|x\|} = +\infty.$$

Then the range of f', $\{f'(x) \mid x \in X\}$, is dense in X^*.

Exercise 2.1.3 As a comparison, show that in Exercise 2.1.2, if X is a finite dimensional Banach space, then f' is onto. (Note also the assumption that f bounded from below on bounded sets is not necessary in finite dimensional spaces).

Exercise 2.1.4 We say a function f is partially lower semicontinuous (plsc) at x provided that, for any $x_i \to x$ with $f(x_i)$ monotone decreasing, one has $f(x) \le \lim f(x_i)$. Prove that in Theorems 2.1.1 and 2.1.2, the assumption that f is lsc can be replaced by the weaker condition that f is plsc.

Exercise 2.1.5 Construct a class of plsc functions that are not lsc.

Exercise 2.1.6 Prove Theorem 2.1.4.

One of the most important—though simple—applications of the Ekeland variational principle is given in the following exercise:

Exercise 2.1.7 (Existence of Approximate Critical Points) Let $U \subset X$ be an open subset of a Banach space and let $f : U \to \mathbb{R}$ be a Gâteaux differentiable function. Suppose that for some $\varepsilon > 0$ we have $\inf_X f > f(\bar{x}) - \varepsilon$. Prove that, for any $\lambda > 0$, there exists a point $x \in B_\lambda(\bar{x})$ where the Gâteaux derivative $f'(x)$ satisfies $\|f'(x)\| \le \varepsilon/\lambda$. Such a point is an *approximate critical point*.

2.2 Geometric Forms Of the Variational Principle

In this section we discuss the Bishop–Phelps Theorem, the flower-petal theorem and the drop theorem. They capture the essence of the Ekeland variational principle from a geometric perspective.

2.2.1 The Bishop–Phelps Theorem

Among the three, the Bishop–Phelps Theorem [24, 25] is the closest to the Ekeland variational principle in its geometric explanation.

Let X be a Banach space. For any $x^* \in X^* \backslash \{0\}$ and any $\varepsilon > 0$ we say that

$$K(x^*, \varepsilon) := \{x \in X \mid \varepsilon\|x^*\|\|x\| \le \langle x^*, x\rangle\}$$

is a *Bishop–Phelps cone* associated with x^* and ε. We illustrate this in Figure 2.2 with the classic "ice cream cone" in three dimensions.

Theorem 2.2.1 (Bishop–Phelps Theorem) *Let X be a Banach space and let S be a closed subset of X. Suppose that $x^* \in X^*$ is bounded on S. Then, for every $\varepsilon > 0$, S has a $K(x^*, \varepsilon)$ support point y, i.e.,*

$$\{y\} = S \cap [K(x^*, \varepsilon) + y].$$

Proof. Apply the Ekeland variational principle of Theorem 2.1.1 to the lsc function $f := -x^*/\|x^*\| + \iota_S$. We leave the details as an exercise.   ●

The geometric picture of the Bishop–Phelps Theorem and that of the Ekeland variational principle are almost the same: the Bishop–Phelps cone

Fig. 2.2. A Bishop–Phelps cone.

$K(x^*, \varepsilon) + y$ in Theorem 2.2.1 plays a role similar to that of $f(y) - \varepsilon d(x, y)$ in Theorem 2.1.1. One can easily derive a Banach space version of the Ekeland variational principle by applying the Bishop–Phelps Theorem to the epigraph of a lsc function bounded from below (Exercise 2.2.2).

If we have additional information, e.g., known points inside and/or outside the given set, then the supporting cone can be replaced by more delicately constructed bounded sets. The flower-petal theorem and the drop theorem discussed in the sequel are of this nature.

2.2.2 The Flower-Petal Theorem

Let X be a Banach space and let $a, b \in X$. We say that

$$P_\gamma(a, b) := \{x \in X \mid \gamma\|a - x\| + \|x - b\| \leq \|b - a\|\}$$

is a *flower petal* associated with $\gamma \in (0, +\infty)$ and $a, b \in X$. A flower petal is always convex, and interesting flower petals are formed when $\gamma \in (0, 1)$ (see Exercises 2.2.3 and 2.2.4).

Figure 2.3 draws the petals $P_\gamma((0, 0), (1, 0))$ for $\gamma = 1/3$, and $\gamma = 1/2$.

Theorem 2.2.2 (Flower Petal Theorem) *Let X be a Banach space and let S be a closed subset of X. Suppose that $a \in S$ and $b \in X \backslash S$ with $r \in (0, d(S; b))$ and $t = \|b - a\|$. Then, for any $\gamma > 0$, there exists $y \in S \cap P_\gamma(a, b)$ satisfying $\|y - a\| \leq (t - r)/\gamma$ such that $P_\gamma(y, b) \cap S = \{y\}$.*

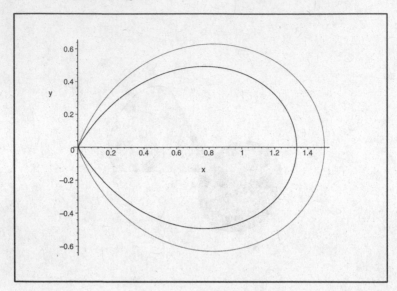

Fig. 2.3. Two flower petals.

Proof. Define $f(x) := \|x - b\| + \iota_S(x)$. Then

$$f(a) < \inf_X f + (t - r).$$

Applying the Ekeland variational principle of Theorem 2.1.2 to the function $f(x)$ with and $\varepsilon = t - r$ and $\lambda = (t - r)/\gamma$, we have that there exists $y \in S$ such that $\|y - a\| < (t - r)/\gamma$ satisfying

$$\|y - b\| + \gamma\|a - y\| \le \|a - b\|$$

and

$$\|x - b\| + \gamma\|x - y\| > \|y - b\|, \text{ for all } x \in S\backslash\{y\}.$$

The first inequality says $y \in P_\gamma(a, b)$ while the second implies that $P_\gamma(y, b) \cap S = \{y\}$. $\qquad\bullet$

2.2.3 The Drop Theorem

Let X be a Banach space, let C be a convex subset of X and let $a \in X$. We say that

$$[a, C] := \text{conv}(\{a\} \cup C) = \{a + t(c - a) \mid c \in C\}$$

is the *drop* associated with a and C.

The following lemma provides useful information on the relationship between drops and flower petals. This is illustrated in Figure 2.4 and the easy proof is left as an exercise.

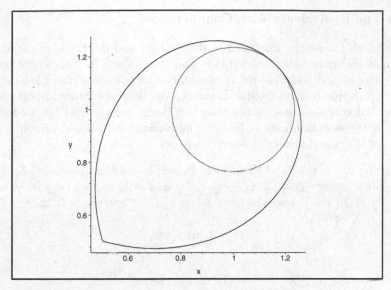

Fig. 2.4. A petal capturing a ball.

Lemma 2.2.3 (Drop and Flower Petal) *Let X be a Banach space, let $a, b \in X$ and let $\gamma \in (0,1)$. Then*

$$B_{\|a-b\|(1-\gamma)/(1+\gamma)}(b) \subset P_\gamma(a, b),$$

so that

$$[a, B_{\|a-b\|(1-\gamma)/(1+\gamma)}(b)] \subset P_\gamma(a, b).$$

Proof.   Exercise 2.2.5.   ●

Now we can deduce the drop theorem from the flower petal theorem.

Theorem 2.2.4 (The Drop Theorem) *Let X be a Banach space and let S be a closed subset of X. Suppose that $b \in X \backslash S$ and $r \in (0, d(S; b))$. Then, for any $\varepsilon > 0$, there exists $y \in \mathrm{bd}(S)$ satisfying $\|y - b\| \leq d(S; b) + \varepsilon$ such that $[y, B_r(b)] \cap S = \{y\}$.*

Proof.   Choose $a \in S$ satisfying $\|a - b\| < d(S; b) + \varepsilon$ and choose

$$\gamma = \frac{\|a-b\| - r}{\|a-b\| + r} \in (0,1).$$

It follows from Theorem 2.2.2 that there exists $y \in S \cap P_\gamma(a, b)$ such that $P_\gamma(y, b) \cap S = \{y\}$. Clearly, $y \in \mathrm{bd}(S)$. Moreover, $y \in P_\gamma(a, b)$ implies that $\|y - b\| < \|a - b\| < d(S; y) + \varepsilon$. Finally, it follows from Lemma 2.2.3 and $r = \frac{1-\gamma}{1+\gamma}\|a-b\|$ that $[y, B_r(b)] \cap S = \{y\}$.   ●

2.2.4 The Equivalence with Completeness

Actually, all the results discussed in this section and the Ekeland variational principle are equivalent provided that one states them in sufficiently general form (see e.g. [135]). In the setting of a general metric space, the Ekeland variational principle is more flexible in various applications. More importantly it shows that completeness, rather than the linear structure of the underlying space, is the essential feature. In fact, the Ekeland variational principle characterizes the completeness of a metric space.

Theorem 2.2.5 (Ekeland Variational Principle and Completeness) *Let (X, d) be a metric space. Then X is complete if and only if for every lsc function $f\colon X \to \mathbb{R} \cup \{+\infty\}$ bounded from below and for every $\varepsilon > 0$ there exists a point $y \in X$ satisfying*

$$f(y) \leq \inf_X f + \varepsilon,$$

and

$$f(x) + \varepsilon d(x, y) \geq f(y), \quad \text{for all } x \in X.$$

Proof. The "if" part follows from Theorem 2.1.4. We prove the "only if" part. Let (x_i) be a Cauchy sequence. Then, the function $f(x) := \lim_{i\to\infty} d(x_i, x)$ is well-defined and nonnegative. Since the distance function is Lipschitz with respect to x we see that f is continuous. Moreover, since (x_i) is a Cauchy sequence we have $f(x_i) \to 0$ as $i \to \infty$ so that $\inf_X f = 0$. For $\varepsilon \in (0, 1)$ choose y such that $f(y) \leq \varepsilon$ and

$$f(y) \leq f(x) + \varepsilon d(x, y), \quad \text{for all } x \in X \tag{2.2.1}$$

Letting $x = x_i$ in (2.2.1) and taking limits as $i \to \infty$ we obtain $f(y) \leq \varepsilon f(y)$ so that $f(y) = 0$. That is to say $\lim_{i\to\infty} x_i = y$. ●

2.2.5 Commentary and Exercises

The Bishop–Phelps theorem is the earliest of this type [24, 25]. In fact, this important result in Banach space geometry is the main inspiration for Ekeland's variational principle (see [107]). The drop theorem was discovered by Danes [95]. The flower-petal theorem was derived by Penot in [217]. The relationship among the Ekeland variational principle, the drop theorem and the flower-petal theorem were discussed in Penot [217] and Rolewicz [238]. The book [141] by Hyers, Isac and Rassias is a nice reference containing many other variations and applications of the Ekeland variational principle.

Exercise 2.2.1 Provide details for the proof of Theorem 2.2.1.

Exercise 2.2.2 Deduce the Ekeland variational principle in a Banach space by applying the Bishop–Phelps Theorem to the epigraph of a lsc function.

Exercise 2.2.3 Show that, for $\gamma > 1$, $P_\gamma(a,b) = \{a\}$ and $P_1(a,b) = \{\lambda a + (1-\lambda)b \mid \lambda \in [0,1]\}$.

Exercise 2.2.4 Prove that $P_\gamma(a,b)$ is convex.

Exercise 2.2.5 Prove Lemma 2.2.3.

2.3 Applications to Fixed Point Theorems

Let X be a set and let f be a map from X to itself. We say x is a *fixed point* of f if $f(x) = x$. Fixed points of a mapping often represent equilibrium states of some underlying system, and they are consequently of great importance. Therefore, conditions ensuring the existence and uniqueness of fixed point(s) are the subject of extensive study in analysis. We now use Ekeland's variational principle to deduce several fixed point theorems.

2.3.1 The Banach Fixed Point Theorem

Let (X,d) be a complete metric space and let ϕ be a map from X to itself. We say that ϕ is a *contraction* provided that there exists $k \in (0,1)$ such that

$$d(\phi(x),\phi(y)) \le kd(x,y), \quad \text{for all } x,y \in X.$$

Theorem 2.3.1 (Banach Fixed Point Theorem) *Let (X,d) be a complete metric space. Suppose that $\phi: X \to X$ is a contraction. Then ϕ has a unique fixed point.*

Proof. Define $f(x) := d(x,\phi(x))$. Applying Theorem 2.1.1 to f with $\varepsilon \in (0,1-k)$, we have $y \in X$ such that

$$f(x) + \varepsilon d(x,y) \ge f(y), \quad \text{for all } x \in X.$$

In particular, setting $x = \phi(y)$ we have

$$d(y,\phi(y)) \le d(\phi(y),\phi^2(y)) + \varepsilon d(y,\phi(y)) \le (k+\varepsilon)d(y,\phi(y)).$$

Thus, y must be a fixed point. The uniqueness follows directly from the fact that ϕ is a contraction and is left as an exercise.   ●

2.3.2 Clarke's Refinement

Clarke observed that the argument in the proof of the Banach fixed point theorem works under weaker conditions. Let (X,d) be a complete metric space. For $x,y \in X$ we define the *segment* between x and y by

$$[x,y] := \{z \in X \mid d(x,z) + d(z,y) = d(x,y)\}. \tag{2.3.1}$$

Definition 2.3.2 (Directional Contraction) *Let (X, d) be a complete metric space and let ϕ be a map from X to itself. We say that ϕ is a* directional contraction *provided that*

(i) *ϕ is continuous, and*
(ii) *there exists $k \in (0, 1)$ such that, for any $x \in X$ with $\phi(x) \neq x$ there exists $z \in [x, \phi(x)] \backslash \{x\}$ such that*

$$d(\phi(x), \phi(z)) \leq kd(x, z).$$

Theorem 2.3.3 *Let (X, d) be a complete metric space. Suppose that $\phi \colon X \to X$ is a directional contraction. Then ϕ admits a fixed point.*

Proof. Define

$$f(x) := d(x, \phi(x)).$$

Then f is continuous and bounded from below (by 0). Applying the Ekeland variational principle of Theorem 2.1.1 to f with $\varepsilon \in (0, 1 - k)$ we conclude that there exists $y \in X$ such that

$$f(y) \leq f(x) + \varepsilon d(x, y), \quad \text{for all } x \in X. \tag{2.3.2}$$

If $\phi(y) = y$, we are done. Otherwise, since ϕ is a directional contraction there exists a point $z \neq y$ with $z \in [y, \phi(y)]$, i.e.,

$$d(y, z) + d(z, \phi(y)) = d(y, \phi(y)) = f(y) \tag{2.3.3}$$

satisfying

$$d(\phi(z), \phi(y)) \leq kd(z, y). \tag{2.3.4}$$

Letting $x = z$ in (2.3.2) and using (2.3.3) we have

$$d(y, z) + d(z, y) \leq d(z, \phi(z)) + \varepsilon d(z, y)$$

or

$$d(y, z) \leq d(z, \phi(z)) - d(z, \phi(y)) + \varepsilon d(z, y) \tag{2.3.5}$$

By the triangle inequality and (2.3.4) we have

$$d(z, \phi(z)) - d(z, \phi(y)) \leq d(\phi(y), \phi(z)) \leq kd(y, z). \tag{2.3.6}$$

Combining (2.3.5) and (2.3.6) we have

$$d(y, z) \leq (k + \varepsilon)d(y, z),$$

a contradiction.   ●

Clearly any contraction is a directional contraction. Therefore, Theorem 2.3.3 generalizes the Banach fixed point theorem. The following is an example where Theorem 2.3.3 applies when the Banach contraction theorem does not.

Example 2.3.4 Consider $X = \mathbb{R}^2$ with a metric induced by the norm $\|x\| = \|(x_1, x_2)\| = |x_1| + |x_2|$. A segment between two points (a_1, a_2) and (b_1, b_2) consists of the closed rectangle having the two points as diagonally opposite corners. Define

$$\phi(x_1, x_2) = \left(\frac{3x_1}{2} - \frac{x_2}{3}, x_1 + \frac{x_2}{3} \right).$$

Then ϕ is a directional contraction. Indeed, if $y = \phi(x) \neq x$. Then $y_2 \neq x_2$ (for otherwise we will also have $y_1 = x_1$). Now the set $[x, y]$ contains points of the form (x_1, t) with t arbitrarily close to x_2 but not equal to x_2. For such points we have

$$d(\phi(x_1, t), \phi(x_1, x_2)) = \frac{2}{3} d((x_1, t), (x_1, x_2)),$$

so that ϕ is a directional contraction. We can directly check that the fixed points of ϕ are all points of the form $(x, 3x/2)$. Since ϕ has more than one fixed point clearly the Banach fixed point theorem does not apply to this mapping.

2.3.3 The Caristi–Kirk Fixed Point Theorem

A similar argument can be used to prove the Caristi–Kirk fixed point theorem for multifunctions. For a multifunction $F: X \to 2^X$, we say that x is a fixed point for F provided that $x \in F(x)$.

Theorem 2.3.5 (Caristi–Kirk Fixed Point Theorem) *Let (X, d) be a complete metric space and let $f: X \to \mathbb{R} \cup \{+\infty\}$ be a proper lsc function bounded below. Suppose $F: X \to 2^X$ is a multifunction with a closed graph satisfying*

$$f(y) \leq f(x) - d(x, y), \quad \text{for all } (x, y) \in \text{graph} \, F. \qquad (2.3.7)$$

Then F has a fixed point.

Proof. Define a metric ρ on $X \times X$ by $\rho((x_1, y_1), (x_2, y_2)) := d(x_1, x_2) + d(y_1, y_2)$ for any $(x_1, y_1), (x_2, y_2) \in X \times X$. Then $(X \times X, \rho)$ is a complete metric space. Let $\varepsilon \in (0, 1/2)$ and define $g: X \times X \to \mathbb{R} \cup \{+\infty\}$ by $g(x, y) := f(x) - (1 - \varepsilon) d(x, y) + \iota_{\text{graph} \, F}(x, y)$. Then g is a lsc function bounded below (exercise). Applying the Ekeland variational principle of Theorem 2.1.1 to g we see that there exists $(x^*, y^*) \in \text{graph} \, F$ such that

$$g(x^*, y^*) \leq g(x, y) + \varepsilon \rho((x, y), (x^*, y^*)), \quad \text{for all } (x, y) \in X \times X.$$

So for all $(x, y) \in \text{graph} \, F$,

$$f(x^*) - (1 - \varepsilon) d(x^*, y^*)$$
$$\leq f(x) - (1 - \varepsilon) d(x, y) + \varepsilon(d(x, x^*) + d(y, y^*)). \qquad (2.3.8)$$

Suppose $z^* \in F(y^*)$. Letting $(x, y) = (y^*, z^*)$ in (2.3.8) we have

$$f(x^*) - (1 - \varepsilon)d(x^*, y^*) \le f(y^*) - (1 - \varepsilon)d(y^*, z^*) + \varepsilon(d(y^*, x^*) + d(z^*, y^*)).$$

It follows that

$$0 \le f(x^*) - f(y^*) - d(x^*, y^*) \le -(1 - 2\varepsilon)d(y^*, z^*),$$

so we must have $y^* = z^*$. That is to say y^* is a fixed point of F.   ●

We observe that it follows from the above proof that $F(y^*) = \{y^*\}$.

2.3.4 Commentary and Exercises

The variational proof of the Banach fixed point theorem appeared in [107]. While the variational argument provides an elegant confirmation of the existence of the fixed point it does not, however, provide an algorithm for finding such a fixed point as Banach's original proof does. For comparison, a proof using an interactive algorithm is outlined in the guided exercises below. Clarke's refinement is taken from [84]. Theorem 2.3.5 is due to Caristi and Kirk [160] and applications of this theorem can be found in [105]. A very nice general reference book for the metric fixed point theory is [127].

Exercise 2.3.1 Let X be a Banach space and let $x, y \in X$. Show that the segment between x and y defined in (2.3.1) has the following representation:

$$[x, y] = \{\lambda x + (1 - \lambda)y \mid \lambda \in [0, 1]\}.$$

Exercise 2.3.2 Prove the uniqueness of the fixed point in Theorem 2.3.1.

Exercise 2.3.3 Let $f \colon \mathbb{R}^N \to \mathbb{R}^N$ be a C^1 mapping. Show that f is a contraction if and only if $\sup\{\|f'(x)\| : x \in \mathbb{R}^N\} < 1$.

Exercise 2.3.4 Prove that Kepler's equation

$$x = a + b\sin(x), \ b \in (0, 1)$$

has a unique solution.

Exercise 2.3.5 (Iteration Method) Let (X, d) be a complete metric space and let $\phi \colon X \to X$ be a contraction. Define for an arbitrarily fixed $x_0 \in X$, $x_1 = \phi(x_0), \ldots, x_i = \phi(x_{i-1})$. Show that (x_i) is a Cauchy sequence and $x = \lim_{i \to \infty} x_i$ is a fixed point for ϕ.

Exercise 2.3.6 (Error Estimate) Let (X, d) be a complete metric space and let $\phi \colon X \to X$ be a contraction with contraction constant $k \in (0, 1)$. Establish the following error estimate for the iteration method in Exercise 2.3.5.

$$\|x_i - x\| \le \frac{k^i}{1 - k}\|x_1 - x_0\|.$$

Exercise 2.3.7 Deduce the Banach fixed point theorem from the Caristi–Kirk fixed point theorem. Hint: Define $f(x) = d(x, \phi(x))/(1 - k)$.

2.4 Variational Principles in Finite Dimensional Spaces

One drawback of the Ekeland variational principle is that the perturbation involved therein is intrinsically nonsmooth. This is largely overcome in the smooth variational principle due to Borwein and Preiss. We discuss a Euclidean space version in this section to illustrate the nature of this result. The general version will be discussed in the next section.

2.4.1 Smooth Variational Principles in Euclidean Spaces

Theorem 2.4.1 (Smooth Variational Principle in a Euclidean Space) *Let* $f\colon \mathbb{R}^N \to \mathbb{R} \cup \{+\infty\}$ *be a lsc function bounded from below, let* $\lambda > 0$ *and let* $p \geq 1$. *Suppose that* $\varepsilon > 0$ *and* $z \in X$ *satisfy*

$$f(z) \leq \inf_X f + \varepsilon.$$

Then, there exists $y \in X$ *such that*

(i) $\|z - y\| \leq \lambda$,
(ii) $f(y) + \frac{\varepsilon}{\lambda^p}\|y - z\|^p < f(z)$, *and*
(iii) $f(x) + \frac{\varepsilon}{\lambda^p}\|x - z\|^p \geq f(y) + \frac{\varepsilon}{\lambda^p}\|y - z\|^p$, *for all* $x \in X$.

Proof. Observing that the function $x \to f(x) + \frac{\varepsilon}{\lambda^p}\|x - z\|^p$ approaches $+\infty$ as $\|x\| \to \infty$, it must attain its minimum at some $y \in X$. It is an easy matter to check that y satisfies the conclusion of the theorem.   ●

This very explicit formulation which is illustrated in Figure 2.5 – for $f(x) = 1/x, z = 1, c = 1, \lambda = 1/2$, with $p = 3/2$ and $p = 2$ – can be mimicked in Hilbert space and many other classical reflexive Banach spaces [58]. It is interesting to compare this result with the Ekeland variational principle geometrically. The Ekeland variational principle says that one can support a lsc function f near its approximate minimum point by a cone with small slope while the Borwein–Preiss variational principle asserts that under stronger conditions this cone can be replaced by a parabolic function with a small derivative at the supporting point. We must caution the readers that although this picture is helpful in understanding the naturalness of the Borwein–Preiss variational principle it is not entirely accurate in the general case, as the support function is usually the sum of an infinite sequence of parabolic functions.

This result can also be stated in the form of an approximate Fermat principle in the Euclidean space $\mathbb{R}^N$.

Lemma 2.4.2 (Approximate Fermat Principle for Smooth Functions) *Let* $f\colon \mathbb{R}^N \to \mathbb{R}$ *be a smooth function bounded from below. Then there exists a sequence* $x_i \in \mathbb{R}^N$ *such that* $f(x_i) \to \inf_{\mathbb{R}^N} f$ *and* $f'(x_i) \to 0$.

Proof. Exercise 2.4.3.   ●

We delay the discussion of the general form of the Borwein–Preiss variational principle until the next section and digress to some applications.

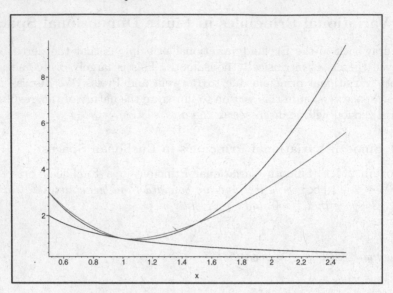

Fig. 2.5. Smooth attained perturbations of $1/x$

2.4.2 Gordan Alternatives

We start with an analytical proof of the Gordan alternative.

Theorem 2.4.3 (Gordan Alternative) *Let $a_1, \ldots, a_M \in \mathbb{R}^N$. Then, exactly one of the following systems has a solution:*

$$\sum_{m=1}^{M} \lambda_m a_m = 0, \sum_{m=1}^{M} \lambda_m = 1, 0 \leq \lambda_m, \; m = 1, \ldots, M, \qquad (2.4.1)$$

$$\langle a_m, x \rangle < 0 \; \text{for } m = 1, \ldots, M, \; x \in \mathbb{R}^N. \qquad (2.4.2)$$

Proof. We need only prove the following statements are equivalent:

(i)   The function

$$f(x) := \ln \Big(\sum_{m=1}^{M} \exp \langle a_m, x \rangle \Big)$$

   is bounded below.

(ii) System (2.4.1) is solvable.

(iii) System (2.4.2) is unsolvable.

The implications (ii)$\Rightarrow$ (iii) $\Rightarrow$ (i) are easy and left as exercises. It remains to show (i) $\Rightarrow$ (ii). Applying the approximate Fermat principle of Lemma 2.4.2 we deduce that there is a sequence (x_i) in $\mathbb{R}^N$ satisfying

$$\|f'(x_i)\| = \Big\| \sum_{m=1}^{M} \lambda_m^i a_m \Big\| \to 0, \qquad (2.4.3)$$

where the scalars

$$\lambda_m^i = \frac{\exp \langle a_m, x_i \rangle}{\sum_{l=0}^{M} \exp \langle a_l, x_i \rangle} > 0, \ m = 1, \dots, M$$

satisfy $\sum_{m=1}^{M} \lambda_m^i = 1$. Without loss of generality we may assume that $\lambda_m^i \to \lambda_m$, $m = 1, \dots, M$. Taking limits in (2.4.3) we see that λ_m, $m = 1, \dots, M$ is a set of solutions of (2.4.1).   ●

2.4.3 Majorization

For a vector $x = (x_1, \dots, x_N) \in \mathbb{R}^N$, we use $x^{\downarrow}$ to denote the vector derived from x by rearranging its components in nonincreasing order. For $x, y \in \mathbb{R}^N$, we say that x is *majorized* by y, denoted by $x \prec y$, provided that $\sum_{n=1}^{N} x_n = \sum_{n=1}^{N} y_n$ and $\sum_{n=1}^{k} x_n^{\downarrow} \leq \sum_{n=1}^{k} y_n^{\downarrow}$ for $k = 1, \dots, N$.

Example 2.4.4 Let $x \in \mathbb{R}^N$ be a vector with nonnegative components satisfying $\sum_{n=1}^{N} x_n = 1$. Then

$$(1/N, 1/N, \dots, 1/N) \prec x \prec (1, 0, \dots, 0).$$

The concept of majorization arises naturally in physics and economics. For example, if we use $x \in \mathbb{R}_+^N$ (the nonnegative orthant of $\mathbb{R}^N$) to represent the distribution of wealth within an economic system, then $x \prec y$ means the distribution represented by x is more even than that of y. Example 2.4.4 then describes the two extremal cases of wealth distribution.

Given a vector $y \in \mathbb{R}^N$ the level set of y with respect to the majorization defined by $l(y) := \{x \in \mathbb{R}^N \mid x \prec y\}$ is often of interest. It turns out that this level set is the convex hull of all the possible vectors derived from permuting the components of y. We will give a variational proof of this fact using a method similar to that of the variational proof of the Gordon alternatives. To do so we will need the following characterization of majorization.

Lemma 2.4.5 *Let $x, y \in \mathbb{R}^N$. Then $x \prec y$ if and only if, for any $z \in \mathbb{R}^N$,* $\langle z^{\downarrow}, x^{\downarrow} \rangle \leq \langle z^{\downarrow}, y^{\downarrow} \rangle$.

Proof. Using Abel's formula we can write

$$\langle z^{\downarrow}, y^{\downarrow} \rangle - \langle z^{\downarrow}, x^{\downarrow} \rangle = \langle z^{\downarrow}, y^{\downarrow} - x^{\downarrow} \rangle$$

$$= \sum_{k=1}^{N-1} \Big((z_k^{\downarrow} - z_{k+1}^{\downarrow}) \times \sum_{n=1}^{k} (y_n^{\downarrow} - x_n^{\downarrow}) \Big) + z_N^{\downarrow} \sum_{n=1}^{N} (y_n^{\downarrow} - x_n^{\downarrow}).$$

Now to see the necessity we observe that $x \prec y$ implies $\sum_{n=1}^{k}(y_n^\downarrow - x_n^\downarrow) \geq 0$ for $k = 1, \ldots, N-1$ and $\sum_{n=1}^{N}(y_n^\downarrow - x_n^\downarrow) = 0$. Thus, the last term in the right hand side of the previous equality is 0. Moreover, in the remaining sum each term is the product of two nonnegative factors, and therefore it is nonnegative. We now prove sufficiency. Suppose that, for any $z \in \mathbb{R}^N$,

$$0 \leq \langle z^\downarrow, y^\downarrow \rangle - \langle z^\downarrow, x^\downarrow \rangle = \sum_{k=1}^{N-1}\left((z_k^\downarrow - z_{k+1}^\downarrow) \times \sum_{n=1}^{k}(y_n^\downarrow - x_n^\downarrow)\right) + z_N^\downarrow \sum_{n=1}^{N}(y_n^\downarrow - x_n^\downarrow).$$

Setting $z = \sum_{n=1}^{k} e_n$ for $k = 1, \ldots, N-1$ (where $\{e_n : n = 1, \ldots, N\}$ is the standard basis of $\mathbb{R}^N$) we have $\sum_{n=1}^{k} y_n^\downarrow \geq \sum_{n=1}^{k} x_n^\downarrow$, and setting $z = \pm\sum_{n=1}^{N} e_n$ we have $\sum_{n=1}^{N} y_n = \sum_{n=1}^{N} x_n$.   ●

Let us denote by $P(N)$ the set of $N \times N$ permutation matrices (those matrices derived by permuting the rows or the columns of the identity matrix). Then we can state the characterization of the level set of a vector with respect to majorization as follows.

Theorem 2.4.6 (Representation of Level Sets of the Majorization) *Let $y \in \mathbb{R}^N$. Then*

$$l(y) = \text{conv}\{Py : P \in P(N)\}.$$

Proof. It is not hard to check that $l(y)$ is convex and, for any $P \in P(N)$, $Py \in l(y)$. Thus, $\text{conv}\{Py : P \in P(N)\} \subset l(y)$ (Exercise 2.4.8).

We now prove the reversed inclusion. For any $x \prec y$, by Lemma 2.4.5 there exists $P = P(z) \in P(N)$ satisfies

$$\langle z, Py \rangle = \langle z^\downarrow, y^\downarrow \rangle \geq \langle z^\downarrow, x^\downarrow \rangle \geq \langle z, x \rangle. \qquad (2.4.4)$$

Observe that $P(N)$ is a finite set (with $N!$ elements to be precise). Thus, the function

$$f(z) := \ln\left(\sum_{P \in P(N)} \exp\langle z, Py - x \rangle\right).$$

is defined for all $z \in \mathbb{R}^N$, is differentiable, and is bounded from below by 0. By the approximate Fermat principle of Lemma 2.4.2 we can select a sequence (z_i) in $\mathbb{R}^N$ such that

$$0 = \lim_{i \to \infty} f'(z_i) = \sum_{P \in P(N)} \lambda_P^i (Py - x). \qquad (2.4.5)$$

where

$$\lambda_P^i = \frac{\exp\langle z_i, Py - x \rangle}{\sum_{P \in P(N)} \exp\langle z_i, Py - x \rangle}.$$

Clearly, $\lambda_P^i > 0$ and $\sum_{P \in P(N)} \lambda_P^i = 1$. Thus, taking a subsequence if necessary we may assume that, for each $P \in P(N)$, $\lim_{i \to \infty} \lambda_P^i = \lambda_P \geq 0$ and $\sum_{P \in P(N)} \lambda_P = 1$. Now taking limits as $i \to \infty$ in (2.4.5) we have

$$\sum_{P\in P(N)} \lambda_P(Py - x) = 0.$$

Thus, $x = \sum_{P\in P(N)} \lambda_P Py$, as was to be shown.   ●

2.4.4 Doubly Stochastic Matrices

We use $E(N)$ to denote the Euclidean space of all real N by N square matrices with inner product

$$\langle A, B \rangle = \text{tr}(B^{\mathsf T}A) = \sum_{n,m=1}^{N} a_{nm}b_{nm}, \quad A, B \in E(N).$$

A matrix $A = (a_{nm}) \in E(N)$ is *doubly stochastic* provided that the entries of A are all nonnegative, $\sum_{n=1}^{N} a_{nm} = 1$ for $m = 1, \ldots, N$ and $\sum_{m=1}^{N} a_{nm} = 1$ for $n = 1, \ldots, N$. Clearly every $P \in P(N)$ is doubly stochastic and they provide the simplest examples of doubly stochastic matrices. Birkhoff's theorem asserts that any doubly stochastic matrix can be represented as a convex combination of permutation matrices. We now apply the method in the previous section to give a variational proof of Birkhoff's theorem.

For $A = (a_{nm}) \in E(N)$, we denote $r_n(A) = \{m \mid a_{nm} \neq 0\}$, the set of indices of columns containing nonzero elements of the nth row of A and we use $\#(S)$ to signal the number of elements in set S. Then a doubly stochastic matrix has the following interesting property.

Lemma 2.4.7 *Let $A \in E(N)$ be a doubly stochastic matrix. Then, for any $1 \leq n_1 < n_2 < \cdots < n_K \leq N$,*

$$\#\left(\bigcup_{k=1}^{K} r_{n_k}(A)\right) \geq K. \tag{2.4.6}$$

Proof. We prove by contradiction. Suppose (2.4.6) is violated for some K. Permuting the rows of A if necessary we may assume that

$$\#\left(\bigcup_{k=1}^{K} r_k(A)\right) < K. \tag{2.4.7}$$

Rearranging the order of the columns of A if needed we may assume

$$A = \begin{pmatrix} O & B \\ C & D \end{pmatrix},$$

where O is a K by L submatrix of A with all entries equal to 0. By (2.4.7) we have $L > N - K$. On the other hand, since A is doubly stochastic, every column of C and every row of B add up to 1. That leads to $L + K \leq N$, a contradiction.   •

Condition (2.4.6) actually ensures a matrix has a diagonal with all elements nonzero which is made precise in the next lemma.

Lemma 2.4.8 *Let $A \in E(N)$. Suppose that A satisfies condition (2.4.6). Then for some $P \in P(N)$, the entries in A corresponding to the 1's in P are all nonzero. In particular, any doubly stochastic matrix has the above property.*

Proof. We use induction on N. The lemma holds trivially when $N = 1$. Now suppose that the lemma holds for any integer less than N. We prove it is true for N. First suppose that, for any $1 \leq n_1 < n_2 < \cdots < n_K \leq N$, $K < N$

$$\#\left(\bigcup_{k=1}^{K} r_{n_k}(A)\right) \geq K + 1. \tag{2.4.8}$$

Then pick a nonzero element of A, say a_{NN} and consider the submatrix A' of A derived by eliminating the Nth row and Nth column of A. Then A' satisfies condition (2.4.6), and therefore there exists $P' \in P(N-1)$ such that the entries in A' corresponding to the 1's in P' are all nonzero. It remains to define $P \in P(N)$ as

$$P = \begin{pmatrix} P' & 0 \\ 0 & 1 \end{pmatrix}.$$

Now consider the case when (2.4.8) fails so that there exist $1 \leq n_1 < n_2 < \cdots < n_K \leq N$, $K < N$ satisfying

$$\#\left(\bigcup_{k=1}^{K} r_{n_k}(A)\right) = K. \tag{2.4.9}$$

By rearranging the rows and columns of A we may assume that $n_k = k, k = 1, \dots, K$ and $\bigcup_{k=1}^{K} r_k(A) = \{1, \dots, K\}$. Then

$$A = \begin{pmatrix} B & O \\ C & D \end{pmatrix},$$

where $B \in E(K)$, $D \in E(N-K)$ and O is a K by $N - K$ submatrix with all entries equal to 0. Observe that for any $1 \leq n_1 < \cdots < n_L \leq K$,

$$\bigcup_{l=1}^{L} r_{n_l}(B) = \bigcup_{l=1}^{L} r_{n_l}(A).$$

Thus,

$$\#\left(\bigcup_{l=1}^{L} r_{n_l}(B)\right) \geq L,$$

and therefore B satisfies condition (2.4.6). On the other hand for any $K+1 \leq n_1 < \cdots < n_L \leq N$,

$$\left[\bigcup_{k=1}^{K} r_k(A)\right] \cup \left[\bigcup_{l=1}^{L} r_{n_l}(A)\right] = \{1,\ldots,K\} \cup \left[\bigcup_{l=1}^{L} r_{n_l}(D)\right].$$

Thus, D also satisfies condition (2.4.6). By the induction hypothesis we have $P_1 \in P(K)$ and $P_2 \in P(N-K)$ such that the elements in B and D corresponding to the 1's in P_1 and P_2, respectively, are all nonzero. It follows that

$$P = \begin{pmatrix} P_1 & O \\ O & P_2 \end{pmatrix} \in P(N),$$

and the elements in A corresponding to the 1's in P are all nonzero.   ●

We now establish the following analogue of (2.4.4).

Lemma 2.4.9 *Let $A \in E(N)$ be a doubly stochastic matrix. Then for any $B \in E(N)$ there exists $P \in P(N)$ such that*

$$\langle B, A - P \rangle \geq 0.$$

Proof. We use an induction argument on the number of nonzero elements of A. Since every row and column of A sums to 1, A has at least N nonzero elements. If A has exactly N nonzero elements then they must all be 1, so that A itself is a permutation matrix and the lemma holds trivially. Suppose now that A has more than N nonzero elements. By Lemma 2.4.8 there exists $P \in P(N)$ such that the entries in A corresponding to the 1's in P are all nonzero. Let $t \in (0,1)$ be the minimum of these N positive elements. Then we can verify that $A_1 = (A - tP)/(1 - t)$ is a doubly stochastic matrix and has at least one fewer nonzero elements than A. Thus, by the induction hypothesis there exists $Q \in P(N)$ such that

$$\langle B, A_1 - Q \rangle \geq 0.$$

Multiplying the above inequality by $1-t$ we have $\langle B, A - tP - (1-t)Q \rangle \geq 0$, and therefore at least one of $\langle B, A - P \rangle$ or $\langle B, A - Q \rangle$ is nonnegative.   ●

Now we are ready to present a variational proof for the Birkhoff theorem.

Theorem 2.4.10 (Birkhoff) *Let $\mathcal{A}(N)$ be the set of all $N \times N$ doubly stochastic matrices. Then*

$$\mathcal{A}(N) = \text{conv}\{P \mid P \in P(N)\}.$$

Proof. It is an easy matter to verify that $\mathcal{A}(N)$ is convex and $P(N) \subset \mathcal{A}(N)$. Thus, conv $P(N) \subset \mathcal{A}(N)$.

To prove the reversed inclusion, define a function f on $E(N)$ by

$$f(B) := \ln\left(\sum_{P \in P(N)} \exp\langle B, A - P \rangle \right).$$

Then f is defined for all $B \in E(N)$, is differentiable and is bounded from below by 0. By the approximate Fermat principle of Theorem 2.4.2 we can select a sequence (B_i) in $E(N)$ such that

$$0 = \lim_{i \to \infty} f'(B_i) = \lim_{i \to \infty} \sum_{P \in P(N)} \lambda_P^i (A - P). \qquad (2.4.10)$$

where

$$\lambda_P^i = \frac{\exp\langle B_i, A - P \rangle}{\sum_{P \in P(N)} \exp\langle B_i, A - P \rangle}.$$

Clearly, $\lambda_P^i > 0$ and $\sum_{P \in P(N)} \lambda_P^i = 1$. Thus, taking a subsequence if necessary we may assume that for each $P \in P(N)$, $\lim_{i \to \infty} \lambda_P^i = \lambda_P \geq 0$ and $\sum_{P \in P(N)} \lambda_P = 1$. Now taking limits as $i \to \infty$ in (2.4.10) we have

$$\sum_{P \in P(N)} \lambda_P (A - P) = 0.$$

It follows that $A = \sum_{P \in P(N)} \lambda_P P$, as was to be shown.   ●

Majorization and doubly stochastic matrices are closely related. Their relationship is described in the next theorem.

Theorem 2.4.11 (Doubly Stochastic Matrices and Majorization) *A nonnegative matrix A is doubly stochastic if and only if $Ax \prec x$ for any vector $x \in \mathbb{R}^N$.*

Proof. We use $e_n, n = 1, \dots, N$, to denote the standard basis of $\mathbb{R}^N$.

Let $Ax \prec x$ for all $x \in \mathbb{R}^N$. Choosing x to be $e_n, n = 1, \dots, N$ we can deduce that the sum of elements of each column of A is 1. Next let $x = \sum_{n=1}^N e_n$; we can conclude that the sum of elements of each row of A is 1. Thus, A is doubly stochastic.

Conversely, let A be doubly stochastic and let $y = Ax$. To prove $y \prec x$ we may assume, without loss of generality, that the coordinates of both x and y are in nonincreasing order. Now note that for any k, $1 \leq k \leq N$, we have

$$\sum_{m=1}^k y_m = \sum_{m=1}^k \sum_{n=1}^N a_{mn} x_n.$$

If we put $t_n = \sum_{m=1}^k a_{mn}$, then $t_n \in [0, 1]$ and $\sum_{n=1}^N t_n = k$. We have

$$\sum_{m=1}^{k} y_m - \sum_{m=1}^{k} x_m = \sum_{n=1}^{N} t_n x_n - \sum_{m=1}^{k} x_m$$

$$= \sum_{n=1}^{N} t_n x_n - \sum_{m=1}^{k} x_m + (k - \sum_{n=1}^{N} t_n) x_k$$

$$= \sum_{n=1}^{k} (t_n - 1)(x_n - x_k) + \sum_{n=k+1}^{N} t_n (x_n - x_k)$$

$$\leq 0.$$

Further, when $k = N$ we must have equality here simply because A is doubly stochastic. Thus, $y \prec x$. $\bullet$

Combining Theorems 2.4.6, 2.4.11 and 2.4.10 we have

Corollary 2.4.12 *Let $y \in \mathbb{R}^N$. Then $l(y) = \{Ay \mid A \in \mathcal{A}(N)\}$.*

2.4.5 Commentary and Exercises

Theorem 2.4.1 is a finite dimensional form of the Borwein–Preiss variational principle [58]. The approximate Fermat principle of Lemma 2.4.2 was suggested by [137]. The variational proof of Gordan's alternative is taken from [56] which can also be used in other related problems (Exercises 2.4.4 and 2.4.5).

Geometrically, Gordan's alternative [129] is clearly a consequence of the separation theorem: it says either 0 is contained in the convex hull of $a_0, \ldots, a_M$ or it can be strictly separated from this convex hull. Thus, the proof of Theorem 2.4.3 shows that with an appropriate auxiliary function variational method can be used in the place of a separation theorem – a fundamental result in analysis.

Majorization and doubly stochastic matrices are import concepts in matrix theory with many applications in physics and economics. Ando [3], Bhatia [22] and Horn and Johnson [138, 139] are excellent sources for the background and preliminaries for these concepts and related topics. Birkhoff's theorem appeared in [23]. Lemma 2.4.8 is a matrix form of Hall's matching condition [134]. Lemma 2.4.7 was established in König [163]. The variational proofs for the representation of the level sets with respect to the majorization and Birkhoff's theorem given here follow [279].

Exercise 2.4.1 Supply the details for the proof of Theorem 2.4.1.

Exercise 2.4.2 Prove the implications (ii) $\Rightarrow$ (iii) $\Rightarrow$ (i) in the proof of the Gordan Alternative of Theorem 2.4.3.

Exercise 2.4.3 Prove Lemma 2.4.2.

*Exercise 2.4.4 (Ville's Theorem) Let $a_1, \ldots, a_M \in \mathbb{R}^N$ and define $f \colon \mathbb{R}^N \to \mathbb{R}$ by

$$f(x) := \ln \Big(\sum_{m=1}^{M} \exp \langle a_m, x \rangle \Big).$$

Consider the optimization problem

$$\inf\{f(x) \mid x \geq 0\} \tag{2.4.11}$$

and its relationship with the two systems

$$\sum_{m=1}^{M} \lambda_m a_m = 0, \sum_{m=1}^{M} \lambda_m = 1, 0 \leq \lambda_m, \ m = 1, \ldots, M, \tag{2.4.12}$$

$$\langle a_m, x \rangle < 0 \text{ for } m = 1, \ldots, M, \ x \in \mathbb{R}^N_+. \tag{2.4.13}$$

Imitate the proof of Gordan's alternatives to prove the following are equivalent:

(i)   Problem (2.4.11) is bounded below.
(ii) System (2.4.12) is solvable.
(iii) System (2.4.13) is unsolvable.

Generalize by considering the problem $\inf\{f(x) \mid x_m \geq 0, m \in K\}$, where K is a subset of $\{1, \ldots, M\}$.

*Exercise 2.4.5 (Stiemke's Theorem) Let $a_1, \ldots, a_M \in \mathbb{R}^N$ and define $f \colon \mathbb{R}^N \to \mathbb{R}$ by

$$f(x) := \ln \Big(\sum_{m=1}^{M} \exp \langle a_m, x \rangle \Big).$$

Consider the optimization problem

$$\inf\{f(x) \mid x \in \mathbb{R}^N\} \tag{2.4.14}$$

and its relationship with the two systems

$$\sum_{m=1}^{M} \lambda_m a_m = 0, 0 < \lambda_m, \ m = 1, \ldots, M, \tag{2.4.15}$$

and

$$\langle a_m, x \rangle \leq 0 \text{ for } m = 1, \ldots, M, \text{ not all } 0, \ x \in \mathbb{R}^N. \tag{2.4.16}$$

Prove the following are equivalent:

(i)   Problem (2.4.14) has an optimal solution.

(ii) System (2.4.15) is solvable.

(iii) System (2.4.16) is unsolvable.

Hint: To prove (iii) implies (i), show that if problem (2.4.14) has no optimal solution then neither does the problem

$$\inf\left\{\sum_{m=1}^{M} \exp y_m \mid y \in K\right\}, \tag{2.4.17}$$

where K is the subspace $\{(\langle a^1, x\rangle, \ldots, \langle a^M, x\rangle) \mid x \in \mathbb{R}^N\} \subset \mathbb{R}^M$. Hence, by considering a minimizing sequence for (2.4.17), deduce system (2.4.16) is solvable.

*Exercise 2.4.6 Prove the following

Lemma 2.4.13 (Farkas Lemma) *Let $a_1, \ldots, a_M$ and let $b \neq 0$ in $\mathbb{R}^N$. Then exactly one of the following systems has a solution:*

$$\sum_{m=1}^{M} \lambda_m a_m = b, \; 0 \leq \lambda_m, \; m = 1, \ldots, M, \tag{2.4.18}$$

$$\langle a_m, x\rangle \leq 0 \text{ for } m = 1, \ldots, M, \; \langle b, x\rangle > 0, \; x \in \mathbb{R}^N \tag{2.4.19}$$

Hint: Use the Gordan alternatives and induction.

Exercise 2.4.7 Verify Example 2.4.4.

Exercise 2.4.8 Let $y \in \mathbb{R}^N$. Verify that $l(y)$ is a convex set and, for any $P \in P(N)$, $Py \in l(y)$.

Exercise 2.4.9 Give an alternative proof of Birkhoff's theorem by going through the following steps.

(i)   Prove $P(N) = \{(a_{mn}) \in \mathcal{A}(N) \mid a_{mn} = 0 \text{ or } 1 \text{ for all } m, n\}$.

(ii) Prove $P(N) \subset \text{ext}(\mathcal{A}(N))$, where $\text{ext}(S)$ signifies *extreme points* of set S.

(iii) Suppose $(a_{mn}) \in \mathcal{A}(N)\backslash P(N)$. Prove there exist sequences of distinct indices $m_1, m_2, \ldots, m_k$ and $n_1, n_2, \ldots, n_k$ such that

$$0 < a_{m_r n_r}, a_{m_{r+1} n_r} < 1 (r = 1, \ldots, k)$$

(where $m_{k+1} = m_1$). For these sequences, show the matrix (a'_{mn}) defined by

$$a'_{mn} - a_{mn} = \begin{cases} \varepsilon & \text{if } (m, n) = (m_r, n_r) \text{ for some } r, \\ -\varepsilon & \text{if } (m, n) = (m_{r+1}, n_r) \text{ for some } r, \\ 0 & \text{otherwise,} \end{cases}$$

is doubly stochastic for all small real ε. Deduce $(a_{mn}) \notin \text{ext}(\mathcal{A}(N))$.

(iv) Deduce $\text{ext}(\mathcal{A}(N)) = P(N)$. Hence prove Birkhoff's theorem.

(v) Use Carathéodory's theorem [77] to bound the number of permutation matrices needed to represent a doubly stochastic matrix in Birkhoff's theorem.

2.5 Borwein–Preiss Variational Principles

Now we turn to a general form of the Borwein–Preiss smooth variational principle and a variation thereof derived by Deville, Godefroy and Zizler with a category proof.

2.5.1 The Borwein–Preiss Principle

Definition 2.5.1 *Let (X, d) be a metric space. We say that a continuous function $\rho\colon X \times X \to [0, \infty]$ is a* gauge-type function *on a complete metric space (X, d) provided that*

(i) $\rho(x, x) = 0$, *for all $x \in X$,*
(ii) *for any $\varepsilon > 0$ there exists $\delta > 0$ such that for all $y, z \in X$ we have $\rho(y, z) \leq \delta$ implies that $d(y, z) < \varepsilon$.*

Theorem 2.5.2 (Borwein–Preiss Variational Principle) *Let (X, d) be a complete metric space and let $f\colon X \to \mathbb{R} \cup \{+\infty\}$ be a lsc function bounded from below. Suppose that ρ is a gauge-type function and $(\delta_i)_{i=0}^{\infty}$ is a sequence of positive numbers, and suppose that $\varepsilon > 0$ and $z \in X$ satisfy*

$$f(z) \leq \inf_X f + \varepsilon.$$

Then there exist y and a sequence $\{x_i\} \subset X$ such that

(i) $\rho(z, y) \leq \varepsilon/\delta_0$, $\rho(x_i, y) \leq \varepsilon/(2^i \delta_0)$,
(ii) $f(y) + \sum_{i=0}^{\infty} \delta_i \rho(y, x_i) \leq f(z)$, *and*
(iii) $f(x) + \sum_{i=0}^{\infty} \delta_i \rho(x, x_i) > f(y) + \sum_{i=0}^{\infty} \delta_i \rho(y, x_i)$, *for all $x \in X \backslash \{y\}$.*

Proof. Define sequences (x_i) and (S_i) inductively starting with $x_0 := z$ and

$$S_0 := \{x \in X \mid f(x) + \delta_0 \rho(x, x_0) \leq f(x_0)\}. \tag{2.5.1}$$

Since $x_0 \in S_0$, S_0 is nonempty. Moreover it is closed because both f and $\rho(\cdot, x_0)$ are lsc functions. We also have that, for all $x \in S_0$,

$$\delta_0 \rho(x, x_0) \leq f(x_0) - f(x) \leq f(z) - \inf_X f \leq \varepsilon. \tag{2.5.2}$$

Take $x_1 \in S_0$ such that

$$f(x_1) + \delta_0 \rho(x_1, x_0) \leq \inf_{x \in S_0} [f(x) + \delta_0 \rho(x, x_0)] + \frac{\delta_1 \varepsilon}{2\delta_0}. \tag{2.5.3}$$

and define similarly

$$S_1 := \left\{ x \in S_0 \,\middle|\, f(x) + \sum_{k=0}^{1} \delta_k \rho(x, x_k) \leq f(x_1) + \delta_0 \rho(x_1, x_0) \right\}. \tag{2.5.4}$$

In general, suppose that we have defined x_j, S_j for $j = 0, 1, \ldots, i-1$ satisfying

$$f(x_j) + \sum_{k=0}^{j-1} \delta_k \rho(x_j, x_k) \le \inf_{x \in S_{j-1}} \left[f(x) + \sum_{k=0}^{j-1} \delta_k \rho(x, x_k) \right] + \frac{\varepsilon \delta_j}{2^j \delta_0} \quad (2.5.5)$$

and

$$S_j := \left\{ x \in S_{j-1} \;\middle|\; f(x) + \sum_{k=0}^{j} \delta_k \rho(x, x_k) \le f(x_j) + \sum_{k=0}^{j-1} \delta_k \rho(x_j, x_k) \right\}. (2.5.6)$$

We choose $x_i \in S_{i-1}$ such that

$$f(x_i) + \sum_{k=0}^{i-1} \delta_k \rho(x_i, x_k) \le \inf_{x \in S_{i-1}} \left[f(x) + \sum_{k=0}^{i-1} \delta_k \rho(x, x_k) \right] + \frac{\varepsilon \delta_i}{2^i \delta_0} \quad (2.5.7)$$

and we define

$$S_i := \left\{ x \in S_{i-1} \;\middle|\; f(x) + \sum_{k=0}^{i} \delta_k \rho(x, x_k) \le f(x_i) + \sum_{k=0}^{i-1} \delta_k \rho(x_i, x_k) \right\}. (2.5.8)$$

We can see that for every $i = 1, 2, \ldots,$ S_i is a closed and nonempty set. It follows from (2.5.7) and (2.5.8) that, for all $x \in S_i$,

$$\delta_i \rho(x, x_i) \le \left[f(x_i) + \sum_{k=0}^{i-1} \delta_k \rho(x_i, x_k) \right] - \left[f(x) + \sum_{k=0}^{i-1} \delta_k \rho(x, x_k) \right]$$

$$\le \left[f(x_i) + \sum_{k=0}^{i-1} \delta_k \rho(x_i, x_k) \right] - \inf_{x \in S_{i-1}} \left[f(x) + \sum_{k=0}^{i-1} \delta_k \rho(x, x_k) \right]$$

$$\le \frac{\varepsilon \delta_i}{2^i \delta_0},$$

which implies that

$$\rho(x, x_i) \le \frac{\varepsilon}{2^i \delta_0}, \quad \text{for all } x \in S_i. \quad (2.5.9)$$

Since ρ is a gauge-type function, inequality (2.5.9) implies that $d(x, x_i) \to 0$ uniformly, and therefore $\operatorname{diam}(S_i) \to 0$. Since X is complete, by Cantor's intersection theorem there exists a unique $y \in \bigcap_{i=0}^{\infty} S_i$, which satisfies (i) by (2.5.2) and (2.5.9). Obviously, we have $x_i \to y$. For any $x \ne y$, we have that $x \notin \bigcap_{i=0}^{\infty} S_i$, and therefore for some j,

$$f(x) + \sum_{k=0}^{\infty} \delta_k \rho(x, x_k) \ge f(x) + \sum_{k=0}^{j} \delta_k \rho(x, x_k)$$

$$> f(x_j) + \sum_{k=0}^{j-1} \delta_k \rho(x_j, x_k). \quad (2.5.10)$$

On the other hand, it follows from (2.5.1), (2.5.8) and $y \in \bigcap_{i=0}^{\infty} S_i$ that, for any $q \geq j$,

$$
\begin{aligned}
f(x_0) &\geq f(x_j) + \sum_{k=0}^{j-1} \delta_k \rho(x_j, x_k) \\
&\geq f(x_q) + \sum_{k=0}^{q-1} \delta_k \rho(x_q, x_k) \\
&\geq f(y) + \sum_{k=0}^{q} \delta_k \rho(y, x_k).
\end{aligned} \tag{2.5.11}
$$

Taking limits in (2.5.11) as $q \to \infty$ we have

$$
\begin{aligned}
f(z) = f(x_0) &\geq f(x_j) + \sum_{k=0}^{j-1} \delta_k \rho(x_j, x_k) \\
&\geq f(y) + \sum_{k=0}^{\infty} \delta_k \rho(y, x_k),
\end{aligned} \tag{2.5.12}
$$

which verifies (ii). Combining (2.5.10) and (2.5.12) yields (iii). ●

We shall frequently use the following normed space form of the Borwein–Preiss variational principle, especially in spaces with a Fréchet smooth renorm, in which case we may deduce first-order (sub)differential information from the conclusion.

Theorem 2.5.3 *Let X be a Banach space with norm $\|\cdot\|$ and let $f: X \to \mathbb{R} \cup \{+\infty\}$ be a lsc function bounded from below, let $\lambda > 0$ and let $p \geq 1$. Suppose that $\varepsilon > 0$ and $z \in X$ satisfy*

$$
f(z) < \inf_X f + \varepsilon.
$$

Then there exist y and a sequence (x_i) in X with $x_1 = z$ and a function $\varphi_p: X \to \mathbb{R}$ of the form

$$
\varphi_p(x) := \sum_{i=1}^{\infty} \mu_i \|x - x_i\|^p,
$$

where $\mu_i > 0$ for all $i = 1, 2, \dots$ and $\sum_{i=1}^{\infty} \mu_i = 1$ such that

(i) $\|x_i - y\| \leq \lambda, n = 1, 2, \dots,$
(ii) $f(y) + (\varepsilon/\lambda^p)\varphi_p(y) \leq f(z),$ *and*
(iii) $f(x) + (\varepsilon/\lambda^p)\varphi_p(x) > f(y) + (\varepsilon/\lambda^p)\varphi_p(y),$ *for all $x \in X \setminus \{y\}$.*

Proof. Exercise 2.5.1. ●

Note that when $\|\cdot\|$ is Fréchet smooth so is φ_p for $p > 1$.

2.5.2 The Deville–Godefroy–Zizler Principle

An important counterpart of the Borwein–Preiss variational principle subsequently found by Deville, Godefroy and Zizler [98] is given below. It is interesting to see how the Baire category theorem is used in the proof. Recall that the Baire category theorem states that in a complete metric space every countable intersection of dense open sets is dense: a set containing such a dense G_δ set is called *generic* or *residual* and the complement of such a set is *meager*. We say a function $f\colon X \to \mathbb{R} \cup \{+\infty\}$ attains a *strong minimum* at $x \in X$ if $f(x) = \inf_X f$ and $\|x_i - x\| \to 0$ whenever $x_i \in X$ and $f(x_i) \to f(x)$. If f is bounded on X, we define $\|f\|_\infty := \sup\{|f(x)| \mid x \in X\}$. We say that $\phi\colon X \to \mathbb{R}$ is a *bump function* if ϕ is bounded and has bounded nonempty *support* $\operatorname{supp}(\phi) := \{x \in X \mid \phi(x) \neq 0\}$.

Theorem 2.5.4 (The Deville–Godefroy–Zizler Variational Principle) *Let X be a Banach space and Y a Banach space of continuous bounded functions g on X such that*

(i)   *$\|g\|_\infty \leq \|g\|_Y$ for all $g \in Y$.*
(ii) *For each $g \in Y$ and $z \in X$, the function $x \to g_z(x) = g(x+z)$ is in Y and $\|g_z\|_Y = \|g\|_Y$.*
(iii) *For each $g \in Y$ and $a \in \mathbb{R}$, the function $x \to g(ax)$ is in Y.*
(iv) *There exists a bump function in Y.*

If $f\colon X \to \mathbb{R} \cup \{+\infty\}$ is a proper lsc function and bounded below, then the set G of all $g \in Y$ such that $f + g$ attains a strong minimum on X is residual (in fact a dense G_δ set).

Proof.   Given $g \in Y$, define $S(g;a) := \{x \in X \mid g(x) \leq \inf_X g + a\}$ and $U_i := \{g \in Y \mid \operatorname{diam} S(f+g;a) < 1/i, \text{ for some } a > 0\}$. We show that each of the sets U_i is dense and open in Y and that their intersection is the desired set G.

To see that U_i is open, suppose that $g \in U_i$ with a corresponding $a > 0$. Then, for any $h \in Y$ such that $\|g - h\|_Y < a/3$, we have $\|g - h\|_\infty < a/3$. Now, for any $x \in S(f + h; a/3)$,

$$(f+h)(x) \leq \inf_X (f+h) + \frac{a}{3}.$$

It is an easy matter to estimate

$$(f+g)(x) \leq (f+h)(x) + \|g-h\|_\infty \leq \inf_X (f+h) + \frac{a}{3} + \|g-h\|_\infty$$

$$\leq \inf_X (f+g) + \frac{a}{3} + 2\|g-h\|_\infty \leq \inf_X (f+g) + a.$$

This shows that $S(f + h; a/3) \subset S(f + g; a)$. Thus, $h \in U_i$.

To see that each U_i is dense in Y, suppose that $g \in Y$ and $\varepsilon > 0$; it suffices to produce $h \in Y$ such that $\|h\|_Y < \varepsilon$ and for some $a > 0$ $\operatorname{diam} S(f + g +$

$h; a) < 1/i$. By hypothesis (iv), Y contains a bump function ϕ. Without loss of generality we may assume that $\|\phi\|_Y < \varepsilon$. By hypothesis (ii) we can assume that $\phi(0) \neq 0$, and therefore that $\phi(0) > 0$. Moreover, by hypothesis (iii) we can assume that $\operatorname{supp}(\phi) \subset B(0, 1/2i)$. Let $a = \phi(0)/2$ and choose $\bar{x} \in X$ such that

$$(f + g)(\bar{x}) < \inf_X(f + g) + \phi(0)/2.$$

Define h by $h(x) := -\phi(x - \bar{x})$; by hypothesis (ii), $h \in Y$ and $\|h\|_Y = \|\phi\|_Y < \varepsilon$ and $h(\bar{x}) = -\phi(0)$. To show that $\operatorname{diam} S(f + g + h; a) < 1/i$, it suffices to show that this set is contained in the ball $B(\bar{x}, 1/2i)$; that is, if $\|x - \bar{x}\| > 1/2i$, then $x \notin S(f + g + h; a)$, the latter being equivalent to

$$(f + g + h)(x) > \inf_X(f + g + h) + a.$$

Now, $\operatorname{supp}(h) \subset B(\bar{x}, 1/2i)$, so $h(x) = 0$ if $\|x - \bar{x}\| > 1/2i$ hence

$$(f + g + h)(x) = (f + g)(x) \geq \inf_X(f + g) > (f + g)(\bar{x}) - a$$
$$= (f + g + h)(\bar{x}) + \phi(0) - \phi(0)/2 \geq \inf_X(f + g + h) + a.$$

as was to be shown.

Finally we show $\bigcap_{i=1}^{\infty} U_i = G$. The easy part of $G \subset \bigcap_{i=1}^{\infty} U_i$ is left as an exercise. Let $g \in \bigcap_{i=1}^{\infty} U_i$. We will show that $g \in G$; that is, $f + g$ attains a strong minimum on X. First, for all i there exists $a_i > 0$ such that $\operatorname{diam} S(f + g; a_i) < 1/i$ and hence there exists a unique point $\bar{x} \in \bigcap_{i=1}^{\infty} S(f + g; a_i)$. Suppose that $x_k \in X$ and that $(f + g)(x_k) \to \inf_X(f + g)$. Given $i > 0$ there exists i_0 such that $(f + g)(x_k) \leq \inf_X(f + g) + a_i$ for all $i \geq i_0$, therefore $x_k \in S(f + g; a_i)$ for all $i \geq i_0$ and hence $\|x_k - \bar{x}\| \leq \operatorname{diam} S(f + g; a_i) < 1/i$ if $k \geq i_0$. Thus, $x_k \to \bar{x}$, and therefore $g \in G$. $\qquad\bullet$

2.5.3 Commentary and Exercises

The Borwein–Preiss smooth variational principle appeared in [58]. The proof here is adapted from Li and Shi [182]. Their original proof leads to a clean generalization of both the Ekeland and Borwein–Preiss variational principle (see Exercises 2.5.2 and 2.5.3). The Deville–Godefroy–Zizler variational principle and its category proof is from [98]. Another very useful variational principle due to Stegall, is given in Section 6.3.

Exercise 2.5.1 Deduce Theorem 2.5.3 from Theorem 2.5.2. Hint: Set $\rho(x, y) = \|x - y\|^p$.

Exercise 2.5.2 Check that, with $\delta_0 := 1$, $\delta_i := 0, i = 1, 2, \ldots$ and $\rho := \varepsilon d$, the procedure in the proof of Theorem 2.5.2 reduces to a proof of the Ekeland variational principle.

If one works harder, the two variational principles can be unified.

*Exercise 2.5.3 Adapt the proof of Theorem 2.5.2 for a nonnegative sequence $(\delta_i)_{i=0}^{\infty}$, $\delta_0 > 0$ to derive the following generalization for both the Ekeland and the Borwein–Preiss variational principles.

Theorem 2.5.5 *Let (X, d) be a complete metric space and let $f: X \to \mathbb{R} \cup \{+\infty\}$ be a lsc function bounded from below. Suppose that ρ is a gauge-type function and $(\delta_i)_{i=0}^{\infty}$ is a sequence of nonnegative numbers with $\delta_0 > 0$. Then, for every $\varepsilon > 0$ and $z \in X$ satisfying*

$$f(z) \leq \inf_X f + \varepsilon,$$

there exists a sequence $\{x_i\} \subset X$ converging to some $y \in X$ such that

(i) $\rho(z, y) \leq \varepsilon / \delta_0$,
(ii) $f(y) + \sum_{i=0}^{\infty} \delta_i \rho(y, x_i) \leq f(z)$, *and*
(iii) $f(x) + \sum_{i=0}^{\infty} \delta_i \rho(x, x_i) > f(y) + \sum_{i=0}^{\infty} \delta_i \rho(y, x_i)$, *for all $x \in X \setminus \{y\}$.*

Moreover, if $\delta_k > 0$ and $\delta_l = 0$ for all $l > k \geq 0$, then (iii) may be replaced by

(iii') *for all $x \in X \setminus \{y\}$, there exists $j \geq k$ such that*

$$f(x) + \sum_{i=0}^{k-1} \delta_i \rho(x, x_i) + \delta_k \rho(x, x_j) > f(y) + \sum_{i=0}^{k-1} \delta_l \rho(y, x_i) + \delta_k \rho(y, x_j).$$

The Ekeland variational principle, the Borwein–Preiss variational principle and the Deville–Godefroy–Zizler variational principle are related in the following exercises.

Exercise 2.5.4 Deduce the following version of Ekeland's variational principle from Theorem 2.5.4.

Theorem 2.5.6 *Let X be a Banach space and let $f: X \to \mathbb{R} \cup \{+\infty\}$ be a proper lsc function and bounded below. Then for all $\varepsilon > 0$ there exists $\bar{x} \in X$ such that*

$$f(\bar{x}) \leq \inf_X f + 2\varepsilon$$

and the perturbed function $x \to f(x) + \varepsilon \|x - \bar{x}\|$ attains a strong minimum at $\bar{x}$.

Hint: Let Y be the space of all bounded Lipschitz continuous functions g on X with norm

$$\|g\|_Y := \|g\|_{\infty} + \sup \left\{ \frac{|g(x) - g(y)|}{\|x - y\|} \;\middle|\; x, y \in X, x \neq y \right\}.$$

Exercise 2.5.5 Deduce the following version of the smooth variational principle from Theorem 2.5.4.

Theorem 2.5.7 *Let X be a Banach space with a Lipschitz Fréchet smooth bump function and let $f\colon X \to \mathbb{R}\cup\{+\infty\}$ be a proper lsc function and bounded below. Then there exists a constant $a > 0$ (depending only on X) such that for all $\varepsilon \in (0,1)$ and for any $y \in X$ satisfying $f(y) < \inf_X f + a\varepsilon^2$, there exist a Lipschitz Fréchet differentiable function g and $x \in X$ such that*

(i) $f + g$ *has a strong minimum at* x,
(ii) $\|g\|_\infty < \varepsilon$ *and* $\|g'\|_\infty < \varepsilon$,
(iii) $\|x - y\| < \varepsilon$.

*∗**Exercise 2.5.6** (Range of Bump Functions) Let $b\colon \mathbb{R}^N \to \mathbb{R}$ be a C^1 bump function.

(i) Show that $0 \in \operatorname{int} \operatorname{range}(b')$ by applying the smooth variational principle.
(ii) Find an example where $\operatorname{range}(b')$ is not simply connected.

Reference: [37].

3

Variational Techniques in Subdifferential Theory

For problems of smooth variation we can usually apply arguments based on Fermat's principle – that a differentiable function has a vanishing derivative at its minima (maxima). However, nonsmooth functions and mappings arise intrinsically in many applications. The following are several such examples of intrinsic nonsmoothness.

Example 3.0.1 (Max Function) Let $f_n \colon X \to \mathbb{R} \cup \{+\infty\}, n = 1, \ldots, N$ be lsc functions. Then so is

$$f = \max(f_1, \ldots, f_N).$$

However, this maximum is often nonsmooth even if all $f_n, n = 1, \ldots, N$ are smooth functions. For example,

$$|x| = \max(x, -x).$$

is nonsmooth at $x = 0$.

Example 3.0.2 (Optimal Value Functions) Consider the simple constrained minimization problem of minimizing $f(x)$ subject to $g(x) = a$, $x \in \mathbb{R}$. Here $a \in \mathbb{R}$ is a parameter allowing for perturbation of the constraint. In practice it is often important to know how the model responds to the perturbation a. For this we need to consider, for example, the *optimal value*

$$v(a) := \inf\{f(x) : g(x) = a\}$$

as a function of a. Consider a concrete example, illustrated in Figure 3.1, of the two smooth functions $f(x) := 1 - \cos x$ and $g(x) := \sin(6x) - 3x$, and $a \in [-\pi/2, \pi/2]$ which corresponds to $x \in [-\pi/6, \pi/6]$. It is easy to show that the optimal value function v is not smooth, in fact, not even continuous.

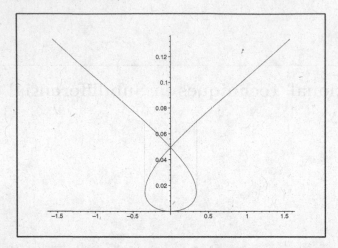

Fig. 3.1. Smooth becomes nonsmooth: g (vertical) plotted against f.

Example 3.0.3 (Penalization Functions) Constrained optimization problems occur naturally in many applications. A simplified form of such a problem is

$$\mathcal{P} \qquad\qquad \text{minimize}\quad f(x)$$

$$\text{subject to}\quad x \in S,$$

where S is a closed subset of X often referred to as the feasible set. One often wishes to convert such a problem to a simpler one without constraint. The use of nonsmooth functions makes this conversion easier. For example, if f is Lipschitz with a Lipschitz constant L then, for any $\mu > L$, problem $\mathcal{P}$ is equivalent to

$$\text{minimize } f + \mu d_S.$$

This is often referred to as *exact penalization*. If f is lsc then $\mathcal{P}$ is equivalent to

$$\text{minimize } f + \iota_S.$$

Example 3.0.4 (Spectral Functions) The maximum eigenvalue of a matrix often plays an important role in problems related to a matrix. When the matrix contains one or more parameters, the maximum eigenvalue then becomes a function of those parameters. This maximum eigenvalue function is often intrinsically nonsmooth. For example, consider the 2 by 2 matrix with a parameter x,

$$\begin{bmatrix} 1 & x \\ x & 1 \end{bmatrix}.$$

Then the maximum eigenvalue is $1 + |x|$, a nonsmooth function.

This intrinsic nonsmoothness motivated the development of nonsmooth analysis. Concepts generalizing that of the derivative for smooth functions have been introduced which enable us to apply the variational technique to nonsmooth functions. There are many competing concepts of subdifferentials; we mainly focus on the Fréchet subdifferential which is a natural fit for the variational technique.

3.1 The Fréchet Subdifferential and Normal Cones

3.1.1 The Fréchet Subdifferential

To generate the Fréchet subdifferential at a nondifferentiable point of a lsc function, we use the collection of all the (Fréchet) derivatives of smooth "osculating" functions (functions lying below and touching at the point in question), if they exist, to replace the missing derivative. More often than not, this simple contrivance is sufficient. Moreover, in the language of analysis, we are led to study a local minimum of the difference of two functions which fits very well with techniques of variational analysis. The geometric concept of the Fréchet normal cone to a closed set is then introduced through the subdifferential of the indicator function of the set – an extended-valued lsc function.

Let X be a Banach space. We say a function f on X is *Fréchet differentiable* at x and $f'(x) \in X^*$ is the *Fréchet derivative* of f at x provided that

$$\lim_{\|h\| \to 0} \frac{|f(x+h) - f(x) - \langle f'(x), h \rangle|}{\|h\|} = 0.$$

We say f is C^1 at x if $f' \colon X \to X^*$ is norm continuous at x. We say a Banach space is *Fréchet smooth* provided that it has an equivalent norm that is differentiable, indeed C^1, for all $x \neq 0$.

Definition 3.1.1 (Fréchet Subdifferential) *Let X be a real Banach space. Let $f \colon X \to \mathbb{R} \cup \{+\infty\}$ be a proper lsc function. We say f is* Fréchet-sub-differentiable *and x^* is a* Fréchet-subderivative *of f at x if $x \in \mathrm{dom} f$ and*

$$\liminf_{\|h\| \to 0} \frac{f(x+h) - f(x) - \langle x^*, h \rangle}{\|h\|} \geq 0. \tag{3.1.1}$$

We denote the set of all Fréchet-subderivatives of f at x by $\partial_F f(x)$ and call this object the Fréchet subdifferential *of f at x. For convenience we define $\partial_F f(x) = \emptyset$ if $x \notin \mathrm{dom} f$.*

Definition 3.1.2 (Viscosity Fréchet Subdifferential) *Let X be a real Banach space. Let $f \colon X \to \mathbb{R} \cup \{+\infty\}$ be a proper lsc function. We say f is* viscosity Fréchet-subdifferentiable *and x^* is a* viscosity Fréchet-subderivative *of f at x if $x \in \mathrm{dom} f$ and there exists a C^1 function g such that $g'(x) = x^*$ and*

$f - g$ attains a local minimum at x. We denote the set of all viscosity Fréchet-subderivatives of f at x by $\partial_{VF} f(x)$ and call this object the viscosity Fréchet subdifferential of f at x. For convenience we define $\partial_{VF} f(x) = \emptyset$ if $x \notin \mathrm{dom} f$.

Since shifting g by a constant does not influence its derivative we can require that $f - g$ attains a local minimum of 0 at x in the above definition.

The following relationship between the Fréchet subdifferential and the viscosity Fréchet subdifferential is easy and useful.

Proposition 3.1.3 Let X be a Banach space and let $f \colon X \to \mathbb{R} \cup \{+\infty\}$ be a lsc function. Then $\partial_{VF} f(x) \subset \partial_F f(x)$.

Proof. Exercise 3.1.1.   ●

In fact, with some additional effort one can show that in a Fréchet-smooth Banach space $\partial_{VF} f(x) = \partial_F f(x)$ [99]. Since we work mostly in Fréchet smooth Banach spaces in this book, we will use ∂_F for both Fréchet and viscosity Fréchet subdifferentials unless pointed out otherwise.

If f is Fréchet differentiable at x then it is not hard to show that $\partial_F f(x) = \{f'(x)\}$. The converse is not true (Exercises 3.1.3). In general, $\partial_F f(x)$ may be empty even if $x \in \mathrm{dom} f$. An easy example is $\partial_F (-\|\cdot\|)(0) = \emptyset$. However, a variational argument leads to the following important result about the existence of the Fréchet subdifferential.

Theorem 3.1.4 Let X be a Fréchet smooth Banach space and let $f \colon X \to \mathbb{R} \cup \{+\infty\}$ be a lsc function. Then $\{x \in X \mid \partial_F f(x) \neq \emptyset\}$ is dense in $\mathrm{dom} f$.

Proof. Let $\bar{x} \in \mathrm{dom} f$ and let ε be an arbitrary positive number. We show f is Fréchet subdifferentiable at some point $y \in B_\varepsilon(\bar{x})$. Since f is lsc at $\bar{x}$ there exists $\delta > 0$ such that $f(x) > f(\bar{x}) - 1$ for all $x \in B_\delta(\bar{x})$. Define $\tilde{f} := f + \iota_{B_\delta(\bar{x})}$. Then, $\tilde{f}$ is lsc and

$$\tilde{f}(\bar{x}) = f(\bar{x}) < \inf_{B_\delta(\bar{x})} f + 1 = \inf_X \tilde{f} + 1.$$

Applying the Borwein–Preiss Variational Principle of Theorem 2.5.3, using the asserted Fréchet smooth renorm with $\lambda < \min(\delta, \varepsilon)$, we conclude that there exists $y \in B_\lambda(\bar{x}) \subset \mathrm{int}(B_\delta(\bar{x}) \cap B_\varepsilon(\bar{x}))$ and $\varphi_2(x) := \sum_{i=1}^{\infty} \mu_i \|x - x_i\|^2$ where (x_i) is a sequence converging to y and (μ_i) is a sequence of positive numbers satisfying $\sum_{i=1}^{\infty} \mu_i = 1$ such that $\tilde{f} + \lambda^{-2} \varphi_2$ attains a minimum at y. Since y is an interior point of $B_\delta(\bar{x})$, $f + \lambda^{-2} \varphi_2$ attains a local minimum at y. After checking that φ_2 is Fréchet differentiable, we see that f is Fréchet subdifferentiable at $y \in B_\varepsilon(\bar{x})$.   ●

We put meat on the bones of the last result by recalling that Hilbert space and $L_p (1 < p < \infty)$ are Fréchet smooth in their original norms while every reflexive space has a Fréchet smooth renorm [58, 99].

Note that the subdifferential is usually a set. The following are subdifferentials of several nonsmooth functions at typical nonsmooth points that can easily be verified.

Example 3.1.5

$$\partial_F |\cdot|(0) = [-1, 1],$$

$$\partial_F \sqrt{|\cdot|}(0) = (-\infty, \infty),$$

$$\partial_F \max(\cdot, 0)(0) = [0, 1],$$

and

$$\partial_F \iota_{[0,1]}(0) = (-\infty, 0].$$

3.1.2 The Fréchet Normal Cone

The central geometric concept of the normal cone to a closed set can now be defined through the indicator function of the set.

Definition 3.1.6 (Fréchet Normal Cone) *Let S be a closed subset of X. We define the Fréchet normal cone of S at x to be $N_F(S; x) := \partial_F \iota_S(x)$.*

Some easy facts directly follow from the definition. It is easy to verify that $N_F(S; x)$ is a cone that always contains $\{0\}$ and when $x \in \operatorname{int} S$, $N_F(S; x) = \{0\}$ (Exercises 3.1.6, 3.1.8 and 3.1.9). Moreover, consider the constrained minimization problem

$$\text{minimize} \quad f(x) \tag{3.1.2}$$

$$\text{subject to} \quad x \in S \subset X.$$

We have an easy and useful necessary optimality condition in terms of the normal cone of S.

Proposition 3.1.7 *Let X be a Fréchet smooth Banach space, let f be a C^1 function on X and let S be a closed subset of X. Suppose that $\bar{x}$ is a solution of the constrained minimization problem (3.1.2). Then*

$$0 \in f'(\bar{x}) + N_F(S; \bar{x}).$$

Proof. Exercise 3.1.13.   ●

Recall that for a C^1 function f, $v = f'(x)$ if and only if $(v, -1)$ is a normal vector for the graph of f at $(x, f(x))$. Our next theorem is a Fréchet subdifferential version of this fact which characterizes the Fréchet subdifferential of a function in terms of the normal cone to its epigraph.

Theorem 3.1.8 *Let X be a Fréchet smooth Banach space and let $f : X \to \mathbb{R} \cup \{+\infty\}$ be a lsc function. Then $x^* \in \partial_F f(x)$ if and only if*

$$(x^*, -1) \in N_F(\operatorname{epi} f; (x, f(x))).$$

Proof. (a) The "only if" part. Let $x^* \in \partial_F f(x)$. Then there exists a C^1 function g such that $g'(x) = x^*$ and $f - g$ attains a minimum at x. Define $h(y, r) := g(y) - r$. We have $h'(x, f(x)) = (x^*, -1)$ and

$$\iota_{\text{epi}f}(y, r) - h(y, r) \geq \iota_{\text{epi}f}(x, f(x)) - h(x, f(x)). \tag{3.1.3}$$

Thus, $(x^*, -1) \in N_F(\text{epi}f; (x, f(x)))$.

(b) The "if" part. Let $(x^*, -1) \in N_F(\text{epi}f; (x, f(x)))$. Then there exists a C^1 function h such that $h'(x, f(x)) = (x^*, -1)$ and $h(y, r) \leq h(x, f(x)) = 0$ for any $(y, r) \in \text{epi}f$. By the implicit function theorem (see e.g. [271]) there exists a C^1 function $g \colon X \to \mathbb{R}$ such that in a neighborhood of x, $h(y, g(y)) = 0$, $g(x) = f(x)$ and $g'(x) = x^*$. Since h is C^1 and the second component of $h'(x, f(x))$ is negative there exists $a > 0$ such that $h(y, r) < h(y, r')$, for any $y \in B_a(x)$ and $f(x) - a < r' < r < f(x) + a$. Take $b \in (0, a)$ such that for any $y \in B_b(x)$, $g(y) \in (f(x) - a, f(x) + a)$ and $f(y) > f(x) - a$. Then, for any $y \in B_b(x)$, we have $f(y) - g(y) \geq 0 = f(x) - g(x)$. In fact, the inequality is obvious when $f(y) \geq f(x) + a$. If $f(y) < f(x) + a$ then it follows from $h(y, f(y)) \leq 0 = h(y, g(y))$. ●

The normal cone to the epigraph of a function has the following special properties.

Lemma 3.1.9 *Let f be a lsc function. Then*

(i) *for any $(x, r) \in \text{epi}f$, $N_F(\text{epi}f; (x, r)) \subset N_F(\text{epi}f; (x, f(x)))$,*
(ii) *if $(x^*, -\lambda) \in N_F(\text{epi}f; (x, f(x)))$ and $\lambda \neq 0$ then $\lambda > 0$ and $x^* \in \lambda\partial_F f(x)$.*

Proof. Exercise 3.1.10. ●

Thus, Theorem 3.1.8 also characterizes $(x^*, \lambda) \in N_F(\text{epi}f; (x, f(x)))$ when $\lambda \neq 0$ in terms of the subdifferentials of f. The characterization of $(x^*, 0) \in N_F(\text{epi}f; (x, f(x)))$ in terms of the subdifferentials of f is more delicate and will be discussed later after we have developed the subdifferential calculus.

3.1.3 The Subdifferential Form of the Variational Principle

We conclude this section with a subdifferential version of the Borwein–Preiss Variational Principle. This is the form most frequently used in applications involving subdifferentials. The easy proof is left as an exercise.

Theorem 3.1.10 *Let X be a Banach space with a Fréchet smooth norm $\|\cdot\|$ and let $f \colon X \to \mathbb{R} \cup \{+\infty\}$ be a lsc function bounded from below, $\lambda > 0$ and $p > 1$. Then, for every $\varepsilon > 0$ and $z \in X$ satisfying*

$$f(z) < \inf_X f + \varepsilon,$$

there exists a point $y \in X$ such that $\|z - y\| \leq \lambda$ and a C^1 function φ with $|\varphi(y)| < \varepsilon/\lambda$ and $\|\varphi'(y)\| < p\varepsilon/\lambda$ such that $f + \varphi$ attains a minimum at y. Consequently,

$$\partial_F f(y) \cap \frac{p\varepsilon}{\lambda} B_{X^*} \neq \emptyset.$$

Proof. Exercise 3.1.12.   •

3.1.4 Commentary and Exercises

Although the use of generalized (one-sided) derivatives dates back explicitly to Dini and before, especially in the context of integration theory, the systematic study of such concepts for variational analysis, especially off the real line, is quite recent. Consistent theory was developed first for certain classes of functions, e.g., the convex subdifferential for convex functions (see [229]) and the quasi-differential for quasi-differentiable functions (see [223]). Clarke's pioneering work [81] on the generalized gradient opened the door to methodical study of general nonsmooth problems. Many competing concepts of generalized derivatives were introduced in the ensuing past several decades. Several frequently used concepts are Halkin's screen [133], the limiting subdifferential developed by Mordukhovich [195, 196, 198], Ioffe's approximate and G-subdifferential [142, 145, 146], Michel and Penot's subdifferential [193], Treiman's linear subdifferential [250, 251], Warga's derivative container [263, 264] and Sussmann's semidifferential [245, 246].

The last decade has witnessed a unification and reconciliation of much of this work in two directions. One is along the ideas pioneered by Warga to study abstract subdifferentials that satisfy a set of axioms so as to provide basic properties of many different subdifferentials alluded to above with a unified framework. The other, which is more relevant to this book, is to turn our attention to the simpler smooth subdifferentials based on the fact that many of the above subdifferentials can be represented by such smooth subdifferentials in spaces with a reasonable geometric property [48, 85, 146, 185]. In this book we primarily consider the Fréchet subdifferential in Fréchet smooth Banach spaces. It was introduced by Bazaraa, Goode and Nashed in finite dimensions [20] and developed in detail in infinite dimensions by Borwein and Strojwas [61], Kruger [164, 165, 166], Kruger and Mordukhovich [167] and others. This allows us to illustrate variational techniques without too many technical assumptions. Most of the results apply to more general bornological smooth subdifferentials or s-Hölder subdifferentials [58, 221] with minor changes. Systematic accounts of nonsmooth analysis and its applications can be found in [8, 84, 85, 91, 185, 150, 198, 208, 237, 263, 264].

Unlike derivatives, subdifferentials do not determine functions up to a constant, even on well connected sets. Thus, we do not have an "integration" theory corresponding to the subdifferentials (see guided Exercises 3.1.19, 3.1.20, 3.1.21 and 3.1.22 for details).

Exercise 3.1.1 Prove Proposition 3.1.3.

Exercise 3.1.2 Verify the Fréchet subdifferentials in Example 3.1.5.

Exercise 3.1.3 Show that

(i) If f is Fréchet differentiable at x then $\partial_F f(x) = \{f'(x)\}$.
(ii) A function can have a unique Fréchet subdifferential without being differentiable.
(iii) There exists a Lipschitz function having the properties described in (ii).

Hint: Consider $f(x) := |x|(\sin(\log(|x|)) + 1), x \neq 0$ and $f(0) := 0$.

Exercise 3.1.4 (Fréchet Superdifferential) Let $f \colon X \to \mathbb{R} \cup \{-\infty\}$ be an upper semicontinuous function (i.e., $-f$ is lsc). We define the Fréchet superdifferential of f at x to be $\partial^F f(x) = -\partial_F(-f)(x)$. Prove that f is Fréchet differentiable at x if and only if $\partial^F f(x) = \partial_F(f)(x) = \{f'(x)\}$. Indeed it suffices that $\partial^F f(x) \cap \partial_F f(x) \neq \emptyset$.

Exercise 3.1.5 Show that for any $\lambda > 0$, $\partial_F(\lambda f)(x) = \lambda \partial_F f(x)$. Care must be taken with zero, when $\partial_F f(x)$ is empty.

Exercise 3.1.6 Verify that for any closed set S and $x \in S$, $N_F(S; x)$ is a cone, i.e., for any $x^* \in N_F(S; x)$ and any $r \geq 0$, $rx^* \in N_F(S; x)$.

Exercise 3.1.7 Show that for any lsc function $f \colon \mathbb{R}^N \to \mathbb{R} \cup \{+\infty\}$ and any $x \in \operatorname{dom} f$. The set $\partial_F f(x)$ is always closed. Deduce that, for any closed subset S in $\mathbb{R}^N$ and any $x \in S$, the normal cone $N_F(S; x)$ is closed. Reference: See Theorem 8.6 in [237].

Exercise 3.1.8 Show that if $s \in \operatorname{int} S$, then $N_F(S; s) = \{0\}$.

Exercise 3.1.9 Let $\{e_i\}$ be the standard orthonormal basis of ℓ_2 and let $S := \overline{\operatorname{conv}}\{\pm e_i/i\}_{i=1}^{\infty}$. Show that $0 \notin \operatorname{int} S$ yet $N_F(S; 0) = \{0\}$.

Exercise 3.1.10 Prove Lemma 3.1.9.

Exercise 3.1.11 Show that in Definition 3.1.2 we can require that $f - g$ attains a local minimum of 0 at x.

Exercise 3.1.12 Suppose that f is a lsc function and that g is a C^1 function. Show that $\partial_F(f + g)(x) = \partial_F f(x) + g'(x)$.

Exercise 3.1.13 Prove Proposition 3.1.13.

Exercise 3.1.14 Prove that if f is a Lipschitz function with rank L then, for any x, $x^* \in \partial_F f(x)$ implies that $\|x^*\| \leq L$.

Exercise 3.1.15 Let X be a Fréchet smooth Banach space and let $f \colon X \to \mathbb{R} \cup \{+\infty\}$ be a lsc function. Prove that f is Lipschitz with rank L if and only if, for any x, $x^* \in \partial_F f(x)$ implies that $\|x^*\| \leq L$.

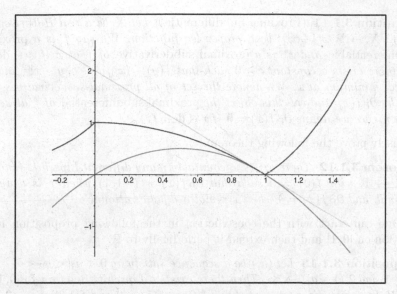

Fig. 3.2. Every Fréchet subdifferential is a "viscosity" subdifferential.

∗Exercise 3.1.16 Let X be a Fréchet smooth Banach space and let $f\colon X \to \mathbb{R} \cup \{+\infty\}$ be a lsc function. Prove that $\partial_{VF} f(x) = \partial_F f(x)$. Reference: [99].

∗Exercise 3.1.17 Let X be a Banach space with a Fréchet smooth equivalent norm and let $f\colon X \to \mathbb{R} \cup \{+\infty\}$ be a lsc function. Prove that $x^* \in \partial_F f(x)$ if and only if there exists a *concave* C^1 function g such that $g'(x) = x^*$ and $f - g$ attains a local minimum at x, as drawn in Figure 3.2. Reference: [71, Remark 1.4].

Exercise 3.1.18 Prove Theorem 3.1.10.

Exercise 3.1.19 Construct two lsc functions on $\mathbb{R}$ with the identical Fréchet subdifferential yet their difference is not a constant. Hint: Consider $f = 1 - \chi_{[0,1]}$ and $2f$ where χ_S is the characteristic function of set S defined by $\chi_S(x) = 1$ for $x \in S$ and $\chi_S(x) = 0$ for $x \notin S$.

Exercise 3.1.20 Construct two continuous functions on $\mathbb{R}$ with the identical Fréchet subdifferential yet their difference is not a constant. Hint: Consider the Cantor function f and $2f$ (see [71] and also Exercise 3.5.5).

Exercise 3.1.21 Prove that if two Lipschitz functions on $\mathbb{R}$ have the identical Fréchet subdifferential then they differ only by a constant.

∗Exercise 3.1.22 The conclusion in Exercise 3.1.21 fails if the Fréchet subdifferential is replaced by the proximal subdifferential. Recall the proximal subdifferential is defined as follows.

Definition 3.1.11 (Proximal Subdifferential) *Let X be a real Hilbert space. Let $f\colon X \to \mathbb{R} \cup \{+\infty\}$ be a proper lsc function. We say f is a proximal subdifferentiable and x^* is a proximal subderivative of f at x if $x \in \mathrm{dom} f$ and there exists a constant $c \geq 0$ such that $f(y) - \langle x^*, y \rangle - c\|y - x\|^2$ attains a local minimum at x. We denote the set of all proximal-subderivatives of f at x by $\partial_P f(x)$ and call this object the proximal subdifferential of f at x. For convenience we define $\partial_P f(x) = \emptyset$ if $x \notin \mathrm{dom} f$.*

Precisely prove the following theorem.

Theorem 3.1.12 *There exists uncountably many different Lipschitz function $f\colon \mathbb{R} \to \mathbb{R}$ with $f(0) = 0$ such that $\partial_P f(x) = (-1,1)$ when x is a dyadic rational, and $\partial_P f(x) = \emptyset$ when x is not a dyadic rational.*

One can start with the construction in the following proposition for a function on $[0,1]$ and then extend it periodically to $\mathbb{R}$.

Proposition 3.1.13 *Let (a_i) be a sequence satisfying $0 < a_1 < a_2 < \cdots < 1$, $a_i \to 1$ and $2^i(1 - a_i) \to \infty$. Then there exists a Lipschitz function $f\colon [0,1] \to \mathbb{R}$ with Lipschitz constant 1 satisfying $f(0) = f(1) = 0$ and $f(1/2) = a_1/2$ such that $\partial_P f(x) = (-1,1)$ when $x \in (0,1)$ is a dyadic rational, and $\partial_P f(x) = \emptyset$ when $x \in (0,1)$ is not a dyadic rational.*

Hint: Define $f = \lim_i f_i$ where f_i are affine on the intervals $[n/2^i, (n+1)/2^i]$ for $n = 0, 1, \ldots, 2^i - 1$. Denote the slope of f_i on this interval by $s_{n,i}$ and define $f_i(0) = 0$ and

$$s_{2n,i} := a_i, \qquad s_{2n+1,i} := 2s_{n,i-1} - a_i, \quad \text{if } s_{n,i-1} \geq 0,$$

$$s_{2n,i} := 2s_{n,i-1} + a_i, \quad s_{2n+1,i} := -a_i, \quad \text{if } s_{n,i-1} \leq 0.$$

Then show that

(i)   For all $i = 1, 2, \ldots$, f_i is defined and Lipschitz on $[0,1]$ and $f_i(2n/2^i) = f_{i-1}(n/2^{i-1})$ for $n = 0, \ldots, 2^{i-1}$ and $i = 2, 3, \ldots$.
(ii) $s_{n,i} \in [-a_i, a_i]$ for all $n = 0, \ldots, 2^{i-1}$ and $i = 1, 2, \ldots$.
(iii)   The sequence (f_i) uniformly converges to a Lipschitz function f with a Lipschitz constant 1.
(iv) $\partial_P f(x) = (-1,1)$ when $x \in (0,1)$ is a dyadic rational.
(v) $\partial_P f(x) = \emptyset$ when $x \in (0,1)$ is not a dyadic rational.
(vi)   Verify that $f(0) = f(1) = 0$ and $f(1/2) = a_1/2$.
(vii)   Extend f periodically to $\mathbb{R}$ and check $\partial_P f(x) = (-1,1)$ when x is an integer.

Reference: see [46] for details and check [21, 66, 92] for related earlier examples.

3.2 Nonlocal Approximate Sum Rule and Viscosity Solutions

To effectively use the Fréchet subdifferential one often needs to decouple subdifferential information in various different settings. Here we discuss a result that combines a variational principle and a decoupling mechanism. It is flexible enough that one can conveniently use it to derive several other results on the calculus of subdifferentials of a similar nature.

3.2.1 The Decoupled Infimum

We start by introducing the concept of a decoupled infimum.

Definition 3.2.1 (Decoupled Infimum) *Let X be a Banach space, $f_n\colon X \to \mathbb{R} \cup \{+\infty\}, n = 1,\dots,N$ extended-value functions and S a subset of X. We define the* decoupled infimum *of $f_1,\dots,f_N$ over S by*

$$\bigwedge[f_1,\dots,f_N](S) := \lim_{\eta \to 0} \inf \left\{ \iota_S(x_0) + \sum_{n=1}^{N} f_n(x_n) : \right.$$

$$\left. \operatorname{diam}(x_0, x_1,\dots,x_N) \le \eta \right\}. \qquad (3.2.1)$$

Some useful elementary properties of the decoupled infimum are discussed in Exercises 3.2.1, 3.2.2 and 3.2.3. The following lemma is often useful.

Lemma 3.2.2 *Let $s_p(y_1,\dots,y_N) := \sum_{n,m=1}^{N} \|y_n - y_m\|^p$ with $p \ge 1$. Suppose that $(x_1^*,\dots,x_N^*) \in \partial_F s_p(x_1,\dots,x_N)$. Then*

$$\sum_{n=1}^{N} x_n^* = 0. \qquad (3.2.2)$$

Moreover, if $p = 1$ and $s_1(x_1,\dots,x_N) > 0$, then

$$\max\{\|x_n^*\| \mid n = 1,\dots,N\} \ge 1. \qquad (3.2.3)$$

Proof. Conclusion (3.2.2) is easy and left as an exercise. To prove (3.2.3) we observe that s_1 is homogeneous. By Proposition 3.1.3 we have

$$\sum_{n=1}^{N} \langle x_n^*, -x_n \rangle \le \liminf_{t \to 0+} \frac{s_1(x_1 - tx_1,\dots,x_N - tx_N) - s_1(x_1,\dots,x_N)}{t}$$

$$= -s_1(x_1,\dots,x_N). \qquad (3.2.4)$$

Combining (3.2.2) and (3.2.4) we have

$$s_1(x_1,\ldots,x_N) \le \sum_{n=1}^{N}\langle x_n^*, x_n\rangle = \sum_{n=1}^{N-1}\langle x_n^*, x_n - x_N\rangle$$

$$\le \max\{\|x_n^*\| \mid n = 1,\ldots,N-1\}\sum_{n=1}^{N-1}\|x_n - x_N\|$$

$$\le \max\{\|x_n^*\| \mid n = 1,\ldots,N\}s_1(x_1,\ldots,x_N), \qquad (3.2.5)$$

which implies (3.2.3) when $s_1(x_1,\ldots,x_N) > 0$.   ●

3.2.2 Nonlocal Approximate Sum Rules

Now we can prove the main result of this section.

Theorem 3.2.3 (Nonlocal Approximate Sum Rule) *Let X be a Fréchet smooth Banach space and let $f_1,\ldots,f_N\colon X \to \mathbb{R}\cup\{+\infty\}$ be lsc functions bounded below. Suppose that $\bigwedge[f_1,\ldots,f_N](X) < +\infty$. Then, for any $\varepsilon > 0$, there exist x_n and $x_n^* \in \partial_F f_n(x_n), n = 1,\ldots,N$ satisfying*

$$\mathrm{diam}(x_1,\ldots,x_N) \times \max(1,\|x_1^*\|,\ldots,\|x_N^*\|) < \varepsilon, \qquad (3.2.6)$$

and

$$\sum_{n=1}^{N} f_n(x_n) < \bigwedge[f_1,\ldots,f_N](X) + \varepsilon \qquad (3.2.7)$$

such that

$$\left\|\sum_{n=1}^{N} x_n^*\right\| < \varepsilon. \qquad (3.2.8)$$

Proof. Without loss of generality, we may assume that $\|\cdot\|$ is C^1 away from 0. Define, for any real number $r > 0$ and s_2 as in Lemma 3.2.2,

$$w_r(y_1,\ldots,y_N) := \sum_{n=1}^{N} f_n(y_n) + rs_2(y_1,\ldots,y_N)$$

and $M_r := \inf w_r$. Then M_r is an increasing function of r and is bounded above by $\bigwedge[f_1,\ldots,f_N](X)$ (Exercise 3.2.4). Let $M := \lim_{r\to\infty} M_r$. Observe that the product space X^N of N copies of the Fréchet smooth Banach space X (with the Euclidean product norm) is also Fréchet smooth. For each r, applying the Borwein–Preiss Variational Principle of Theorem 3.1.10 to the function w_r, we obtain a C^1 function ϕ_r and $x_{n,r}, n = 1,\ldots,N$ such that

$$\sum_{n=1}^{N} f_n(x_{n,r}) \le w_r(x_{1,r},\ldots,x_{N,r}) < \inf w_r + \frac{1}{r} \le M + \frac{1}{r}, \qquad (3.2.9)$$

$\|\phi'_r(x_{1,r}, \ldots, x_{N,r})\| < \varepsilon/N$, and

$$\sum_{n=1}^{N} f_n(y_n) + rs_2(y_1, \ldots, y_N) + \phi_r(y_1, \ldots, y_N)$$

attains a local minimum at $(x_{1,r}, \ldots, x_{N,r})$. Thus,

$$(x^*_{1,r}, \ldots, x^*_{N,r}) := -\phi'_r(x_{1,r}, \ldots, x_{N,r}) - rs'_2(x_{1,r}, \ldots, x_{N,r})$$
$$\in \partial_F f_1(x_{1,r}) \times \cdots \times \partial_F f_N(x_{N,r}).$$

Summing the N components of the above inclusion and using Lemma 3.2.2 we obtain (3.2.8). By the definition of M_r we have

$$M_{r/2} \leq w_{r/2}(x_{1,r}, \ldots, x_{N,r})$$
$$= w_r(x_{1,r}, \ldots, x_{N,r}) - \frac{r}{2}s_2(x_{1,r}, \ldots, x_{N,r})$$
$$\leq M_r + \frac{1}{r} - \frac{r}{2}s_2(x_{1,r}, \ldots, x_{N,r}). \tag{3.2.10}$$

Rewriting (3.2.10) as $rs_2(x_{1,r}, \ldots, x_{N,r}) \leq 2(M_r - M_{r/2} + \frac{1}{r})$ yields

$$\lim_{r\to\infty} rs_2(x_{1,r}, \ldots, x_{N,r}) = 0. \tag{3.2.11}$$

Therefore,

$$\lim_{r\to\infty} \text{diam}(x_{1,r}, \ldots, x_{N,r}) = 0. \tag{3.2.12}$$

Moreover,

$$\lim_{r\to\infty} \text{diam}(x_{1,r}, \ldots, x_{N,r}) \times \max(\|x^*_{1,r}\|, \ldots, \|x^*_{N,r}\|) = 0. \tag{3.2.13}$$

Also,

$$M \leq \bigwedge[f_1, \ldots, f_N](X)$$
$$\leq \liminf_{r\to\infty} \sum_{n=1}^{N} f_n(x_{n,r}) = \liminf_{r\to\infty} w_r(x_{1,r}, \ldots, x_{N,r}) \leq M$$

which yields

$$M = \bigwedge[f_1, \ldots, f_N](X). \tag{3.2.14}$$

For r sufficiently large, set $x_n := x_{n,r}$ and $x^*_n := x^*_{n,r}$, $n = 1, \ldots, N$. Then (3.2.6) follows from (3.2.12) and (3.2.13) and (3.2.7) follows from (3.2.9) and (3.2.14). $\bullet$

Theorem 3.2.3 is a powerful result. The guided Exercise 3.2.7 illustrates how to use it to deduce Theorem 3.1.4. The conditions in the nonlocal approximate sum rule are minimum in a certain sense (see examples in Exercises 3.2.5 and 3.2.6).

3.2.3 The Uniqueness of Viscosity Solutions

We now use the nonlocal approximate sum rule to prove a uniqueness theorem for the viscosity solution of the following Hamilton–Jacobi equation

$$u + H(x, u') = 0. \tag{3.2.15}$$

This equation is closely related to the optimal value function of certain optimal control problems.

Consider the value function u of the optimal control problem

$$u(x) := \inf\left\{ \int_0^\infty e^{-t} f(x(t), c(t))\, dt : x'(t) = g(x(t), c(t)), \right.$$

$$\left. c(t) \in C, x(0) = x \right\} \tag{3.2.16}$$

where f and g are Lipschitz functions, c is a measurable function modeling the control and C is a compact set modeling the admissible range of the control function. We assume that for any given x, an optimal control for the above problem exists. When u is smooth it satisfies equation (3.2.15) with

$$H(x, p) := \sup\{\langle -g(x, c), p\rangle - f(x, c) : c \in C\} \tag{3.2.17}$$

The proof is outlined in the guided exercises 3.2.11 and 3.2.12.

In general such a value function is not necessarily smooth and (3.2.15) does not necessarily have a classical solution. Viscosity solutions were introduced to replace classical solutions. We recall the definition below. First, let $f \colon X \to \mathbb{R} \cup \{-\infty\}$ be an upper semicontinuous function. We define the *Fréchet superdifferential* of f at x, $\partial^F f(x)$, by

$$\partial^F f(x) := -\partial_F(-f)(x).$$

Definition 3.2.4 (Viscosity Solutions) *A function $u\colon X \to \mathbb{R}$ is a* viscosity supersolution (viscosity subsolution) *of (3.2.15) if u is lower (upper) semicontinuous and, for every $x \in X$ and every $x^* \in \partial_F(u)(x)$ $(x^* \in \partial^F(u)(x))$,*

$$u(x) + H(x, x^*) \geq 0 \quad (u(x) + H(x, x^*) \leq 0).$$

A continuous function u is called a viscosity solution *if u is both a viscosity subsolution and a viscosity supersolution.*

One can using an argument similar to that in Exercises 3.2.11 and 3.2.12 to show that when u is continuous it is a viscosity solution of (3.2.15). The uniqueness of viscosity solutions to the Hamilton–Jacobi equation follows readily from the following comparison theorem.

Theorem 3.2.5 (Comparison Theorem) *Let u be an upper semicontinuous function bounded above and v be a lower semicontinuous function bounded below. Suppose $H\colon X \times X^* \to \mathbb{R}$ satisfies the following assumption:*

(A) *for any $x_1, x_2 \in X$ and $x_1^*, x_2^* \in X^*$,*

$$|H(x_1, x_1^*) - H(x_2, x_2^*)| \leq \omega(x_1 - x_2, x_1^* - x_2^*)$$
$$+ M \max(\|x_1^*\|, \|x_2^*\|)\|x_1 - x_2\|$$

where $M > 0$ is a constant and $\omega \colon X \times X^* \to \mathbb{R}$ is a continuous function with $\omega(0,0) = 0$.

Suppose furthermore that u is a viscosity subsolution of (3.2.15) *and v is a viscosity supersolution of* (3.2.15). *Then $u \leq v$.*

Proof. Let ε be an arbitrary positive number. Applying the nonlocal approximate sum rule of Theorem 3.2.3 with $f_1 = v$ and $f_2 = -u$, there exist $x_1, x_2 \in X$, $x_1^* \in \partial_F v(x_1)$ and $x_2^* \in \partial^F u(x_2)$ satisfying

- $\|x_1 - x_2\| < \varepsilon$, $\|x_1^*\|\|x_1 - x_2\| < \varepsilon$ and $\|x_2^*\|\|x_1 - x_2\| < \varepsilon$;
- $v(x_1) - u(x_2) < \inf_X(v - u) + \varepsilon$; and
- $\|x_1^* - x_2^*\| \leq \varepsilon$.

Since the function v is a viscosity supersolution of (3.2.15) we have

$$v(x_1) + H(x_1, x_1^*) \geq 0.$$

Similarly

$$u(x_2) + H(x_2, x_2^*) \leq 0.$$

Therefore,

$$\inf_X(v - u) > v(x_1) - u(x_2) - \varepsilon$$
$$\geq [H(x_2, x_2^*) - H(x_1, x_1^*)] - \varepsilon$$
$$\geq -[\omega(x_2 - x_1, x_2^* - x_1^*)$$
$$+ M \max(\|x_1^*\|, \|x_2^*\|)\|x_2 - x_1\|] - \varepsilon.$$

As $\varepsilon \to 0$ the right hand side converges to 0 which yields $\inf_X(v - u) \geq 0$. ●

Corollary 3.2.6 (Uniqueness of Viscosity Solutions) *Under the assumptions of Theorem 3.2.5 any continuous bounded viscosity solution to* (3.2.15) *is unique.*

We note that the function H defined in (3.2.17) satisfies condition (A) in Theorem 3.2.5 (Exercise 3.2.13).

3.2.4 Commentary and Exercises

The nonlocal approximate sum rule was derived by Zhu in [272]. In essence, it is a combination of the Borwein-Preiss smooth variational principle and a decoupling method. Decoupling techniques are important parts of calculus of variations and are behind many important results in functional, convex and nonsmooth analysis. The decoupling method we use here with a smooth symmetric penalty function was inspired by the proof of uniqueness of the viscosity solution in [93]. Many other calculus results for subdifferentials in this chapter, such as approximate local sum rules, mean value theorems, multidirectional mean value inequalities and extremal principles, are also of this nature and can be deduced from the nonlocal approximate sum rule. Actually they are equivalent (see [273] and Section 6.1). We note that approximate local sum rules, mean value theorems and extremal principles are local in nature, and therefore also accompanied with corresponding limiting forms under suitable conditions for the limiting process to be justified. We will discuss these limiting forms in Chapter 5 after giving the definition of limiting subdifferentials and their associated normal cones. The importance of the decoupled infimum is emphasized by Lassonde in [170], where it is called a uniform infimum. He points out that many conditions associated with decoupling methods (see [48, 68, 88, 148, 272, 273]) involve this quantity. The assumptions in Theorem 3.2.3 cannot be dispensed with as shown by examples in Exercises 3.2.5 and 3.2.6. The name arises since, in contrast to earlier more conventional (local) rules, one cannot guarantee the placement of the approximate minimizers. Nonetheless, Exercise 3.2.7 shows that conclusion (3.2.7) in Theorem 3.2.3 can often provide information on the location of points x_n, $n = 1, \ldots, N$, indirectly. The fact that a smooth value function of the optimal control problem (3.2.16) satisfies the Hamilton–Jacobi equation (3.2.15) is a classical result in dynamic programming theory (see e.g., [122]). That a continuous value function satisfies the same equation in the sense of viscosity solutions is the seminal work of Crandall and Lions [94]. Early forms of the uniqueness theorem of the viscosity solution are discussed in [94, 93]. The relationship between the uniqueness of viscosity solutions and the Fréchet subdifferential calculus is discussed by Deville and Haddad in [108]. However, the condition in [108] is stronger and does not apply to the control problem (3.2.16). Theorem 3.2.5 with the more realistic condition (A) is derived in [68, 71]. The improvement is achieved through better estimates on the size of the subderivative in the subdifferential calculus such as (3.2.6) in Theorem 3.2.3.

Exercise 3.2.1 Show that $\underline{f}(x) := \bigwedge [f](\{x\})$ is the *lsc closure* of $f(x)$, i.e., the largest lsc function dominated by f.

Exercise 3.2.2 Verify that

$$\bigwedge [f_1, \ldots, f_N](S) = \bigwedge [f_1, \ldots, f_N](\bar{S}) = \bigwedge [f_1, \ldots, f_N, \iota_S](X).$$

Exercise 3.2.3 Prove the following.

(i)   If $S \subset T$ then $\bigwedge [f_1, \ldots, f_N](S) \geq \bigwedge [f_1, \ldots, f_N](T)$.
(ii) $\bigwedge [f_1, \ldots, f_N](S) \leq \inf_{x \in S} \bigwedge [f_1, \ldots, f_N](\{x\})$ and equality holds when S is compact.

(iii) $\inf_{x \in S} \bigwedge[f_1, \ldots, f_N](\{x\}) \leq \inf_{x \in S} \sum_{n=1}^{N} f_n(x)$.

(iv) Show that the inequality in (iii) can be strict. Can you do it for $N = 1$?

Exercise 3.2.4 Show that M_r defined in the proof of Theorem 3.2.3 is an increasing function of r and is bounded above by $\bigwedge[f_1, \ldots, f_N](X)$.

Exercise 3.2.5 Show that functions $f_1(x) = x$ and $f_2(x) = 0$ defined on $\mathbb{R}$ do not satisfy the nonlocal approximate sum rule and explain why.

Exercise 3.2.6 Show that functions $f_1(x) = \iota_{\{0\}}(x)$ and $f_2(x) = \iota_{\{1\}}(x)$ defined on $\mathbb{R}$ do not satisfy the nonlocal approximate sum rule and explain why.

Exercise 3.2.7 Deduce Theorem 3.1.4 from Theorem 3.2.3. Hint: Let f be a lsc function bounded below. For any $x \in \text{dom} f$, define $f_1 := f$ and $f_2 := \iota_{\{x\}}$ and apply Theorem 3.2.3 to f_1 and f_2 to conclude that there exist x_1, x_2 with $\partial_F f_n(x_n) \neq \emptyset$, $n = 1, 2$, satisfying $\|x_1 - x_2\| < \varepsilon$. Note that we must have $x_2 = x$. Thus, $x_1 \in B_\varepsilon(x)$ and $\partial_F f(x_1) \neq \emptyset$.

Exercise 3.2.8 Prove (3.2.2) in Lemma 3.2.2.

Exercise 3.2.9 Show that if all but one of $f_1, \ldots, f_N$ are uniformly continuous around S then $\bigwedge[f_1, \ldots, f_N](S) = \bigwedge[\sum_{n=1}^{N} f_n](S)$.

Exercise 3.2.10 Let $\bar{x}$ minimize f over a closed set S. Prove that if f is C^1 then $-f'(\bar{x}) \in N_F(S; \bar{x})$.

*Exercise 3.2.11 Consider the optimal value function v of the following optimal control problem

$$v(t, x) := \inf \left\{ \int_t^\infty f(s, x(s), c(s)) \, ds : x'(s) = g(s, x(s), c(s)), \right.$$

$$\left. c(s) \in C, x(t) = x \right\}$$

where f and g are Lipschitz functions, c is a measurable function and C is a compact set. We assume that for any given (t, x), an optimal control for the above problem always exists.

(i) Prove the optimal principle: for any solution pair (x, c) of the control system $x'(s) = g(s, x(s), c(s))$, $c(s) \in C, x(t) = x$, we have

$$v(t, x) \leq \int_t^r f(s, x(s), c(s)) \, ds + v(r, x(r))$$

and equality holds when (x, c) is an optimal control pair.

(ii) If in addition v is a C^1 function then it satisfies the following Hamilton–Jacobi equation

$$v_t(t, x) = \hat{H}(t, x, v_x(t, x)),$$

where $\hat{H}(t, x, p) := \sup\{\langle -g(t, x, c), p \rangle - f(t, x, c) : c \in C\}$.

*Exercise 3.2.12 Show that if the optimal value function u defined in (3.2.16) is C^1 then it satisfies the Hamilton–Jacobi equation (3.2.15).

Exercise 3.2.13 Show that the function H defined in (3.2.17) satisfies condition (A) in Theorem 3.2.5.

3.3 Local Approximate Sum Rules and Constrained Minimization

Local approximate sum rules are generalizations of the sum rule for derivatives of smooth functions. They are important in studying constrained optimization problems and other problems involving local properties of subdifferentials. Consider lsc functions f_1 and f_2 such that $f_1 + f_2$ attains a minimum at x. Then $0 \in \partial_F(f_1 + f_2)(x)$. We would hope that we could conclude that

$$0 \in \partial_F f_1(x) + \partial_F f_2(x) \tag{3.3.1}$$

as is the case when both f_1 and f_2 are differentiable or convex continuous functions. Unfortunately, (3.3.1) is false in general. For example, if $f_1(x) := -f_2(x) := |x| \colon \mathbb{R} \to \mathbb{R}$ then $f_1 + f_2$ attains a minimum at 0. Yet $\partial_F f_2(0) = \emptyset$ so that $0 \notin \partial_F f_1(0) + \partial_F f_2(0) = \emptyset$. Thus, one has to settle for an approximate form of (3.3.1) in terms of subdifferentials at points near x. Such a result is referred to as a *local approximate sum rule*. In infinite dimensional Banach spaces there are two basic types of local approximate sum rules: strong and weak, corresponding to the approximation's accuracy up to an arbitrary strong- or weak-star neighborhood, respectively.

3.3.1 Strong Approximate Sum Rules

Theorem 3.3.1 (Strong Local Approximate Sum Rule) *Let X be a Fréchet smooth Banach space and let $f_1, \ldots, f_N \colon X \to \mathbb{R} \cup \{+\infty\}$ be lsc functions. Suppose that $\bar{x} \in \bigcap_{n=1}^{N} \operatorname{dom} f_n$ and there exists an $h > 0$ such that*

$$\sum_{n=1}^{N} f_n(\bar{x}) \le \bigwedge [f_1, \ldots, f_N](B_h(\bar{x})). \tag{3.3.2}$$

Then, for any $\varepsilon > 0$, there exist x_n and $x_n^* \in \partial_F f_n(x_n)$, $n = 1, \ldots, N$ satisfying

$$\operatorname{diam}(x_1, \ldots, x_N) \times \max(1, \|x_1^*\|, \ldots, \|x_N^*\|) < \varepsilon, \tag{3.3.3}$$

and

$$(x_n, f_n(x_n)) \in B_\varepsilon((\bar{x}, f_n(\bar{x}))) \tag{3.3.4}$$

such that

$$0 \in \sum_{n=1}^{N} x_n^* + \varepsilon B_{X^*}. \tag{3.3.5}$$

Proof. Since $f_n, n = 1, \ldots, N$ are lsc and since (3.3.2) implies that for any $0 < h' < h$,

$$\sum_{n=1}^{N} f_n(\bar{x}) \leq \bigwedge[f_1, \ldots, f_N](B_h(\bar{x}))$$

$$\leq \bigwedge[f_1, \ldots, f_N](B_{h'}(\bar{x})) \leq \sum_{n=1}^{N} f_n(\bar{x}), \tag{3.3.6}$$

decreasing h if necessary we may assume that $h \in (0, \min(1, \varepsilon))$ and that for any $x \in B_h(\bar{x})$,

$$f_n(x) > f_n(\bar{x}) - \varepsilon/N, n = 1, \ldots, N. \tag{3.3.7}$$

Moreover, it follows from condition (3.3.2) that for $\varepsilon_1 = h^2/32N^2$, we can choose $\eta \in (0, h)$ satisfying

$$\sum_{n=1}^{N} f_n(\bar{x}) \leq \inf \Big\{ \sum_{n-1}^{N} f_n(y_n) + \iota_{B_h(\bar{x})}(y_0) \ \Big| \ \|y_n - y_m\| \leq \eta,$$

$$n, m = 0, 1, \ldots, N \Big\} + \varepsilon_1. \tag{3.3.8}$$

Define, for $n = 1, \ldots, N$,

$$g_n := f_n + \| \cdot - \bar{x} \|^2 + \iota_{B_h(\bar{x})}.$$

Then g_n are lsc and bounded from below with

$$\bigwedge[g_1, \ldots, g_N](X) \leq \sum_{n=1}^{N} f_n(\bar{x}) < \infty.$$

Applying the nonlocal approximate sum rule of Theorem 3.2.3 to $g_n, n = 1, \ldots, N$ with $\varepsilon_2 \in (0, \min(\eta, \varepsilon_1))$ yields x_n and $y_n^* \in \partial_F g_n(x_n), n = 1, \ldots, N$ satisfying

$$\mathrm{diam}(x_1, \ldots, x_N) \times \max(1, \|y_1^*\|, \ldots, \|y_N^*\|) < \varepsilon_2, \tag{3.3.9}$$

and

$$\sum_{n=1}^{N} g_n(x_n) < \bigwedge[g_1, \ldots, g_N](X) + \varepsilon_2 \tag{3.3.10}$$

such that

$$\left\|\sum_{n=1}^{N} y_n^*\right\| < \varepsilon_2. \tag{3.3.11}$$

It follows from (3.3.10) that $x_n \in B_h(\bar{x})$, $n = 1, \ldots, N$ and from (3.3.8) and (3.3.9) that

$$\sum_{n=1}^{N}[f_n(\bar{x}) + \|x_n - \bar{x}\|^2] - \varepsilon_1 \leq \sum_{n=1}^{N} g_n(x_n) < \bigwedge[g_1, \ldots, g_N](X) + \varepsilon_2$$

$$\leq \sum_{n=1}^{N} g_n(\bar{x}) + \varepsilon_2 = \sum_{n=1}^{N} f_n(\bar{x}) + \varepsilon_2. \tag{3.3.12}$$

Thus,

$$\sum_{n=1}^{N} \|x_n - \bar{x}\|^2 \leq \varepsilon_1 + \varepsilon_2 < 2\varepsilon_1 \tag{3.3.13}$$

which implies that $x_n \in \operatorname{int} B_h(\bar{x})$, and each $(\|\cdot\|^2)'(x_n - \bar{x})$, $n = 1, \ldots, N$ is bounded by $\varepsilon/2N$. We can check that

$$x_n^* := y_n^* - (\|\cdot\|^2)'(x_n - \bar{x}) \in \partial_F f_n(x_n), \tag{3.3.14}$$

and from (3.3.9) and (3.3.11) that x_n and x_n^* satisfy (3.3.3) and (3.3.5). It remains to verify (3.3.4) which follows from (3.3.13), (3.3.7) and the following estimate:

$$f_n(x_n) \leq f_n(\bar{x}) + \sum_{m \neq n}[f_m(\bar{x}) - f_m(x_m)] + \varepsilon_2$$

$$< f_n(\bar{x}) + \frac{(N-1)\varepsilon}{N} + \varepsilon_2 < f_n(\bar{x}) + \varepsilon. \tag{3.3.15}$$

●

Condition (3.3.2) in the strong approximate sum rule cannot be replaced by a usual infimum in general. Examples are discussed in [100, 255]. The following proposition provides two useful sufficient conditions for (3.3.2). The proof is elementary and is left as an exercise.

Proposition 3.3.2 *Let $f_n \colon X \to \mathbb{R} \cup \{+\infty\}$, $n = 1, \ldots, N$ be lsc functions and let $\bar{x} \in \bigcap_{n=1}^{N} \operatorname{dom} f_n$. Then there exists a number $h > 0$ such that*

$$\sum_{n=1}^{N} f_n(\bar{x}) \leq \bigwedge[f_1, \ldots, f_N](B_h(\bar{x}))$$

if $\bar{x}$ is a local minimum of $\sum_{n=1}^{N} f_n$ and either

(i) *all but one of f_n are uniformly continuous in a neighborhood of $\bar{x}$, or*

(ii) *at least one of f_n has compact lower level sets in a neighborhood of $\bar{x}$.*

Proof. Exercise 3.3.2.   ●

3.3.2 Weak Approximate Sum Rules

If we are willing to weaken the conclusion of Theorem 3.3.1 in replacing εB_{X^*} in (3.3.5) by an arbitrary weak-star neighborhood, then condition (3.3.2) can be eliminated. Such a result is often called a weak local approximate sum rule.

Theorem 3.3.3 (Weak Local Approximate Sum Rule) *Let X be a Fréchet smooth Banach space and let $f_1, \dots, f_N \colon X \to \mathbb{R} \cup \{+\infty\}$ be lsc functions. Suppose that $\bar{x} \in \bigcap_{n=1}^N \mathrm{dom} f_n$ and $x^* \in \partial_F(\sum_{n=1}^N f_n)(\bar{x})$. Then, for any $\varepsilon > 0$ and any weak-star neighborhood V of 0 in X^*, there exist x_n and $x_n^* \in \partial_F f_n(x_n), n = 1, \dots, N$ satisfying*

$$\mathrm{diam}(x_1, \dots, x_N) \times \max(1, \|x_1^*\|, \dots, \|x_N^*\|) < \varepsilon, \tag{3.3.16}$$

and

$$(x_n, f_n(x_n)) \in B_\varepsilon((\bar{x}, f_n(\bar{x}))) \tag{3.3.17}$$

such that

$$x^* \in \sum_{n=1}^N x_n^* + V. \tag{3.3.18}$$

Proof. Let $\varepsilon > 0$ be a positive number and let V be a weak-star neighborhood of 0 in X^*. Fix $r > 0$ and a finite dimensional subspace L of X such that $L^\perp + 2r B_{X^*} \subset V$. Since $x^* \in \partial_F(\sum_{n=1}^N f_n)(\bar{x})$ there exists a C^1 function g such that $g'(\bar{x}) = x^*$ and $\sum_{n=1}^N f_n - g$ attains a local minimum at $\bar{x}$. Choose $0 < \eta < \min(\varepsilon, r)$ such that $\|y - \bar{x}\| < \eta < \varepsilon$ implies that $\|g'(x) - g(\bar{x})\| < r$. Then $\sum_{n=1}^N f_n - g + \iota_{\bar{x}+L}$ attains a local minimum at $\bar{x}$. Since $\iota_{\bar{x}+L}$ has locally compact sublevel sets, by Proposition 3.3.2, the functions $f_1, \dots, f_N, -g, \iota_{\bar{x}+L}$ satisfy the condition of Theorem 3.3.1.

Applying the strong local approximate sum rule of Theorem 3.3.1 yields the existence of $x_n, n = 1, \dots, N + 2$ such that $\|x_n - \bar{x}\| < \eta < \varepsilon, n = 1, \dots, N + 2$, $x_n^* \in \partial_F f(x_n), n = 1, \dots, N$, $x_{N+1}^* = -g'(x_{N+1})$ and $x_{N+2}^* \in \partial_F \iota_{\bar{x}+L}(x_{N+2})$ satisfying the conclusion of Theorem 3.3.1. That is, for $n = 1, \dots, N$,

$$|f_n(x_n) - f_n(\bar{x})| < \eta < \varepsilon,$$

$$\|x_n^*\| \times \mathrm{diam}(\{x_1, \dots, x_N\}) \leq \|x_n^*\| \times \mathrm{diam}(\{x_1, \dots, x_{N+2}\}) < \eta < \varepsilon$$

and

$$|\iota_{\bar{x}+L}(x_{N+2}) - \iota_{\bar{x}+L}(\bar{x})| < \eta.$$

Thus $x_{N+2} \in \bar{x} + L$, and

$$\left\| \sum_{n=1}^{N} x_n^* - g'(x_{N+1}) + x_{N+2}^* \right\| < r.$$

Note that $\partial_F \iota_{\bar{x}+L}(x_{N+2}) = L^\perp$ and $\|x^* - g'(x_{N+1})\| < r$. Therefore,

$$x^* \in \sum_{n=1}^{N} x_n^* + L^\perp + 2rB_{X^*} \subset \sum_{n=1}^{N} x_n^* + V.$$

$\bullet$

Note that in finite dimensional Banach spaces the strong and weak(star) topologies coincide, and therefore the strong approximate sum rule holds in finite dimensional Banach spaces without condition (3.3.2).

3.3.3 Normal Vectors to (Sub)Level Sets

When f is a C^1 function it is well known that if $f'(\bar{x}) \neq 0$ then it generates the normal cone of $f^{-1}((-\infty, f(\bar{x})])$ at $\bar{x}$. We now discuss nonsmooth versions of this fact as an application of the approximate local sum rules. They are closely related to the approximation of the singular normal vectors to the epigraph and graph of functions and are important in studying necessary optimality conditions for constrained optimization problems.

Theorem 3.3.4 (Representation of Normal Vectors of Sublevel Sets) *Let X be a Fréchet smooth Banach space and let $f: X \to \mathbb{R} \cup \{+\infty\}$ be a lsc function. Suppose that $\liminf_{x \to \bar{x}} d(\partial_F f(x) \; ; \; 0) > 0$ and $\xi \in N_F(f^{-1}((-\infty, a]); \bar{x})$. Then, for any $\varepsilon > 0$, there exist $\lambda > 0$, $(x, f(x)) \in B_\varepsilon((\bar{x}, f(\bar{x})))$ and $x^* \in \partial_F f(x)$ such that*

$$\|\lambda x^* - \xi\| < \varepsilon.$$

Proof. Assume that

$$\liminf_{x \to \bar{x}} d(\partial_F f(x), 0) > c > 0. \tag{3.3.19}$$

We consider the nontrivial case when $\xi \neq 0$. Moreover, since for any $(\bar{x}, a) \in \text{epi} f$, $N_F(f^{-1}((-\infty, a]); \bar{x}) \subset N_F(f^{-1}((-\infty, f(\bar{x})]); \bar{x})$ we may assume that $a = f(\bar{x})$. Choose $\eta \in (0, \varepsilon)$ satisfying

$$2\eta\|\xi\| + \frac{2\eta\|\xi\|(1 + 2\eta)}{c} < \varepsilon/2. \tag{3.3.20}$$

and choose $\delta \in (0, \varepsilon)$ such that

$$\langle \xi, h \rangle < \eta \|\xi\| \|h\|, \text{ for all } \bar{x} + h \in f^{-1}((-\infty, a]) \cap B_\delta(\bar{x}), \ h \neq 0, \text{(3.3.21)}$$

f is bounded below on $B_\delta(\bar{x})$ and

$$\inf_{x \in B_\delta(\bar{x})} d(\partial_F f(x) ; 0) > c.$$

Then

$$[\bar{x} + K(\xi, \eta)] \cap f^{-1}((-\infty, a]) \cap B_\delta(\bar{x}) = \{\bar{x}\}, \qquad (3.3.22)$$

where $K(\xi, \eta)$ is the Bishop–Phelps cone. Define, for each natural number i,

$$g_i := f - a + id_{\bar{x} + K(\xi, 2\eta)}.$$

Then (g_i) is bounded below on $B_\delta(\bar{x})$ and $g_i(\bar{x}) = 0$. We consider two possible cases: (A) $\inf_{B_\delta(\bar{x})} g_i < 0$ and (B) $\inf_{B_\delta(\bar{x})} g_i = 0$. For case (A), by the Ekeland variational principle of Theorem 2.1.1 there exists $y_i \in B_\delta(\bar{x})$ such that $g_i(y_i) < 0$ and

$$g_i + \frac{1}{i} \| \cdot - y_i \|$$

attains a (local) minimum at y_i over $B_\delta(\bar{x})$. We must have $y_i \notin \bar{x} + K(\xi, \eta)$ for otherwise (3.3.22) implies that $y_i \notin f^{-1}((-\infty, a])$ and $g_i(y_i) \geq 0$, a contradiction. We claim that

$$d_{\bar{x} + K(\xi, 2\eta)}(y_i) \geq \frac{\eta}{(2\eta + 1)} \|y_i - \bar{x}\|. \qquad (3.3.23)$$

Indeed, if

$$\|h\| < \frac{\eta}{(2\eta + 1)} \|y_i - \bar{x}\|$$

or

$$\|h\| < \eta \|y_i - \bar{x}\| - 2\eta \|h\|,$$

then

$$\begin{aligned}
\langle \xi, y_i - \bar{x} + h \rangle &= \langle \xi, y_i - \bar{x} \rangle + \langle \xi, h \rangle \\
&< \eta \|\xi\| \|y_i - \bar{x}\| + \|\xi\| \|h\| \\
&< 2\eta \|\xi\| [\|y_i - \bar{x}\| - \|h\|] \\
&\leq 2\eta \|\xi\| \|y_i - \bar{x} + h\|.
\end{aligned}$$

That is to say $y_i - \bar{x} + h \notin K(\xi, 2\eta)$. Since $(f(y_i))$ is bounded from below we have

$$\lim_{i \to \infty} d_{\bar{x} + K(\xi, 2\eta)}(y_i) = 0. \qquad (3.3.24)$$

Combining (3.3.23) and (3.3.24) we have $y_i \to \bar{x}$ as $i \to \infty$. Therefore, for i sufficiently large we have $y_i \in \text{int } B_\delta(\bar{x})$. In case (B) we set $y_i = \bar{x}$. Thus, in both cases (A) and (B),

$$g_i + \frac{1}{i}\|\cdot - y_i\|$$

attains a local minimum at y_i when i is sufficiently large. By the strong approximate sum rule of Theorem 3.3.1 there exists $x_i, z_i \in \text{int } B_\delta(\bar{x})$, $x_i^* \in \partial_F f(x_i)$ and $z_i^* \in \partial_F d_{\bar{x}+K(\xi,2\eta)}(z_i)$ such that

$$\|x_i^* + iz_i^*\| < \eta + 1/i. \tag{3.3.25}$$

Since $K(\xi, 2\eta)$ is convex, so is the function $d_{\bar{x}+K(\xi,2\eta)}$. Moreover, this function is a Lipschitz function with Lipschitz constant 1. Using the convex separation theorem as an exercise (see Exercises 4.2.8 and 4.3.13) we can show that

$$\partial_F d_{\bar{x}+K(\xi,2\eta)}(\cdot) \subset \{\alpha(-\xi + 2\eta\|\xi\|B_{X^*}) \mid \alpha > 0\} \cap B_{X^*} \tag{3.3.26}$$

Let $z_i^* = \alpha_i(-\xi + 2\eta\|\xi\|b^*)$ for some $b^* \in B_{X^*}$. It follows from (3.3.25) that

$$\|x_i^* - i\alpha_i\xi\| < 2i\alpha_i\|\xi\|\eta + \eta + 1/i. \tag{3.3.27}$$

We must have $i\alpha_i > c/2\|\xi\|(1 + 2\eta)$ for otherwise we would have $\|x_i^*\| < c$, a contradiction. Now letting $\lambda_i = 1/i\alpha_i$ and multiplying (3.3.27) by λ_i we have

$$\|\lambda_i x_i^* - \xi\| < 2\eta\|\xi\| + \frac{2\eta\|\xi\|(1 + 2\eta)}{c} + \frac{2\|\xi\|(1 + 2\eta)}{ic}.$$

For

$$i > \frac{4\|\xi\|(1 + 2\eta)}{c\varepsilon},$$

setting $\lambda = \lambda_i$, $x = x_i$ and $x^* = x_i^*$ we have

$$\|\lambda x^* - \xi\| < \varepsilon.$$

A corresponding approximation can be derived for the normal cone of $f^{-1}(a)$ when f is a continuous function.

Theorem 3.3.5 (Representation of Normal Vectors to the Level Sets) *Let X be a Fréchet smooth Banach space and let $f: X \to \mathbb{R}$ be a continuous function. Suppose that $\liminf_{x \to \bar{x}} d(\partial_F f(x) \cup \partial_F(-f)(x), 0) > 0$ and $\xi \in N_F(f^{-1}(a); \bar{x})$. Then, for any $\varepsilon > 0$, there exists $\lambda > 0$, $(x, f(x)) \in B_\varepsilon((\bar{x}, f(\bar{x})))$ and $x^* \in \partial_F f(x) \cup \partial_F(-f)(x)$ such that*

$$\|\lambda x^* - \xi\| < \varepsilon.$$

Proof. We give a sketch of the proof here and the details are left as an exercise. As in the proof of Theorem 3.3.4 we may assume that

$$\liminf_{x \to \bar{x}} d(\partial_F f(x) \cup \partial_F(-f)(x), 0) > c > 0 \tag{3.3.28}$$

and $\xi \neq 0$. Note that we must have $a = f(\bar{x})$. As in the proof of Theorem 3.3.4, choose $\eta, \delta \in (0, \varepsilon)$ small enough so that

$$\langle \xi, h \rangle < \eta \|\xi\| \|h\|, \text{ for all } \bar{x} + h \in f^{-1}(a) \cap B_\delta(\bar{x}), \ h \neq 0, \quad (3.3.29)$$

and

$$\inf_{x \in B_\delta(\bar{x})} d(\partial_F f(x), 0) > c.$$

Then

$$[\bar{x} + K(\xi, \eta)] \cap f^{-1}(a) \cap B_\delta(\bar{x}) = \{\bar{x}\}, \quad (3.3.30)$$

where $K(\xi, \eta)$ is the Bishop–Phelps cone defined in Section 2.2.1. We have that either

(a) $f(x) \geq a$ for all $x \in [\bar{x} + K(\xi, \eta)] \cap B_\delta(\bar{x})$, or
(b) $f(x) \leq a$ for all $x \in [\bar{x} + K(\xi, \eta)] \cap B_\delta(\bar{x})$.

In fact, suppose on the contrary that there exist $x_1, x_2 \in [\bar{x} + K(\xi, \eta)] \cap B_\delta(\bar{x})$ such that $f(x_1) > a$ and $f(x_2) < a$. Then $x_1, x_2 \neq \bar{x}$. Since f is continuous there exists $r \in (0, 1)$ such that $z := rx_1 + (1-r)x_2$ satisfies $f(z) = a$. Clearly $z \in [\bar{x} + K(\xi, \eta)] \cap B_\delta(\bar{x})$, and therefore $z = \bar{x}$ by (3.3.30). However, this leads to $0 = r(x_1 - \bar{x}) + (1-r)(x_2 - \bar{x})$ or $r(x_1 - \bar{x}) = -(1-r)(x_2 - \bar{x}) \in K(\xi, \eta) \cap [-K(\xi, \eta)] = \{0\}$, a contradiction. Define

$$g_i := \begin{cases} f - a + id_{\bar{x} + K(\xi, 2\eta)} & \text{in case (a)} \\ -f + a + id_{\bar{x} + K(\xi, 2\eta)} & \text{in case (b)}. \end{cases}$$

The rest of the proof is similar to that of Theorem 3.3.4.   ●

Using Theorems 3.3.4 and 3.3.5 we can also easily derive the following subdifferential characterizations for the singular normal vectors of the epigraph or the graph of a function.

Theorem 3.3.6 (Subdifferential Approximation of the Singular Normal Cone to the Epigraph and Graph of a Function) *Let X be a Fréchet smooth Banach space and let $f: X \to \mathbb{R} \cup \{+\infty\}$ be a lower semicontinuous (resp. continuous) function. Suppose that $(\bar{x}^*, 0) \in N_F(\text{epi} f; (\bar{x}, a))$ (resp. $(\bar{x}^*, 0) \in N_F(\text{graph} f; (\bar{x}, f(\bar{x})))$). Then, for any $\varepsilon > 0$, there exist x, x^* and $\lambda \in (0, \varepsilon)$, such that $x^* \in \partial_F f(x)$ (resp. $x^* \in \partial_F f(x) \cup \partial_F(-f)(x)$), $(x, f(x)) \in B_\varepsilon(\bar{x}, f(\bar{x}))$ and*

$$\|\lambda x^* - \bar{x}^*\| < \varepsilon.$$

Proof. We sketch the proof for the case of epigraph and leave the details as an exercise. Without loss of generality, we may assume that $a = f(\bar{x})$. Set $F(x, t) = f(x) - t$. Then $\text{epi} f = F^{-1}((-\infty, 0])$. Applying Theorem 3.3.4 to the function F yields the conclusion.   ●

3.3.4 Constrained Minimization Problems

Constrained optimization problems provide important models in many different applications. Finding good first order necessary conditions for solutions to such problems is a prerequisite. For smooth finite dimensional problems the Lagrange multiplier theorem and Karush–Kuhn–Tucker conditions are the canonical results. There is a vast literature on the generalizations of these "critical point" conditions to nonsmooth and infinite dimensional settings. We now present a form of such necessary conditions in terms of Fréchet subdifferentials as an application of the local approximate sum rule which enables us to impose minimal assumptions on the data: lower semicontinuity for the inequality constraints, continuity for the equality constraints and closedness for the feasible set. We consider the following minimization problem:

$$\mathcal{P} \qquad \text{minimize} \quad f_0(x)$$
$$\text{subject to} \quad f_n(x) \le 0, \quad n = 1, \ldots, N,$$
$$f_n(x) = 0, \quad n = N+1, \ldots, M,$$
$$x \in S \subset X.$$

Here X is a Fréchet smooth Banach space, $f_n, n = 0, 1, \ldots, N$ are lsc functions, f_n, $n = N+1, \ldots, M$ are continuous functions and S is a closed subset of X.

We will derive necessary optimality conditions for $\mathcal{P}$. Observe that if $\bar{x}$ is a solution to the constrained optimization problem $\mathcal{P}$ then it is a local minimum of the following function

$$f_0 + \sum_{n=1}^{N} \iota_{f_n^{-1}((-\infty,0])} + \sum_{n=N+1}^{M} \iota_{f_n^{-1}(0)} + \iota_S.$$

Applying the weak approximate sum rule of Theorem 3.3.3 yields a necessary condition in terms of the subdifferential of f_0 and the normal cones to the level sets of the f_n and S. Then we relate the normal cones to the level sets of the functions f_n to their subdifferentials by Theorems 3.3.4 and 3.3.5.

To simplify notation we introduce $\tau_n = 1$ for $n = 0, 1, \ldots, N$ and $\tau_n \in \{-1, 1\}$ for $n = N+1, \ldots, M$ corresponding, respectively, to inequality and equality constraints.

Theorem 3.3.7 (Approximate Multiplier Rule) *Let X be a Fréchet smooth Banach space, let S be a closed subset of X, let f_n be lsc for $n = 0, 1, \ldots, N$ and f_n be continuous for $n = N+1, \ldots, M$ and let $\bar{x}$ be a local solution of $\mathcal{P}$. Suppose that $\liminf_{x \to \bar{x}} d(\partial_F f_n(x), 0) > 0$, for $n = 1, \ldots, N$ and $\liminf_{x \to \bar{x}} d(\partial_F f_n(x) \cup \partial_F(-f_n)(x), 0) > 0$, for $n = N+1, \ldots, M$. Then, for any positive number $\varepsilon > 0$ and any weak-star neighborhood U of 0 in X^*, there exist $(x_n, f_n(x_n)) \in (\bar{x}, f_n(\bar{x})) + \varepsilon B_{X \times \mathbb{R}}, n = 0, 1, \ldots, M$ and $x_{M+1} \in \bar{x} + \varepsilon B_X$ such that*

$$0 \in \partial_F f_0(x_0) + \sum_{n=1}^{M} \mu_n \partial_F(\tau_n f_n)(x_n) + N_F(S, x_{M+1}) + U$$

where $\mu_n > 0, n = 1, \ldots, M$.

Proof. Let V be a weak-star neighborhood of 0 in X^* and assume that $(M + 1)V \subset U$. Decreasing ε if necessary we may assume that for any $y \in B_\varepsilon(\bar{x})$,

$$\liminf_{x \to y} d(\partial_F f_n(x), 0) > 0, \quad \text{for } n = 1, \ldots, N, \text{ and}$$

$$\liminf_{x \to y} d(\partial_F f_n(x) \cup \partial_F(-f_n)(x), 0) > 0, \quad \text{for } n = N+1, \ldots, M.$$

Observe that $\bar{x}$ is a local minimum of the following function

$$f_0 + \sum_{n=1}^{N} \iota_{f_n^{-1}((-\infty,0])} + \sum_{n=N+1}^{M} \iota_{f_n^{-1}(0)} + \iota_S.$$

Since f_n, $n = 1, \ldots, N$ are lower semicontinuous and f_n, $n = N+1, \ldots, M$ are continuous we can choose $\eta \in (0, \varepsilon/2)$ such that $y \in B_\eta(\bar{x})$ implies $f_n(y) > f_n(\bar{x}) - \varepsilon/2$ for $n = 1, \ldots, N$ and $f_n(y) \in (f_n(\bar{x}) - \varepsilon/2, f_n(\bar{x}) + \varepsilon/2)$ for $n = N+1, \ldots, M$. By the weak approximate sum rule of Theorem 3.3.3 there exist $(x_0, f_0(x_0)) \in B_\eta((x, f_0(\bar{x})))$, $x_{M+1} \in B_\eta(\bar{x}) \cap S$, $y_n \in B_\eta(\bar{x}), n = 1, \ldots, M$ with $|\iota_{f_n^{-1}((-\infty, f_n(\bar{x})])}(y_n) - \iota_{f_n^{-1}((-\infty, f_n(\bar{x})])}(\bar{x})| < \eta$, $n = 1, \ldots, M$ and $|\iota_{f_n^{-1}(0)}(y_n) - \iota_{f_n^{-1}(0)}(\bar{x})| < \eta$, $n = N+1, \ldots, M$, (and therefore $(y_n, f_n(y_n)) \in B_{\varepsilon/2}((\bar{x}, f_n(\bar{x}))))$, $x_0^* \in \partial_F f(x_0)$, $x_{M+1}^* \in N_F(C; x_{M+1})$, $y_n^* \in N_F(f_n^{-1}((-\infty, f_n(\bar{x})]); y_n)$ for $n = 1, \ldots, N$ and $y_n^* \in N_F(f_n^{-1}(0); y_n)$ for $n = N+1, \ldots, M$ such that

$$0 \in x_0^* + \sum_{n=1}^{M} y_n^* + x_{M+1}^* + V. \tag{3.3.31}$$

Theorems 3.3.4 and 3.3.5 imply that there exist

$$(x_n, f_n(x_n)) \in B_{\varepsilon/2}((y_n, f_n(y_n))) \subset B_\varepsilon((\bar{x}, f_n(\bar{x}))),$$

$\mu_n > 0$ and $x_n^* \in \partial_F f_n(x_n)$ ($x_n^* \in \partial_F f_n(x_n) \cup \partial_F(-f_n)(x_n)$) for $n = 1, \ldots, N$ ($n = N+1, \ldots, M$) such that

$$y_n^* \in \mu_n x_n^* + V. \tag{3.3.32}$$

Combining (3.3.31) and (3.3.32) completes the proof.   ●

Conditions $\liminf_{x \to \bar{x}} d(\partial_F f_n(x), 0) > 0$, for $n = 1, \ldots, N$ and

$$\liminf_{x \to \bar{x}} d(\partial_F f_n(x) \cup \partial_F(-f_n)(x), 0) > 0,$$

for $n = N + 1, \ldots, M$ serve as "constraint qualifications" to force the coefficient μ_0 of $\partial_F f_0$ to be one. However, since our necessary conditions are in an approximate form they are less stringent than the usual constraint qualifications such as the Mangasarian–Fromovitz condition. These conditions are not necessary if we do not insist μ_0 to be nonzero. Indeed, if the above condition fails for one of the f_n's then we can assign the multiplier corresponding to that f_n to be 1 and the rest of the multipliers to be 0. Thus, the following form of the multiplier rule holds without any constraint qualification.

Theorem 3.3.8 (Weak Approximate Multiplier Rule) *Let X be a Fréchet smooth Banach space, let S be a closed subset of X, let f_n be lsc for $n = 0, 1, \ldots, N$ and let f_n be continuous for $n = N + 1, \ldots, M$. Assume that $\bar{x}$ is a local solution of $\mathcal{P}$. Then, for any positive number $\varepsilon > 0$ and any weak-star neighborhood U of 0 in X^*, there exist $(x_n, f_n(x_n)) \in (\bar{x}, f_n(\bar{x})) + \varepsilon B_{X \times R}, n = 0, 1, \ldots, M$ and $x_{M+1} \in \bar{x} + \varepsilon B_X$ such that*

$$0 \in \sum_{n=0}^{M} \mu_n \partial_F(\tau_n f_n)(x_n) + N_F(S, x_{M+1}) + U$$

where $\mu_n \geq 0, n = 0, 1, \ldots, M$ and $\sum_{n=0}^{M} \mu_n = 1$.

Proof. Exercise 3.3.16.   ●

When X is a finite dimensional space and $f_n, n = 0, \ldots, M$ are C^1 functions and $S = X$ we can recover the Fritz John condition from the weak approximate multiplier rule by taking limits.

Theorem 3.3.9 (Fritz John Condition) *Let X be a finite dimensional Banach space, let $S = X$, let f_n be C^1 functions for $n = 0, 1, \ldots, M$. Assume that $\bar{x}$ is a local solution of $\mathcal{P}$. Then,*

$$0 \in \sum_{n=0}^{M} \mu_n \tau_n f_n'(\bar{x}).$$

where $\mu_n \geq 0, n = 0, 1, \ldots, M$ and $\sum_{n=0}^{M} \mu_n = 1$. Moreover, $\mu_n f_n(\bar{x}) = 0$ for $n = 1, \ldots, N$.

Proof. Exercise 3.3.17   ●

3.3.5 Sensitivity Analysis

In almost all practical problems, inaccuracies occur in both the modeling of a problem and collecting data for a given mathematical model. Thus, it is important to be able to gauge the influence of such inaccuracies on the outcome of the underlying mathematical model. This is often referred to as sensitivity analysis. The basic pattern is well illustrated by the following example.

Example 3.3.10 Consider the optimization problem $\mathcal{P}_a$ of minimizing $f(x)$ subject to $h(x) = a$ and define the optimal value or marginal function $v(a) := \inf\{f(x) : h(x) = a\}$. Then it is not hard to see that, for any x, $v(h(x)) \leq f(x)$. On the other hand, if $\bar{x}$ is a solution to $\mathcal{P}_0$ then $v(h(\bar{x})) = f(\bar{x})$. Thus, $\bar{x}$ is a minimum point for the function,

$$x \to f(x) - v(h(x)).$$

Assuming all the functions involved are smooth then

$$f'(\bar{x}) - v'(0)h'(\bar{x}) = 0.$$

In other words, $-v'(0)$ is a Lagrange multiplier (*shadow price*) of the problem $\mathcal{P}_0$.

We have seen that v is rarely a smooth function. Therefore, the above argument will not apply in general. Nevertheless the general pattern does persist and it turns out that the Fréchet subdifferential provides a convenient language to describe it.

Consider the following family of constrained minimization problems without the set constraint.

$$\mathcal{P}_a \qquad \text{minimize} \quad f_0(x)$$
$$\text{subject to} \quad f_n(x) \leq a_n, \quad n = 1, \ldots, M,$$
$$f_n(x) = a_n, \quad n = M+1, \ldots, N.$$

We denote $a = (a_1, \ldots, a_N)$ and the infimum of f_0 over the feasible set of $\mathcal{P}_a$ by $v(a)$.

Motivated by Theorem 3.3.7 we define the *multiplier set* of problem $\mathcal{P}_a$ as follows.

Definition 3.3.11 *Let ε be a positive number and let U be a weak-star neighborhood of 0 in X^*. We say $\nu = (\nu_1, \ldots, \nu_N)$ is a multiplier of problem $\mathcal{P}_b$ corresponding to $(\bar{x}, \varepsilon, U)$ if $\nu_n, n = 1, \ldots, N$ are not all 0, $\tau_n \nu_n \geq 0$ and there exist $(x_n, f_n(x_n)) \in (\bar{x}, f_n(\bar{x})) + \varepsilon B_{X \times \mathbb{R}}, n = 0, 1, \ldots, N$ such that*

$$0 \in \partial_F f_0(x_0) + \sum_{n=1}^{N} \tau_n \nu_n \partial_F(\tau_n f_n)(x_n) + U.$$

We denote the set of all such multipliers by $M_{\varepsilon,U}(\bar{x})$. Here τ_n are as in Theorem 3.3.7.

Theorem 3.3.12 (Sensitivity) *Let x_a be a solution to problem $\mathcal{P}_a$. Then, for any $\varepsilon > 0$ and any weak-star neighborhood U of 0 in X^*,*

$$-\partial_F v(a) \subset M_{\varepsilon,U}(x_a) + \varepsilon B_{\mathbb{R}^N}.$$

Proof.   There is nothing to prove if $\partial_F v(a) = \emptyset$. Let $\lambda \in -\partial_F v(a) \neq \emptyset$. Then there exists a Fréchet smooth function g such that $v + g$ attains a local minimum 0 at a and $g'(a) = \lambda$. Note that for any x satisfying the constraint, $f_n(x) \leq b_n, n = 1, \ldots, M$ and $f_n(x) = b_n, n = M + 1, \ldots, N$, we have $f_0(x) \geq v(b)$ so that

$$f_0(x) + g(b) \geq v(b) + g(b) \geq v(a) + g(a) = f_0(x_a) + g(a).$$

Thus, (x_a, a) is a solution to the minimization problem

$$\text{minimize} \quad f_0(x) + g(b)$$
$$\text{subject to} \quad f_n(x) - b_n \leq 0, \quad n = 1, \ldots, M,$$
$$f_n(x) - b_n = 0, \quad n = M + 1, \ldots, N.$$

Choose $\varepsilon' > 0$ smaller than $\varepsilon/2$ such that $a' \in \varepsilon' B_{\mathbb{R}^N}$ implies that $\|g'(a') - g'(a)\| < \varepsilon/2$. Note that $U \times \varepsilon' B_{\mathbb{R}^N}$ is a weak-star neighborhood in $X^* \times \mathbb{R}^N$. By Theorem 3.3.7, there exists $x_n \in x_a + \varepsilon' B_X \subset x_a + \varepsilon B_X$ and $a' \in a + \varepsilon' B_{\mathbb{R}^N}$ such that

$$0 \in \partial_F f_0(x_0) \times \{g'(a')\} + \sum_{n=1}^{N} \nu_n [\partial_F(\tau_n f_n)(x_n) \times (-\tau_n e_n)] + U \times \varepsilon' B_{\mathbb{R}^N}$$

where $\{e_n, n = 1, \ldots, N\}$ is the standard basis of $\mathbb{R}^N$. We can rewrite this relation as

$$0 \in \partial_F f_0(x_0) + \sum_{n=1}^{N} \nu_n \partial_F(\tau_n f_n)(x_n) + U$$

and

$$0 \in g'(a') - \nu + \varepsilon' B_{\mathbb{R}^N}.$$

That is $\lambda \in M_{\varepsilon,U}(x_a) + \varepsilon B_{\mathbb{R}^N}$.   ●

3.3.6 Constraint Qualifications

If in a multiplier $\mu = (\mu_0, \ldots, \mu_N)$ for the necessary conditions of Theorem 3.3.8, μ_0 is zero or can be arbitrarily close to zero, then this multiplier is detached from the cost function f_0, and therefore does not provide useful information. We call such multipliers *degenerate*. Thus, it is important to find conditions ensuring the existence of nondegenerate multipliers in the necessary conditions. These conditions are called *constraint qualifications*. We will illustrate the ideas in finite dimensional spaces.

In general, there are two different types of constraint qualifications. The first makes μ_0 near 0 impossible in Theorem 3.3.8; or all μ_n's uniformly bounded in Theorem 3.3.7. The following is a natural qualification.

(CQ1) There exist constants $\eta, c > 0$ such that, for any $\varepsilon \in (0, \eta)$ and

$$(x_n, f_n(x_n)) \in (\bar{x}, f_n(\bar{x})) + \varepsilon B_{X \times \mathbb{R}}, n = 0, 1, \ldots, N$$

and $x_{N+1} \in B_\varepsilon(\bar{x})$ such that

$$d\left(0, \sum_{n=1}^{N} \mu_n \partial_F(\tau_n f_n)(x_n) + N_F(S, x_{N+1})\right) \geq c$$

where $\tau_n = 1$ for $n = 1, \ldots, M$ and $\tau_n \in \{-1, 1\}$ for $n = M+1, \ldots, N$, $\mu_n \geq 0, n = 1, \ldots, N$ and $\sum_{n=1}^{N} \mu_n = 1$.

With this constraint qualification we have the following generalization of the Karush–Kuhn–Tucker necessary conditions.

Theorem 3.3.13 *Let X be a finite dimensional Banach space, let S be a closed subset of X, let f_n be lsc for $n = 0, 1, \ldots, M$ and let f_n be continuous for $n = M + 1, \ldots, N$. Assume that $\bar{x}$ is a local solution of $\mathcal{P}$ and suppose (CQ1) holds. Then, for any positive number $\varepsilon > 0$, there exist $(x_n, f_n(x_n)) \in (\bar{x}, f_n(\bar{x})) + \varepsilon B_{X \times \mathbb{R}}, n = 0, 1, \ldots, N$ and $x_{N+1} \in \bar{x} + \varepsilon B_X$ and a positive constant K such that*

$$0 \in \partial_F f_0(x_0) + \sum_{n=1}^{N} \mu_n \partial_F(\tau_n f_n)(x_n) + N_F(S, x_{N+1}) + \varepsilon B_{X^*}.$$

where $\mu_n \in [0, K]$.

Proof. Without loss of generality we may assume that $\varepsilon < c/2$. If there exists $(x_0, f_0(x_0)) \in (\bar{x}, f_n(\bar{x})) + \varepsilon B_{X \times \mathbb{R}}$ such that $d(\partial_F f_0(x_0), 0) < \varepsilon$ then we need only to take $\mu_0 = 1$ and $\mu_n = 0$ for $n = 1, \ldots, N$. Otherwise, applying Theorem 3.3.8 we have $(x_n, f_n(x_n)) \in (\bar{x}, f_n(\bar{x})) + \varepsilon B_{X \times \mathbb{R}}, n = 0, 1, \ldots, N$ and $x_{N+1} \in \bar{x} + \varepsilon B_X$ such that

$$0 \in \sum_{n=0}^{N} \lambda_n \partial_F(\tau_n f_n)(x_n) + N_F(S, x_{N+1}) + (\varepsilon/2) B_{X^*}, \qquad (3.3.33)$$

where $\lambda_n \geq 0$ and $\sum_{n=0}^{N} \lambda_n = 1$. We claim that $\lambda_0 \geq 1/2$. In fact, if $\lambda_0 < 1/2$ then, by (CQ1) we have

$$\lambda_0 \varepsilon/2 \geq \lambda_0 d(\partial_F f_0(x_0), 0) = d(\lambda_0 \partial_F f_0(x_0), 0) \geq (1 - \lambda_0)c - \varepsilon/2 > \varepsilon/2,$$

a contradiction. It remains to multiply (3.3.33) by $1/\lambda_0$ and let $\mu_n = \lambda_n/\lambda_0, n = 1, \ldots, N$ and $K = 2$. ●

For problems with smooth data and without the set constraint, the constraint qualification (CQ1) becomes the following Mangasarian–Fromovitz condition.

(MF) The derivatives $f_n'(\bar{x}), n = 1, \ldots, N$ are linearly independent.

Theorem 3.3.14 (Karush–Kuhn–Tucker Conditions) *Let X be a finite dimensional Banach space, let $S = X$ and let f_n be C^1 functions for $n = 0, \ldots, N$. Assume that $\bar{x}$ is a local solution of $\mathcal{P}$ and suppose the constraint qualification condition* (MF) *holds at $\bar{x}$. Then,*

$$0 \in f_0'(\bar{x}) + \sum_{n=1}^{N} \mu_n \tau_n f_n'(\bar{x}),$$

where $\mu_n \geq 0$ are not all zero and τ_n are as in (CQ1).

Proof. Exercise 3.3.19.   ●

In Theorem 3.3.14, the assumption that $f_n, n = 0, \ldots, N$ are C^1 cannot be replaced by the assumption that they are Fréchet differentiable at $\bar{x}$. A counterexample is given in Exercise 3.3.20.

The second type of constraint qualification condition directly ensures the existence of a multiplier such that μ_0 is bounded away from 0 for Theorem 3.3.8, or the μ_n's are all bounded for Theorem 3.3.7. To state such a condition let us consider the following perturbation of problem $\mathcal{P}$ where $a = (a_1, \ldots, a_N)$.

$$\mathcal{P}_a \qquad \text{minimize} \quad f_0(x)$$
$$\text{subject to} \quad f_n(x) \leq a_n, n = 1, \ldots, M,$$
$$f_n(x) = a_n, n = M + 1, \ldots, N,$$
$$x \in S \subset X.$$

We denote the infimum of f_0 over the feasible set of $\mathcal{P}_a$ by $v(a)$. Now we can state our second constraint qualification condition, which is often referred to as the calmness condition.

(CQ2) $\partial_F v(0) \neq \emptyset$.

Theorem 3.3.15 *Let X be a finite dimensional Banach space, let S be a closed subset of X, let f_n be lsc for $n = 0, 1, \ldots, M$ and let f_n be continuous for $n = M + 1, \ldots, N$. Assume that $\bar{x}$ is a local solution of $\mathcal{P}$ and suppose* (CQ2) *holds. Then, for any positive number $\varepsilon > 0$, there exist $(x_n, f_n(x_n)) \in (\bar{x}, f_n(\bar{x})) + \varepsilon B_{X \times \mathbb{R}}, n = 0, 1, \ldots, N$ and $x_{N+1} \in \bar{x} + \varepsilon B_X$ and a positive constant K such that*

$$0 \in \partial_F f_0(x_0) + \sum_{n=1}^{N} \mu_n \partial_F (\tau_n f_n)(x_n) + N_F(S, x_{N+1}) + \varepsilon B_{X^*}$$

where $\mu_n \in [0, K]$.

Proof. Suppose that $p \in \partial_F v(0) \neq \emptyset$. Then there exists a C^1 function g on R^N with $g'(0) = p$ such that $v - g$ attains a local minimum at $a = 0$. It follows that $(\bar{x}, 0)$ is a solution of the following minimization problem on space $X \times \mathbb{R}^N$

$$\text{minimize} \quad f_0(x) - g(a)$$
$$\text{subject to} \quad f_n(x) - a_n \leq 0, n = 1, \dots, M,$$
$$f_n(x) - a_n = 0, n = M+1, \dots, N,$$
$$(x, a) \in S \times \mathbb{R}^N.$$

Choose $\eta < \min(1, \varepsilon/2)$ such that $\|a'\| < \eta$ implies that $\|g'(a') - p\| < 1$. Applying Theorem 3.3.7 we have $a_0 \in \eta B_{\mathbb{R}^N}$, $(x_n, f_n(x_n)) \in (\bar{x}, f_n(\bar{x})) + \eta B_{X \times \mathbb{R}}, n = 0, 1, \dots, N$, $x_{N+1} \in \bar{x} + \eta B_X$ and $\mu_n \geq 0, n = 1, \dots, N$ such that

$$0 \in (\partial_F f_0(x_0), g'(a_0)) + \sum_{n=1}^{N} \mu_n(\partial_F(\tau_n f_n)(x_n), -e_n)$$
$$+ N_F(S, x_{N+1}) \times \{0_{\mathbb{R}^N}\} + \eta B_{X^* \times \mathbb{R}^N}, \quad (3.3.34)$$

Now the first component of (3.3.34) provides the necessary condition and the second component of (3.3.34) shows that $\mu_n \leq K := \|p\| + 1$. ●

3.3.7 Constrained Optimization and Inequalities

Necessary optimality conditions for constrained optimization problems are often effective tools for discovering and proving inequalities. We illustrate by proving the following inequality.

Theorem 3.3.16 Let $x_n \in \mathbb{R}, n = 1, \dots, N$ satisfy $\sum_{n=1}^{N} x_n \geq 0$. Then

$$\sum_{n=1}^{N} x_n e^{x_n} \geq \frac{2}{N} \sum_{n=1}^{N} x_n^2.$$

Proof. The strategy is to consider the constrained minimization problem of minimizing $\sum_{n=1}^{N} x_n e^{x_n} - \frac{2}{N} \sum_{n=1}^{N} x_n^2$, subject to the constraint $\sum_{n=1}^{N} x_n \geq 0$ and show that the minimum is nonnegative.

First we show that the minimum is attained. For $x = (x_1, \dots, x_N) \in \mathbb{R}^N$ we denote the p norm of x by

$$\|x\|_p := \left(\sum_{n=1}^{N} |x_n|^p \right)^{1/p}.$$

We will also use the notation $x^+ := (x_1^+, \dots, x_N^+)$ and $x^- := (x_1^-, \dots, x_N^-)$ where, for any $t \in \mathbb{R}$, $t^+ := \max(0, t)$ and $t^- := \max(0, -t)$. It is not hard to check $\|x\|_2 \leq \|x\|_1$, for any $x \in \mathbb{R}^N$. Moreover, for $x \in \mathbb{R}^N$ satisfying $\sum_{n=1}^{N} x_n \geq 0$ we have $\|x\|_1 \leq 2\|x^+\|_1$ (Exercise 3.3.25). For $x = (x_1, \ldots, x_N) \in \mathbb{R}^N$, define

$$f(x) := \sum_{n=1}^{N} x_n e^{x_n} - \frac{2}{N} \sum_{n=1}^{N} x_n^2.$$

Since $\psi(t) = te^t$ is convex on $[0, \infty)$ and $te^{-t} \leq 1/e$ for all $t \in [0, \infty)$ (Exercise 3.3.26), we have

$$f(x) = \sum_{n=1}^{N} x_n^+ e^{x_n^+} - \sum_{n=1}^{N} x_n^- e^{-x_n^-} - \frac{2}{N} \sum_{n=1}^{N} x_n^2$$

$$\geq N\psi\left(\frac{1}{N} \sum_{n=1}^{N} x_n^+ e^{x_n^+}\right) - \frac{N}{e} - \frac{2}{N}\|x\|_2^2$$

$$\geq N\psi\left(\frac{1}{N}\|x^+\|_1\right) - \frac{N}{e} - \frac{2}{N}\|x\|_1^2$$

$$\geq \|x^+\|_1 \exp\left(\frac{1}{N}\|x^+\|_1\right) - \frac{N}{e} - \frac{8}{N}\|x^+\|_1^2.$$

Thus, for $x \in \{x \in \mathbb{R}^N \mid \sum_{n=1}^{N} x_n \geq 0\}$ as $\|x\|_1 \to \infty$ we have $f(x) \to \infty$. It follows that the set $\{x \in \mathbb{R}^N \mid f(x) \leq c, \sum_{n=1}^{N} x_n \geq 0\}$ is compact. Therefore, the constrained minimization problem

$$\text{minimize} \quad f(x)$$

$$\text{subject to} \quad -\sum_{n=1}^{N} x_n \leq 0,$$

has a solution, say $\bar{x}$. Applying the Karush–Kuhn–Tucker necessary optimality condition in Theorem 3.3.14 to the above minimization problem we have that there exists a $\mu \geq 0$ such that, for $n = 1, \ldots, N$,

$$(\bar{x}_n + 1)e^{\bar{x}_n} - \frac{4}{N}\bar{x}_n = \mu, \tag{3.3.35}$$

where moreover $\mu = 0$ if $\sum_{n=1}^{N} \bar{x}_n = 0$. This condition is satisfied when $\bar{x} = 0$; then $f(\bar{x}) = 0$. From now on we assume that $\bar{x} \neq 0$ which implies that $\bar{x}$ has at least one strictly positive component.

Consider the function $g(t) = (t+1)e^t - 4t/N$. We can check that g has a unique critical point $\bar{t} \in (-2, 0]$ and is strictly decreasing to the left of $\bar{t}$ and strictly increasing to the right of $\bar{t}$ (see Exercise 3.3.27).

Thus, equation (3.3.35) has at most two solutions. Therefore, the components of $\bar{x}$ can achieve at most only two different values. Since $\bar{x} \neq 0$ and $\sum_{n=1}^{N} \bar{x}_n \geq 0$, the components of $\bar{x}$ take exactly two values, one positive and the other negative. Now suppose there are M positive components all equal to a and $N - M$ negative components all equal to b with $1 \leq M \leq N - 1$. Moreover, one can directly check that for $N \geq 2$ and $t \geq 0$ we have $g(-t) \leq g(t)$ (Exercise 3.3.27). This implies that $|b| \geq a$, and so we can write $b = -sa$ with some $s > 1$. Also the condition $Ma + (N - M)b \geq 0$ implies $(N - M)s \leq M$, which in turn implies that $s \leq N - 1$.

Thus, we have

$$f(\bar{x}) = Mae^a - (N - M)sae^{-sa} - \frac{2}{N}(Ma^2 + (N - M)s^2a^2)$$

$$\leq Mae^a - Mae^{-sa} - \frac{2}{N}(Ma^2 + Msa^2)$$

$$\leq Ma^2(1 + s)\left(\frac{e^a - e^{-sa}}{a(1 + s)} - \frac{2}{N}\right) \geq 0$$

because, for all $a > 0$ and $s \in (1, N - 1]$,

$$\frac{e^a - e^{-sa}}{a(1 + s)} \geq \frac{e^a - e^{-a}}{a(1 + s)} > \frac{2}{(1 + s)} \geq \frac{2}{N}.$$

•

Necessary optimality conditions can be used to derive many other inequalities. Some examples are given in Exercises 3.3.28, 3.3.29 and 3.3.30.

3.3.8 Commentary and Exercises

The prototype of the local approximate sum rules appeared in Ioffe [143]. Generalizations and refinements can be found in [48, 68, 108, 146, 148, 170, 209]. Limiting forms of the subdifferential sum rule in finite dimensional spaces are discussed in Section 5.2 along with counterexamples showing that they fail in infinite dimensional spaces without additional assumptions. The two approximate sum rules in this section are taken from [68]. Theorem 3.3.1 fails without condition (3.3.2) (see examples in Exercises 3.3.11 and 3.3.12). On the other hand, condition (3.3.2) is not tight either (study the example in Exercise 3.3.10). A version of Theorem 3.3.1 with a condition weaker than (3.3.2) is discussed in [170], where one may also find additional examples. Condition (3.3.2) has the drawback that it is not stable in terms of adding new "nice" functions. An example is given in Exercise 3.3.4. In this respect the stronger *sequential uniform lower semicontinuity* condition introduced in [48, 148] is more convenient (see Exercises 3.3.5–3.3.9).

The approximate sum rule is particularly useful in discussing necessary optimality conditions for constrained minimization problems. It allows us to use penalization functions involving indicator functions [251] and make the proof of the necessary condition simple and natural. Theorem 3.3.7 was proved in Borwein, Treiman and Zhu [63] for problems in reflexive Banach spaces and generalized in [213, 276]. Limiting forms of these necessary optimality conditions can be derived by taking limits of the approximate form. An example is Theorem 5.2.25 in Section 5.2. This result is quite general in that it allows lsc inequality constraints and continuous equality constraints. Some more general constraints such as general equilibrium constraints [189, 214] and variational inequality constraints [268] can be conveniently converted to the equality and inequality constraints in $\mathcal{P}$. Such conversions are discussed in [276] and the essence is given in guided Exercises 3.3.24 and 3.3.23. The history of these kinds of necessary conditions can be traced back to Lagrange, whose original work led to the Fritz John [155] and the Karush–Kuhn–Tucker [158, 168] conditions for smooth problems with equality and inequality constraints. Generalizations to nonsmooth problems with set constraints were developed with the advance of nonsmooth analysis; related literature can be found in [84, 144, 185, 198, 223, 229, 265, 266].

Research in sensitivity analysis for nonsmooth mathematical programming problems can be found in [9, 84, 123, 124, 201, 236, 262]. A variational argument naturally reduces the discussion of the sensitivity to an appropriate necessary condition for the corresponding optimization problem. Note that to calculate the value of $v(a)$ one has to solve an optimization problem which usually is costly. By contrast the multiplier set is defined through the original data and is easier to get hold of. Hence, Theorem 3.3.12 provides a useful estimate for the value function v (see Exercise 3.3.18 for an example).

Constraint qualification condition (CQ1) can be viewed as a nonsmooth version of the Mangasarian–Fromovitz condition (MF) [190, 268] while (CQ2) is the calmness condition [84] stated in terms of the Fréchet subdifferential of the optimal value function. Note that, unlike (CQ1), condition (CQ2) does not exclude the possibility of existing degenerate multipliers and is therefore more interesting. However, this condition is in terms of the optimal value function for a perturbed problem rather than the data of the original problem and is harder to verify.

The inequality in Theorem 3.3.16 is from [47] where one can find the more accurate inequality with the constant $2/N$ improved to C_N/N where $C_N = \max\{2, e(1 - 1/N)\}$.

Exercise 3.3.1 Prove that

$$\sum_{n=1}^{N} f_n(\bar{x}) \leq \bigwedge[f_1, \ldots, f_N](B_h(\bar{x}))$$

if and only if, for any $r \leq h$,

$$\sum_{n=1}^{N} f_n(\bar{x}) = \bigwedge[f_1, \ldots, f_N](B_r(\bar{x}))$$

Exercise 3.3.2 Prove Proposition 3.3.2 and show that condition (i) or (ii) in Proposition 3.3.2 cannot be dispensed with.

Exercise 3.3.3 Show that in Theorem 3.3.1, if any f_n is a C^1 function then we can replace x_n^* by $f_n'(\bar{x})$.

Exercise 3.3.4 Define $A := \{0\} \cup \{1/i \mid i = \pm 1, \pm 2, \cdots\}$, $f_1(x) := \iota_A(x)$ and $f_2(x) := x$. Verify that, for any $h > 0$, $f_1(0) \leq \bigwedge [f_1]([-h, h])$ yet

$$\sum_{n=0}^{2} f_n(0) > \bigwedge [f_1, f_2]([-h, h]).$$

*Exercise 3.3.5 (Sequential Uniform Lower Semicontinuity Condition)

Definition 3.3.17 *Let X be a Banach space, let $f_1, \ldots, f_N : X \to \mathbb{R} \cup \{+\infty\}$ be lsc functions and let $\bar{x} \in \bigcap_{n=1}^{N} \mathrm{dom} f_n$. We say that $(f_1, \ldots, f_N)$ is sequentially uniformly lower semicontinuous at $\bar{x}$ if there exists $h > 0$ such that for any sequences $(x_{n,i}), n = 1, \ldots, N$ belonging to $B_h(\bar{x})$ and such that $\lim_{i \to \infty} \mathrm{diam}(x_{1,i}, \ldots, x_{N,i}) = 0$, there is a sequence (u_i) in $B_h(\bar{x})$ such that, for all $n = 1, \ldots, N$, $\lim_{i \to \infty} \|x_{n,i} - u_i\| = 0$ and*

$$\liminf_{i \to \infty} \sum_{n=1}^{N} (f_n(x_{n,i}) - f_n(u_i)) \geq 0.$$

(i)   Show that if $(f_1, \ldots, f_N)$ is sequentially uniformly lower semicontinuous at $\bar{x}$ and $\bar{x}$ is a local minimum of $\sum_{n=1}^{N} f_n$, then condition (3.3.2) holds.

(ii)   Construct two functions f_1, f_2 such that (f_1, f_2) is not sequentially uniformly lower semicontinuous at a point $\bar{x}$ yet they satisfy condition (3.3.2). Reference: [71, Example 2.10].

(iii) Suppose that f_0 is uniformly continuous in a neighborhood of $\bar{x}$. Prove that if $(f_1, \ldots, f_N)$ is sequentially uniformly lower semicontinuous at $\bar{x}$ then so is $(f_0, f_1, \ldots, f_N)$.

Exercise 3.3.6 Prove the following sufficient conditions for sequential uniform lower semicontinuity.

Proposition 3.3.18 *Let $f_n : X \to \mathbb{R} \cup \{+\infty\}, n = 1, \ldots, N$ be lsc functions and let $\bar{x} \in \bigcap_{n=1}^{N} \mathrm{dom} f_n$. Then $(f_1, \ldots, f_N)$ is sequentially uniformly lower semicontinuous at $\bar{x}$ if $\bar{x}$ is a local minimum of $\sum_{n=1}^{N} f_n$ and either*

(i)   *all but one of f_n are uniformly continuous in a neighborhood of $\bar{x}$, or*

(ii) *at least one of f_n has compact lower level sets in a neighborhood of $\bar{x}$.*

Exercise 3.3.7 (Sequential Weak Lower Semicontinuity) We say a function $f : X \to \mathbb{R} \cup \{+\infty\}$ is *sequentially weak lower semicontinuous* (w-lsc) at $\bar{x}$ if for any sequence (x_i) weak convergent to $\bar{x}$, $\liminf_{i \to \infty} f(x_i) \geq f(\bar{x})$. Show that if f is a convex function and is lsc at $\bar{x}$, then it is w-lsc at $\bar{x}$.

Exercise 3.3.8 Prove that if $f_1, \ldots, f_N$ are functions in a reflexive Banach space and are w-lsc at $\bar{x}$ then $(f_1, \ldots, f_N)$ is sequentially uniformly lower semicontinuous at $\bar{x}$.

Exercise 3.3.9 Prove the following general form of the strong local approximate sum rule.

Theorem 3.3.19 *Let X be a Fréchet smooth Banach space and let $f_1, \ldots, f_N \colon X \to \mathbb{R} \cup \{+\infty\}$ be lsc functions. Let $\bar{x} \in \bigcap_{n=1}^{N} \operatorname{dom} f_n$ and suppose that $(f_1, \ldots, f_N)$ is sequentially uniformly lower semicontinuous at $\bar{x}$ and $x^* \in \partial_F(\sum_{n=1}^{N} f_n)(\bar{x})$. Then, for any $\varepsilon > 0$, there exist x_n and $x_n^* \in \partial_F f_n(x_n), n = 1, \ldots, N$ satisfying*

$$\operatorname{diam}(x_1, \ldots, x_N) \times \max(1, \|x_1^*\|, \ldots, \|x_N^*\|) < \varepsilon, \qquad (3.3.36)$$

and

$$(x_n, f_n(x_n)) \in B_\varepsilon((\bar{x}, f_n(\bar{x}))) \qquad (3.3.37)$$

such that

$$\left\| x^* - \sum_{n=1}^{N} x_n^* \right\| < \varepsilon. \qquad (3.3.38)$$

Exercise 3.3.10 Let $X := \ell_2$ and let $\{e_i\}$ be the standard basis. Define

$$f_1(x) := \begin{cases} 0 & \text{if } x = 0, \\ -1/j & \text{if } x = e_i/j, \\ \infty & \text{otherwise}, \end{cases}$$

and

$$f_2(x) := \begin{cases} 0 & \text{if } x = 0, \\ -1/j & \text{if } x = (e_i + e_1/i)/j, \\ \infty & \text{otherwise}, \end{cases}$$

where $i, j = 1, \ldots$. Verify that for $\bar{x} = 0$, $f_n, n = 1, 2$ do not satisfy condition (3.3.2) yet the conclusion of Theorem 3.3.1 holds. Reference: [274].

*Exercise 3.3.11 Let $X := \ell_2$ and let $\{e_i\}$ be the standard basis. Then $x \in X$ can be uniquely represented as $x = \sum_{i=1}^{\infty} x_i e_i$. Moreover, $x_i \to 0$ as $i \to \infty$ and so $\|x\|_\infty := \max\{|x_i| \mid 1 \le i < \infty\}$ exists. Define $F_j := \{x \mid \|x\| \le 3, \ x_i \ge 0$ and $x_i = 0$ if $i \bmod 3 \ne 0$ or $i < 3j\}$. Now we construct two functions

$$f_1(x) := \begin{cases} 0 & \text{if } x = 0, \\ -\frac{1}{\sqrt{n}} - \|y\|_\infty & \text{if } x = \frac{1}{j}e_{3j-1} + y, y \in F_j, \\ +\infty & \text{otherwise}, \end{cases}$$

and

$$f_2(x) := \begin{cases} 0 & \text{if } x = 0, \\ -\frac{1}{\sqrt{j}} - \|y\|_\infty & \text{if } x = \frac{1}{j}e_{3j-2} + y, y \in F_j, \\ +\infty & \text{otherwise}. \end{cases}$$

Prove that f_1 and f_2 are lsc functions and $f_1 + f_2$ attains a minimum at $x = 0$. Yet, for any $\|x_n\| \le 1$ and $x_n^* \in \partial_F f_n(x_n), n = 1, 2, \|x_1^* + x_2^*\| \ge 1$. Reference: [255].

*Exercise 3.3.12 Let $X := \ell_2$ and let $\{e_i\}, i = 1, 2, \ldots$, be the standard basis. Denote $Y := \{x \in X \mid \langle e_1, x \rangle = 0\}$.

(i) Construct a sequence of even functions $\alpha_i \colon \mathbb{R} \to \mathbb{R}$, $i = 2, 3, \ldots$, with support in $[-(i-1)^{-1/2}, -(i+1)^{-1/2}] \cup [(i+1)^{-1/2}, (i-1)^{-1/2}]$ for $i \ge 3$ and with support $(-\infty, -1] \cup [1, +\infty)$ for α_2 such that, for each $t \ne 0$, there exists $i \ge 2$ with $\alpha_i(t) > 1$ and such that for each t and each n, $\alpha_i(t) \in [0, 2]$ and $\alpha_i(i^{-1/2}) = 2$.

(ii) Define
$$f(x) := -\sum_{i=2}^{\infty} \alpha_i(x_1)x_i - |x_1|,$$
where $x_i = \langle x, e_i \rangle$. Show that f is continuous.

(iii) Show that f is C^1 on $X \backslash Y$ and for any $x \in X \backslash Y$, $\|f'(x)\|_{Y^*} \ge 1$.

(iv) Show that for any $x \in Y$, $\partial_F f(x) = \emptyset$.

(v) Verify that the strong fuzzy sum rule fails for $f_1 := f$ and $f_2 := \iota_Y$ at any $x \in Y$ (reference [100]).

Exercise 3.3.13 Consider the problem of minimizing $f(x)$ subject to $x \in C$ where C is a closed nonempty subset of a Fréchet smooth Banach space X. Use appropriate subdifferential sum rules to derive necessary conditions for a solution $\bar{x}$ of this problem in terms of the Fréchet subdifferential of f and the Fréchet normal cone of C for the following three cases: (i) f is C^1, (ii) f is Lipschitz and (iii) f is lsc.

Exercise 3.3.14 Prove inclusion 3.3.26.

Exercise 3.3.15 Provide details for the proof of Theorem 3.3.5.

Exercise 3.3.16 Prove Theorem 3.3.8.

Exercise 3.3.17 Prove Theorem 3.3.9. Hint: Take limits in Theorem 3.3.8. The conditions $\mu_n f_n(\bar{x}) = 0$ for $n = 1, \ldots, N$ follow from the fact that if $f_n(\bar{x}) < 0$ then the corresponding constraint is not binding and can be taken away so that one can let $\mu_n = 0$.

Exercise 3.3.18 Consider the optimal value
$$v(a) := \inf\{f(x) : g(x) = a\}$$
as a function of a, where $f(x) := 1 - \cos x$ and $g(x) := \sin(6x) - 3x$ and $a \in [-\pi/2, \pi/2]$ which corresponds to $x \in [-\pi/6, \pi/6]$.

(i) Use Theorem 3.3.12 to show that
$$\partial_F v(a) \subset \{\lambda : f'(x) - \lambda g'(x) = 0, g(x) = a\}.$$

(ii) Show that v is a C^1 function in the three open intervals $(-\pi/2, \pi/6 - \sqrt{3}/2)$, $(\pi/6 - \sqrt{3}/2, \sqrt{3}/2 - \pi/6)$ and $(\sqrt{3}/2 - \pi/6, \pi)$.

Exercise 3.3.19 Prove Theorem 3.3.14. Hint: Use condition (MF) to show that $\mu_0 > 0$ (and therefore can be scaled to 1) in Theorem 3.3.9.

Exercise 3.3.20 (Nonexistence of Multipliers [118]) Define the sign function $\operatorname{sgn} \colon \mathbb{R} \to \mathbb{R}$ by

$$\operatorname{sgn}(x) = \begin{cases} 1 & \text{if } x > 0, \\ 0 & \text{if } x = 0, \\ -1 & \text{if } x < 0. \end{cases}$$

Consider functions $f(x,y) := x$ and $h(x,y) := y - \operatorname{sgn}(y)(x^+)^2$. Show that h is Fréchet differentiable at $(0,0)$ with $h'(0,0) = (0,1)$ and $(0,0)$ is an optimal solution for the problem

$$\begin{aligned} \text{minimize} \quad & f(x,y) \\ \text{subject to} \quad & h(x,y) = 0 \end{aligned}$$

yet the Karush–Kuhn–Tucker condition is not satisfied at $(0,0)$.

Exercise 3.3.21 Find conditions in terms of f_n and S that ensure that the optimal value function $v(a)$ is lsc.

Exercise 3.3.22 Give an example of a minimization problem that satisfies (CQ2) but not (CQ1).

*$*$**Exercise 3.3.23** Derive necessary optimality conditions for the following optimization problems in finite dimensional Banach spaces:

(i) (Multifunction Constraint)

$$\begin{aligned} \text{minimize} \quad & f(x) \\ \text{subject to} \quad & 0 \in F(x). \end{aligned}$$

Here $f \colon X \to \mathbb{R} \cup \{+\infty\}$ is a lsc function and $F \colon X \to 2^Y$ is a closed multifunction.

(ii) (Variational Inequality Constraints)

$$\begin{aligned} \text{minimize} \quad & f(x) \\ \text{subject to} \quad & \langle h(x), y - x \rangle \geq 0, \text{ for all } y \in C \\ & x \in C. \end{aligned}$$

Here $f \colon X \to \mathbb{R} \cup \{+\infty\}$ is a lsc function, $h \colon X \to \mathbb{R}^N$ is a continuous function and C is a closed convex subset of X.

Hint: Convert the constraint to $0 \in h(x) + N_F(C; x)$. Reference: [276].

*Exercise 3.3.24 Consider a general model of optimization problems with equilibrium constraints.

$\mathcal{MPEC}$ 　　　　　　　minimize 　$f_0(x)$

　　　　　　　　subject to 　$0 \in h(x) + F(g(x))$,

　　　　　　　　　　　$x \in S$,

where $f_0: X \to \mathbb{R} \cup \{+\infty\}$ is a lower semicontinuous function, S is a closed subset of X, $h: X \to \mathbb{R}^{N-M}$ and $g: X \to \mathbb{R}^M$ are continuous functions and $F: \mathbb{R}^M \to 2^{\mathbb{R}^{N-M}}$ is a multifunction with a closed graph. Derive a necessary optimality condition by converting problem $\mathcal{MPEC}$ to a problem with the form of $\mathcal{P}$. Hint: Denote the components of g and h by $g = (f_1, \ldots, f_M)$ and $h = (f_{M+1}, \ldots, f_N)$ and show that if $\bar{x}$ is a solution to problem $\mathcal{MPEC}$ then $(\bar{x}, g(\bar{x}), -h(\bar{x}))$ is a solution to the following optimization problem.

$\mathcal{AP}$ 　　　　　　　minimize 　$f_0(x)$

　　　　　　subject to 　$f_n(x) - u_n = 0, n = 1, \ldots, M$,

　　　　　　　　　　$f_n(x) + u_n = 0, n = M+1, \ldots, N$,

　　　　　　　　　　$(x, u) \in X \times \operatorname{graph} F$.

Reference: [276].

Exercise 3.3.25 Let $x \in \mathbb{R}^N$.

(i) Show that $\|x\|_2 \leq \|x\|_1$.
(ii) Show that if $\sum_{n=1}^N x_n \geq 0$, then $\|x\|_1 \leq 2\|x^+\|_1$.

Exercise 3.3.26 Show that

(i) $\psi(t) = te^t$ is convex on $[0, \infty)$; and
(ii) $te^{-t} \leq 1/e$ for all $t \in [0, \infty)$.

Exercise 3.3.27 Let $g(t) = (t+1)e^t - 4t/N$, where $N \geq 2$.

(i) Show that g has a unique critical point $\bar{t} \in (-2, 0]$ and is strictly decreasing to the left of $\bar{t}$ and strictly increasing to the right of $\bar{t}$.
(ii) Show that for any $t \geq 0$, $g(-t) \leq g(t)$.

Exercise 3.3.28 (Largest Eigenvalue [56, p. 162]) Let A be an $N \times N$ symmetric matrix. Use the Karush–Kuhn–Tucker necessary optimality conditions to calculate

$$\max\{\langle x, Ax \rangle \mid \|x\| = 1, x \in \mathbb{R}^N\}.$$

*Exercise 3.3.29 (Largest Eigenvalue [139, p. 135]) Let A be an $M \times N$ matrix. Consider the optimization problem

$$\alpha = \sup\{\langle x, Ay \rangle \mid \|x\|^2 = 1, \|y\|^2 = 1, x \in \mathbb{R}^M, y \in \mathbb{R}^N\}. \quad (3.3.39)$$

and the matrix

$$\tilde{A} = \begin{bmatrix} 0 & A \\ A^\top & 0 \end{bmatrix}.$$

Here $A^\top$ is the transpose of A.

(i)   If μ is an eigenvalue of $\tilde{A}$, prove $-\mu$ is also.
(ii)   If μ is an eigenvalue of $\tilde{A}$, use a corresponding eigenvector to construct a feasible solution to problem (3.3.39) with objective value μ.
(iii) Prove problem (3.3.39) has an optimal solution.
(iv)   Use the Karush–Kuhn–Tucker necessary optimality conditions to prove any optimal solution to problem (3.3.39) corresponds to an eigenvector of $\tilde{A}$.
(v)   (Jordan [156]) Deduce α is the largest eigenvalue of $\tilde{A}$ (This number is called largest singular value of A.)

Exercise 3.3.30 (Hadamard's Inequality [132]) Let $(x^1, \ldots, x^N)$ be an $N \times N$ matrix with columns $x^1, \ldots, x^N$ in $\mathbb{R}^N$. Prove $(\bar{x}^1, \ldots, \bar{x}^N)$ solves the problem

$$\text{minimize} \quad -\det(x^1, \ldots, x^N)$$

$$\text{subject to} \quad \|x^n\|^2 = 1 \text{ for } x^n \in \mathbb{R}^N, n = 1, \ldots, N$$

if and only if $\det(\bar{x}^1, \ldots, \bar{x}^N) = 1$ and $\bar{x}^1, \ldots, \bar{x}^N$ forming an orthonormal basis of $\mathbb{R}^N$. Deduce the inequality

$$\det(x^1, \ldots, x^N) \leq \Pi_{n=1}^N \|x^n\|.$$

3.4 Approximate Mean Value Theorems and Applications

The mean value theorems are fundamental results in calculus. They have numerous applications. The proofs of mean value theorems are classical examples of variational arguments. This section is devoted to the discussion of mean value theorems, their generalizations to nonsmooth functions and applications.

3.4.1 Mean Value Theorems

We start with Rolle's Theorem which illustrates the pattern and provides a foundation for developing other more general mean value theorems.

Theorem 3.4.1 (Rolle's Mean Value Theorem) *Let $f\colon \mathbb{R} \to \mathbb{R}$ be a function and let $a < b$ be two real numbers. Suppose that f is continuous on $[a, b]$, differentiable on (a, b) and $f(a) = f(b)$. Then there exists a point $c \in (a, b)$ such that $f'(c) = 0$.*

Proof. We consider the nontrivial case when f is not a constant on $[a, b]$. Since f is continuous on $[a, b]$, f or $-f$ attains its minimum at some point $c \in (a, b)$. Thus, $f'(c) = 0$. ●

In applications the following Lagrange mean value theorem is often more flexible.

Theorem 3.4.2 (Lagrange Mean Value Theorem) *Let $f\colon \mathbb{R} \to \mathbb{R}$ be a function and let $a < b$ be two real numbers. Suppose that f is continuous on $[a, b]$ and differentiable on (a, b). Then there exists a point $c \in (a, b)$ such that*

$$f(b) - f(a) = f'(c)(b - a).$$

Proof. Apply Rolle's Theorem to $h(x) := f(x) - \frac{f(b)-f(a)}{b-a}(x - a)$. ●

A similar trick can be used to derive the even more general Cauchy mean value theorem. We leave that as an exercise (Exercise 3.4.1) and turn to applications of differential characterizations of Lipschitz property, monotonicity and convexity. We say a function $f\colon \mathbb{R} \to \mathbb{R}$ is Lipschitz with a Lipschitz constant L provided that for all $x, y \in \mathbb{R}$, $|f(y) - f(x)| \leq L|y - x|$.

Theorem 3.4.3 (Characterization of Lipschitz Property) *Let $f\colon \mathbb{R} \to \mathbb{R}$ be a differentiable function. Then f is Lipschitz with a Lipschitz constant L if and only if, for all $x \in \mathbb{R}$, $|f'(x)| \leq L$.*

Proof. Exercise 3.4.2. ●

We say a function $f\colon \mathbb{R} \to \mathbb{R}$ is (strictly) increasing provided that for any $x < y$, $(f(x) < f(y))\ f(x) \leq f(y)$. We say f is (strictly) decreasing if $-f$ is (strictly) increasing. A function is (strictly) monotone if it is either (strictly) increasing or (strictly) decreasing. Monotonicity of a differentiable function is characterized by the fact that its derivative preserves sign. This can also be proven easily using the mean value theorem.

Theorem 3.4.4 (Characterization for Monotonicity) *Let $f\colon \mathbb{R} \to \mathbb{R}$ be a differentiable function. Then f is increasing if and only if for all $x \in \mathbb{R}$, $f'(x) \geq 0$.*

Proof. Exercise 3.4.3. ●

Note that $f'(x) > 0$ for all $x \in \mathbb{R}$ is a sufficient condition for f to be strictly increasing (Exercise 3.4.4) but is not a necessary condition (examining $f(x) = x^3$).

Recall that a function $f \colon \mathbb{R} \to \mathbb{R}$ is convex provided that for any $x, y \in \mathbb{R}$ and $\lambda \in [0, 1]$,

$$f(\lambda x + (1 - \lambda)y) \leq \lambda f(x) + (1 - \lambda)f(y).$$

For a differentiable function the convex property is characterized by the increasing of its derivative.

Theorem 3.4.5 (Characterization for Convex Property) *Let $f \colon \mathbb{R} \to \mathbb{R}$ be a differentiable function. Then f is convex if and only if f' is increasing.*

Proof. Necessity: Let $x < y$ and let $\lambda \in (0, 1)$. It follows from

$$f(\lambda x + (1 - \lambda)y) \leq \lambda f(x) + (1 - \lambda)f(y)$$

that

$$\frac{f(y + \lambda(x - y)) - f(y)}{\lambda} \leq f(y) - f(x).$$

Taking limits as $\lambda \to 0$ we have $f'(y)(x - y) \leq f(x) - f(y)$ or

$$f'(y) \geq \frac{f(y) - f(x)}{y - x}. \tag{3.4.1}$$

Switch the position of x and y and λ and $1 - \lambda$ and taking limits as $\lambda \to 1$ we have

$$f'(x) \leq \frac{f(y) - f(x)}{y - x}. \tag{3.4.2}$$

Combining (3.4.1) and (3.4.2) we have $f'(y) \geq f'(x)$.

Sufficiency: Suppose that f' is increasing. Let $x < y$ and let $\lambda \in [0, 1]$. Applying the Lagrange mean value theorem on intervals $[x, \lambda x + (1 - \lambda)y]$ and $[\lambda x + (1 - \lambda)y, y]$, respectively, we have that there exist $c_1 \in (x, \lambda x + (1 - \lambda)y)$ and $c_2 \in (\lambda x + (1 - \lambda)y, y)$ such that $f(x) - f(\lambda x + (1 - \lambda)y) = f'(c_1)(1 - \lambda)(x - y)$ and $f(y) - f(\lambda x + (1 - \lambda)y) = f'(c_2)\lambda(y - x)$. Clearly $f(c_2) \geq f(c_1)$. It follows that

$$
\begin{aligned}
\lambda f(x) &+ (1 - \lambda)f(y) - f(\lambda x + (1 - \lambda)y) \\
&= \lambda(f(x) - f(\lambda x + (1 - \lambda)y)) + (1 - \lambda)(f(y) - f(\lambda x + (1 - \lambda)y)) \\
&= \lambda f'(c_1)(1 - \lambda)(x - y) + (1 - \lambda)f'(c_2)\lambda(y - x) \\
&= \lambda(1 - \lambda)(f'(c_2) - f'(c_1))(y - x) \geq 0.
\end{aligned}
$$

Therefore, f is convex.   ●

The above differential characterizations for the Lipschitz property, the monotonicity and the convexity lead to many useful inequalities. Some examples are given as exercises in the end of this section.

3.4.2 Approximate Mean Value Theorems

A closer look at the applications of the mean value theorems will reveal that (a) only one direction of the inequality is important, and (b) the variational arguments proving these mean value theorems are in fact valid for lsc functions. We now make the above observations precise.

Theorem 3.4.6 (Limiting Approximate Mean Value Theorem) *Let X be a Fréchet smooth Banach space, let $f \colon X \to \mathbb{R} \cup \{+\infty\}$ be a lsc function, let $a, b \in X$ be two distinct points with $f(a) < \infty$ and let $r \in \mathbb{R}$ be such that $r \leq f(b) - f(a)$. Then there exist $c \in [a, b)$ and a sequence x_i with $(x_i, f(x_i)) \to (c, f(c))$ and $x_i^* \in \partial_F f(x_i)$ such that*

(i) $\liminf_{i \to \infty} \langle x_i^*, c - x_i \rangle \geq 0$;
(ii) $\liminf_{i \to \infty} \langle x_i^*, b - a \rangle \geq r$;
(iii) $f(c) \leq f(a) + |r|$.

Proof. Take $v \in X^*$ such that $\langle v, a - b \rangle = r$. Then

$$g(x) := f(x) + \langle v, x \rangle + \iota_{[a,b]}(x)$$

attains its minimum at some $c \in [a, b)$ because $g(b) \geq g(a)$. Applying the local approximate sum rule of Theorem 3.3.1, there exist sequences (x_i), (y_i), (x_i^*) and (y_i^*) satisfying $(x_i, f(x_i)) \to (c, f(c))$, $x_i^* \in \partial_F f(x_i)$, $[a, b] \ni y_i \to c$ and $y_i^* \in N_F([a, b], y_i)$ such that $\|x_i^*\| \times \|x_i - y_i\| < 1/i$, $\|y_i^*\| \times \|x_i - y_i\| < 1/i$ and

$$\|x_i^* + y_i^* + v\| < 1/i.$$

Then (i) can be derived directly via:

$$\liminf_{i \to \infty} \langle x_i^*, c - x_i \rangle = \liminf_{i \to \infty} \langle x_i^* + v, c - x_i \rangle$$
$$= \liminf_{i \to \infty} \langle -y_i^*, c - y_i \rangle \geq 0.$$

To show (ii) note that $c \in [a, b)$ implies that $y_i \in [a, b)$ for i sufficiently large. Then

$$\langle x_i^* + v, b - a \rangle = \langle x_i^* + v, b - y_i \rangle \frac{\|b - a\|}{\|b - y_i\|}.$$

Taking limits we obtain

$$\liminf_{i \to \infty} \langle x_i^* + v, b - a \rangle = \liminf_{i \to \infty} \langle x_i^* + v, b - y_i \rangle \frac{\|b - a\|}{\|b - y_i\|}$$
$$= \liminf_{i \to \infty} \langle -y_i^*, b - y_i \rangle \frac{\|b - a\|}{\|b - c\|} \geq 0.$$

This is (ii) in disguise. Clearly $f(c)$ satisfies (iii). ●

By passing to a subsequence one can replace the limit inferior in Theorem 3.4.6 by a limit. One can also write the theorem in the following approximate form.

Theorem 3.4.7 (Approximate Mean Value Theorem) *Let X be a Fréchet smooth Banach space; let $f\colon X \to \mathbb{R} \cup \{+\infty\}$ be a lower semicontinuous function; let $a, b \in X$ be two distinct points with $f(a) < \infty$ and let $r \in \mathbb{R}$ be such that $r \leq f(b) - f(a)$. Then there exist $c \in [a, b)$ such that for any $\varepsilon > 0$, there exist $(x, f(x)) \in B_\varepsilon((c, f(c)))$ and $x^* \in \partial_F f(x)$ satisfying*

(i) $\langle x^*, c - x \rangle > -\varepsilon$;
(ii) $\langle x^*, b - a \rangle > r$;
(iii) $f(x) \leq f(a) + |r| + \varepsilon$.

Proof. Exercises 3.4.10.   ●

As in calculus we can use this approximate mean value theorem to derive subdifferential criteria for various properties of functions such as monotonicity, Lipschitzness, convexity, etc.

3.4.3 A Lipschitz Criterion

Theorem 3.4.8 *Let X be a Fréchet smooth Banach space, let $U \subset X$ be an open convex set with $U \cap \mathrm{dom}(f) \neq \emptyset$ and let $L > 0$. Then f is Lipschitz with a Lipschitz constant L on U if and only if for all $x \in U$, $\sup\{\|x^*\| : x^* \in \partial_F f(x)\} \leq L$.*

Proof. The "only if" part is straightforward. We prove the "if" part. Let $a, b \in U$ with $a \in \mathrm{dom(f)}$ and $a \neq b$, let $r \in \mathbb{R}$ such that $r \leq f(b) - f(a)$, and let $\varepsilon > 0$. It follows from Theorem 3.4.6 (ii) that there exist $x \in U$ and $x^* \in \partial_F f(x)$ such that

$$r \leq \langle x^*, b - a \rangle + \varepsilon \leq L\|b - a\| + \varepsilon.$$

Since $r \leq f(b) - f(a)$ and $\varepsilon > 0$ are arbitrary, we derive that $f(b) - f(a) \leq L\|b - a\|$. Therefore, $f(b) < \infty$. Exchanging the roles of a and b we can conclude that f is Lipschitz of rank L on U.   ●

Corollary 3.4.9 *Let X be a Fréchet smooth Banach space, let $f\colon X \to \mathbb{R} \cup \{+\infty\}$ be a lsc function and let $U \subset X$ be a path connected open set with $U \cap \mathrm{dom}(f) \neq \emptyset$. Then f is a constant function on U if and only if for all $x \in U$, $\partial_F f(x) \subset \{0\}$.*

Note that unlike the smooth counterpart of this result one needs only check where the subdifferential $\partial_F f$ is nonempty.

3.4.4 Cone Monotonicity

The next result generalizes the monotonicity criteria in calculus.

Let X be a Banach space, let K be a cone in X. We define the *polar* of K by

$$K^o := \{x^* \in X^* : \langle x^*, x \rangle \leq 0, \text{ for all } x \in K\}$$

Let f be a function on X. We say that f is K-*nonincreasing* provided that $y \in x + K$ implies $f(y) \leq f(x)$.

Theorem 3.4.10 *Let X be a Fréchet smooth Banach space, let K be a cone in X and let $f : X \to \mathbb{R} \cup \{+\infty\}$ be a lsc function. Suppose that for all x, $\partial_F f(x) \subset K^o$. Then f is K-nonincreasing.*

Proof. Let $x, y \in X$ such that $f(x) < f(y)$. It follows from the Approximate Mean Value Theorem that there exist $z \in \text{dom}(f)$ and $z^* \in \partial_F f(z)$ with $\langle z^*, y - x \rangle > 0$. Therefore $y - x$ does not belong to K.   ●

In particular, we have the following corollary.

Corollary 3.4.11 *Let $f : \mathbb{R} \to \mathbb{R} \cup \{+\infty\}$ be a lsc function. Suppose that for all x, $\partial_F f(x) \subset (-\infty, 0]$. Then f is nonincreasing.*

Proof. Exercise 3.4.12.   ●

3.4.5 Quasi-Convexity

Let X be a Banach space. We recall that a function $f : X \to \mathbb{R} \cup \{+\infty\}$ is called *quasi-convex* provided, for any $x, y \in \text{dom} f$ and $z \in [x, y]$, $f(z) \leq \max\{f(x), f(y)\}$ and that a multifunction $F : X \to X^*$ is *quasi-monotone* if

$$x^* \in F(x), y^* \in F(y) \text{ and } \langle x^*, y - x \rangle > 0 \Rightarrow \langle y^*, y - x \rangle \geq 0.$$

Theorem 3.4.12 *Let X be a Fréchet smooth Banach space and let $f : X \to \mathbb{R} \cup \{+\infty\}$ be a lsc function. Suppose that $\partial_F f$ is quasi-monotone. Then f is quasi-convex.*

Proof. We work by way of contradiction. Assume that there exist some $x, y, z \in X$ such that $z \in [x, y]$ and $f(z) > \max\{f(x), f(y)\}$. Applying Theorem 3.4.6 with $a = x$ and $b = z$, there exist sequences x_i and $x_i^* \in \partial_F f(x_i)$ such that $x_i \to \bar{x} \in [x, z]$, $\liminf_{i \to \infty} \langle x_i^*, \bar{x} - x_i \rangle \geq 0$ and $\liminf_{i \to \infty} \langle x_i^*, z - x \rangle > 0$. Combining with $y - \bar{x} = \frac{\|y - \bar{x}\|}{\|z - x\|}(z - x)$ we have

$$\liminf_{i \to \infty} \langle x_i^*, y - x_i \rangle > 0. \tag{3.4.3}$$

Let $\lambda \in (0, 1)$ be such that $z = \bar{x} + \lambda(y - \bar{x})$ and set $z_i := x_i + \lambda(y - x_i)$. Then $z_i \to z$. Since f is lower semicontinuous, in considering relation (3.4.3) we can pick an integer i such that $f(z_i) > f(y)$ and

$$\langle x_i^*, y - x_i \rangle > 0. \tag{3.4.4}$$

Applying Theorem 3.4.6 again with $a := y$ and $b := z_i$, there exist sequences (y_j) and (y_j^*) satisfying $y_j^* \in \partial_F f(y_j)$ such that $y_j \to \bar{y} \in [y, z_i)$, $\liminf_{j \to \infty} \langle y_j^*, \bar{y} - y_j \rangle \geq 0$ and $\liminf_{j \to \infty} \langle y_j^*, z_i - y \rangle > 0$. Noting that $z_i - y$ and $x_i - \bar{y}$ lie in the same direction, we obtain

$$\liminf_{j \to \infty} \langle y_j^*, x_i - y_j \rangle > 0. \tag{3.4.5}$$

Since $\bar{y} \in [x_i, y)$, inequality (3.4.4) yields

$$\liminf_{j \to \infty} \langle x_i^*, y_j - x_i \rangle = \langle x_i^*, \bar{y} - x_i \rangle > 0. \tag{3.4.6}$$

Inequalities (3.4.5) and (3.4.6) imply that for j sufficiently large, we have both $\langle y_j^*, x_i - y_j \rangle > 0$ and $\langle x_i^*, y_j - x_i \rangle > 0$, i.e., $\partial_F f$ is not quasi-monotone, a contradiction.   ●

3.4.6 Commentary and Exercises

Mean value theorems for differentiable functions are classical results in analysis. Their extension to nonsmooth functions can be traced back to Lebourg [171]. The limiting approximate mean value theorem of Theorems 3.4.6 appeared in Zagrodny [270] in terms of the Clarke-Rockafellar subdifferential, see also Penot [218]. The Fréchet subdifferential forms were given in [186, 208, 220] along with some applications. The proof given here by using the local approximate sum rule is taken from [274]. It is worth pointing out that although Theorems 3.4.6 and 3.4.7 are stated in (possibly) infinite dimensional Banach spaces, they are actually restricted to a line segment. In fact in the guided Exercise 3.4.9 one can find a Fréchet differentiable function on ℓ_2 constructed by Ferrer [119] that equals to 0 on the unit sphere yet whose derivative does not vanish inside the unit ball. Recently, Borwein, Kortezov and Wiersma constructed in [50] a delicate C^1 function on $\mathbb{R}^2$ that is even on the unit circle yet has no critical points in the unit ball showing that the Rolle's theorem is essentially a one dimensional result.

The prototypes of the Lipschitz criterion appeared in Rockafellar [231] for functions on finite dimensional spaces and in Treiman [249] for functions on infinite dimensional spaces. The proof adopted here is from [248, 186].

Characterization of quasi-convexity for lsc functions was discussed in [187]. The short proofs here follow a more general version in [10]. It is easy to deduce a subdifferential characterization for convex functions from this result that extends Theorem 3.4.5 to nonsmooth functions. We will discuss it later.

Exercise 3.4.1 Prove the Cauchy mean value theorem:

Theorem 3.4.13 (Cauchy Mean Value Theorem) *Let $f, g \colon \mathbb{R} \to \mathbb{R}$ be two functions and let $a < b$ be two real numbers. Suppose that f and g are continuous on $[a, b]$, differentiable on (a, b) and suppose g is strictly increasing. Then there exists a point $c \in (a, b)$ such that*

$$g'(c)(f(b) - f(a)) = f'(c)(g(b) - g(a)).$$

Hint: Apply Rolle's Theorem to $h(x) := f(x) - \frac{f(b)-f(a)}{g(b)-g(a)}(g(x) - g(a))$.

Exercise 3.4.2 Prove Theorem 3.4.3. Hint: Necessity follows directly from the definitions of derivative and the Lipschitz property. For sufficiency use the Lagrange mean value theorem and prove by contradiction.

Exercise 3.4.3 Prove Theorem 3.4.4. Hint: Necessity follows directly from the definitions of derivative. For sufficiency use the Lagrange mean value theorem and prove by contradiction.

Exercise 3.4.4 Show that if $f \colon \mathbb{R} \to \mathbb{R}$ is a differentiable function with $f'(x) > 0$ for all $x \in \mathbb{R}$ then f is strictly increasing.

Exercise 3.4.5 Prove that for $x, y \in \mathbb{R}$, $|\sin y - \sin x| \le |y - x|$.

Exercise 3.4.6 Use the trigonometric identity

$$\cos y - \cos x = -2\sin\frac{x+y}{2}\sin\frac{y-x}{2}$$

to establish the inequality, for $0 \le a < b$

$$\cos a - \cos b < \frac{b^2 - a^2}{2}.$$

Then prove that for $0 \le a < b$ and $x \neq 0$,

$$\frac{\cos ax - \cos bx}{x^2} < \frac{b^2 - a^2}{2}.$$

Exercise 3.4.7 Prove that for $h > 0$ and $p > 1$, $(1 + h)^p > 1 + ph$.

Exercise 3.4.8 Let $f \colon \mathbb{R} \to \mathbb{R}$ be a twice differentiable function. Show that f is convex if and only if $f''(x) \ge 0$ for all $x \in \mathbb{R}$.

*Exercise 3.4.9** (Ferrer's Example) In this exercise we break into several steps Ferrer's example of a Fréchet differentiable function on ℓ_2 that vanishes on the unit sphere yet has no critical point inside the unit ball.

We will denote by $\{e_i\}$ the standard basis of ℓ_2. Let $L, R \colon \ell_2 \to \ell_2$ be the linear left and right shift operators on ℓ_2 defined by

$$Lx = (x_2, x_3, x_4, \dots) \quad Rx = (0, x_1, x_2, x_3, \dots).$$

(i) Show that $\|Lx\| \le \|x\|$ and $\|Rx\| = \|x\|$, and L and R are adjoint to each other, that is

$$\langle x, Ru \rangle = \langle Lx, u \rangle.$$

(ii)   Define operator $T \colon \ell_2 \to \ell_2$ by

$$T(x) = (1/2 - \|x\|^2)e_1 + Rx.$$

Show that T is a continuous operator and $Tx \neq x$ for any $x \in \ell_2$.

(iii)   Define Ferrer's function

$$f(x) = \frac{1 - \|x\|^2}{\|x - T(x)\|^2}. \tag{3.4.7}$$

Verify that f is a Fréchet differentiable function that vanishes on the unit sphere.

(iv)   Use the quotient rule to compute

$$\langle f'(x), u \rangle = -\frac{2\langle x, u \rangle \|x - T(x)\|^2 + 2\langle x - T(x), u - T'(x)u \rangle}{\|x - T(x)\|^4}$$

for any $x, u \in \ell_2$.

(v)   Verify that

$$f'(x) = \frac{-2}{\|x - T(x)\|^4} \times \big[(\|x - T(x)\|^2 \\ + (1 - \|x\|^2)(1 + 2x_1 + 2\|x\|^2))x - (1 - \|x\|^2)(Lx + T(x))\big].$$

(vi)   Prove by contradiction that $f'(x) \neq 0$ in the interior of the unit ball of ℓ_2. Hint: Suppose the contrary; then $f'(x) = 0$ for $\|x\| < 1$. It follows that

$$Lx + T(x) = sx,$$

where

$$s = \frac{\|x - T(x)\|^2}{1 - \|x\|^2} + 1 + 2x_1 + 2\|x\|^2.$$

Applying L to both sides gives

$$L^2 x - sLx + x = 0,$$

so x satisfies the second-order linear recurrence relation

$$x_{i+2} - sx_{i+1} + x_i = 0, \quad i \geq 1,$$

whose characteristic equation is

$$t^2 - st + 1 = 0.$$

Discuss case by case the three types of solutions to the recurrence, depending on the discriminant of the above characteristic equation.

Reference: [119, 267].

Exercise 3.4.10   Prove Theorem 3.4.7.

Exercise 3.4.11   Deduce the Lipschitz criterion for the cone monotonicity criterion. Hint: consider the monotonicity of $f(x) \pm Lx$.

Exercise 3.4.12   Prove Corollary 3.4.11.

3.5 Chain Rules and Lyapunov Functions

3.5.1 Approximate Chain Rules

We now introduce an approximate chain rule for the Fréchet subdifferential that estimates the subdifferential of $f \circ F$. Observe that if $f \circ F$ attains its minimum at $\bar{x}$ then $(z, y) \to f(y) + \iota_{\mathrm{graph}\, F}(z, y)$ attains a minimum at $(\bar{x}, F(\bar{x}))$. This suggests that we can use subdifferential sum rules to deduce chain rules. This is the method we adopt. Similar to the subdifferential sum rules there are strong and weak versions of approximate chain rules. First we discuss a strong approximate chain rule which needs some qualification conditions on F.

Let X and Y be Banach spaces. Recall that a map $F \colon X \to Y$ is locally compact at x provided that there is a neighborhood U of x such that for any closed subset $S \subset U$, $F(S)$ is compact.

Theorem 3.5.1 (Strong Approximate Chain Rule) *Let X and Y be Fréchet smooth Banach spaces, let $f \colon Y \to \mathbb{R} \cup \{+\infty\}$ be a lsc function and let $F \colon X \to Y$ be a locally Lipschitz and locally compact mapping at $\bar{x}$. Suppose that $x^* \in \partial_F (f \circ F)(\bar{x})$. Then, for any $\varepsilon > 0$, there exist $x \in B_\varepsilon(\bar{x})$, $y \in B_\varepsilon(F(\bar{x}))$, $y^* \in \partial_F f(y)$, $\|\lambda - y^*\| < \varepsilon$ and $z^* \in \partial_F \langle \lambda, F \rangle(x)$ such that $|f(y) - f(F(\bar{x}))| < \varepsilon$,*

$$\max(\|\lambda\|, \|y^*\|, \|z^*\|)\|y - F(x)\| < \varepsilon \tag{3.5.1}$$

and

$$\|x^* - z^*\| < \varepsilon.$$

Proof. Let g be a C^1 function such that $f \circ F - g$ attains a minimum at $\bar{x}$ and $g'(\bar{x}) = x^*$. Then

$$(z, y) \to f(y) + \iota_{\mathrm{graph}\, F}(z, y) - g(z)$$

attains a minimum at $(\bar{x}, F(\bar{x}))$. Define $f_1(z, y) := f(y) - g(z)$ and $f_2(z, y) = \iota_{\mathrm{graph}\, F}(z, y)$. As an exercise one can verify that f_1, f_2 satisfy condition (3.3.2) in the local approximate sum rule of Theorem 3.3.1. Applying Theorem 3.3.1 we have $x \in B_\varepsilon(\bar{x})$, $z \in B_\varepsilon(\bar{x})$ close enough to $\bar{x}$ so that $\|g'(z) - x^*\| < \varepsilon/2$, $y \in B_\varepsilon(F(\bar{x}))$, $y^* \in \partial_F f(y)$ and

$$(z^*, -\lambda) \in \partial_F \iota_{\mathrm{graph}\, F}(x, F(x)) \tag{3.5.2}$$

satisfying (3.5.1) and

$$\|(-g'(z), y^*) + (z^*, -\lambda)\| < \frac{\varepsilon}{2}. \tag{3.5.3}$$

By (3.5.2) and the definition of the Fréchet subdifferential there exists a C^1 function $h \colon X \times Y \to \mathbb{R}$ such that $h'(x, F(x)) = (z^*, -\lambda)$ and, for y in a neighborhood of x,

$$0 \geq h(y, F(y)) - h(x, F(x)) = \langle z^*, y - x \rangle - \langle \lambda, F(y) - F(x) \rangle + o(\|y - x\|).$$

It follows that $z^* \in \partial_F \langle \lambda, F \rangle(x)$. The rest of the conclusions in the theorem follow from (3.5.3). ●

The same argument can be used to prove the following weak approximate chain rule if we replace Theorem 3.3.1 by Theorem 3.3.3.

Theorem 3.5.2 (Weak Approximate Chain Rule) *Let X and Y be Fréchet smooth Banach spaces, let $f: Y \to \mathbb{R} \cup \{+\infty\}$ be a lsc function and let $F: X \to Y$ be a locally Lipschitz mapping at $\bar{x}$. Suppose that $x^* \in \partial_F(f \circ F)(\bar{x})$. Then, for any $\varepsilon > 0$ and any weak-star neighborhood U of 0 in X^*, there exist $x \in B_\varepsilon(\bar{x})$, $y \in B_\varepsilon(F(\bar{x}))$, $y^* \in \partial_F f(y)$, $\|\lambda - y^*\| < \varepsilon$ and $z^* \in \partial_F \langle \lambda, F \rangle(x)$ such that $|f(y) - f(F(\bar{x}))| < \varepsilon$,*

$$\max(\|\lambda\|, \|y^*\|, \|z^*\|)\|y - F(x)\| < \varepsilon,$$

and

$$x^* \in z^* + U.$$

Proof. Exercise 3.5.1. ●

3.5.2 Lyapunov Functions and Stability

We now use the subdifferential chain rule derived above to obtain an interesting generalization of the Lyapunov indirect method in the stability theory of dynamical systems.

Consider the differential equation

$$x' = f(x), \tag{3.5.4}$$

where $f: \mathbb{R}^N \to \mathbb{R}^N$ is a locally Lipschitz mapping, that is, for any $x \in \mathbb{R}^N$, there exists a neighborhood U of x and a constant $L = L_U > 0$ such that for any $y, z \in U$,

$$\|f(y) - f(z)\| \leq L\|y - z\|.$$

It is well known that for any initial condition $x(0) = x_0$, differential equation (3.5.4) has a unique solution $x(t, x_0)$ defined on $t \in [0, \tau)$ for some $\tau > 0$. We further assume that all the solutions of (3.5.4) are defined on $[0, +\infty)$. A simple sufficient condition ensuring this property is $\|f(x)\| \leq a\|x\| + b$ for some constants a, b.

The idea of the Lyapunov indirect method is rather simple. Suppose that $f(0) = 0$ so that $\{0\}$ is an equilibrium point of (3.5.4). This method tells us that if there is a C^1 positive definite function V with $V(0) = 0$ (a function satisfying $V(x) > 0$ for $x \neq 0$ and $V(x) \to 0$ if and only if $x \to 0$) such that for some $a \geq 0$,

$$\langle V'(x), f(x)\rangle + aV(x) \le 0, \tag{3.5.5}$$

then the equilibrium solution 0 is stable. Moreover, if $a > 0$ then it is asymptotically stable.

In fact, for any trajectory $x(\cdot)$ of (3.5.4), using calculus rules for derivatives and inequality (3.5.5) we have $\frac{d}{dt}(e^{at}V(x(t))) \le 0$, and therefore the function $t \to e^{at}V(x(t))$ is decreasing. The stability and asymptotic stability (in the case when $a > 0$) of the equilibrium $\{0\}$ then follows from the positive definite property of V.

We show that this line of reasoning essentially remains valid for the more general stability concept related to stability sets defined below when the derivatives involved are replaced by the Fréchet subdifferential for lower semicontinuous functions.

Definition 3.5.3 (Stable and Attractive Sets) *Let S be a closed subset of $\mathbb{R}^N$. We say that S is stable with respect to differential equation (3.5.4) provided that for any $\varepsilon > 0$, there exists a $\delta > 0$ such that for any $x_0 \in B_\delta(S)$,*

$$x(t, x_0) \in B_\varepsilon(S), \quad \text{for all } t \in [0, +\infty).$$

We say that S is attractive with respect to differential equation (3.5.4) provided that S is stable and there exists a $\delta > 0$ such that for any $x_0 \in B_\delta(S)$,

$$\lim_{t \to +\infty} d(x(t, x_0); S) = 0.$$

Note that if $S = \{s\}$ for an equilibrium solution s of (3.5.4) then Definition 3.5.3 recovers the classical concepts of Lyapunov stability and asymptotic stability for equilibrium solutions. However Definition 3.5.3 encompasses many useful situations beyond the stability of equilibriums. For example, consider the simple linear system $x_1' = x_2, x_2' = -x_1$. It is easy to see that the phase-portrait of any trajectory of this system is a stable set. Another example is that the phase-portrait of any stable limiting circle of a plane autonomous system is attractive.

Next we define lsc Lyapunov functions and the related critical sets that model the potential stable and attractive sets.

Definition 3.5.4 *Let $V: \mathbb{R}^N \to [0, +\infty]$ be an extended-valued lsc function. We say that a closed set $S \subset \mathbb{R}^N$ is a critical set of V provided that,*

(i) $S \subset \{x \in \mathbb{R}^N : V(x) = 0\}$;
(ii) *for any $\varepsilon > 0$ there exists a $\delta > 0$ such that $B_\delta(S) \subset \{x \in \mathbb{R}^N : V(x) < \varepsilon\}$; and*
(iii) *for any $\varepsilon > 0$ there exist $\delta, \eta > 0$ such that*

$$\{x \in \mathbb{R}^N : V(x) < \delta\} \cap B_{\varepsilon+\eta}(S) \subset B_\varepsilon(S).$$

Now we can state our generalization of the Lyapunov indirect method.

Theorem 3.5.5 (Stability) *Let* $V : \mathbb{R}^N \to [0, +\infty]$ *be a lsc function and let* S *be a compact critical set of* V. *Suppose that there exists a constant* $a \geq 0$, *for any* $x^* \in \partial_F V(x)$ *with* $V(x) > 0$,

$$\langle x^*, f(x) \rangle + aV(x) \leq 0. \tag{3.5.6}$$

Then S *is a stable set of the differential equation* (3.5.4). *Moreover,* S *is attractive when* $a > 0$.

Proof. We need only show that, for any x_0 sufficiently close to S, the function $t \to e^{at}V(x(t))$ is decreasing for the solution $x(t) = x(t, x_0)$ of differential equation (3.5.4). Indeed to see that S is stable, let ε be an arbitrary positive number. Since S is a critical set of V by Definition 3.5.4 (iii) there exist $\varepsilon', \eta > 0$ such that $\{x \in \mathbb{R}^N : V(x) < \varepsilon'\} \cap B_{\varepsilon+\eta}(S) \subset B_\varepsilon(S)$ and by (ii) there exists $\delta > 0$ such that $B_\delta(S) \subset \{x \in R^n : V(x) < \varepsilon'\}$. Now, for any $x_0 \in B_\delta(S)$ and any $t \in [0, +\infty)$,

$$V(x(t, x_0)) \leq e^{-at}V(x_0) \leq \varepsilon' \tag{3.5.7}$$

so that $x(t, x_0) \in B_\varepsilon(S)$.

When $a > 0$, inequality (3.5.7) implies that $\lim_{t \to +\infty} V(x(t, x_0)) = 0$. Thus we must have $\lim_{t \to +\infty} d(S; x(t, x_0)) = 0$, by property (iii) of Definition 3.5.4.

Now we turn to prove that the function $t \to e^{at}V(x(t))$ is decreasing. In the above argument we see that only points in a small neighborhood of S are relevant. So we may consider such a neighborhood and assume without loss of generality that on this neighborhood f is Lipschitz with a Lipschitz constant L and $\|f\|$ is bounded by some constant M. By virtue of Corollary 3.4.11 we need only show that, for any $\xi \in \partial_F(e^{at}V(x(t)))$, $\xi \leq 0$. Define $g(x, y) = yV(x)$ and $G(t) = (x(t), e^{at})$. Then $\xi \in \partial_F(g \circ G)(t)$. Applying the approximate chain rule of Theorem 3.5.1 we have that, for any $\varepsilon > 0$, there exist $t' \in (t - \varepsilon, t + \varepsilon)$, $|x - x(t')| < \varepsilon$, $|y - e^{at'}| < \varepsilon$, $(yx^*, y^*) \in \partial_F g(x, y)$, i.e., $y^* = V(x)$, $x^* \in \partial_F V(x)$, $\|(\lambda, \mu) - (yx^*, y^*)\| < \varepsilon$ and

$$z^* \in \partial_F \langle (\lambda, \mu), (x(\cdot), e^{a\cdot}) \rangle (t') = \{\langle \lambda, x'(t') \rangle + a\mu e^{at'}\}$$

satisfying $\|\xi - z^*\| < \varepsilon$ and

$$\max(\|\lambda\|, \|\mu\|, \|x^*y\|, \|y^*\|, \|z^*\|) \times (\|x - x(t')\| + |y - e^{at'}|) < \varepsilon.$$

It follows that

$$\xi \leq z^* + \varepsilon = \langle \lambda, x'(t') \rangle + a\mu e^{at'} + \varepsilon = \langle \lambda, f(x(t')) \rangle + a\mu e^{at'} + \varepsilon$$
$$\leq \langle yx^*, f(x(t')) \rangle + aV(x)e^{at'} + \varepsilon + M\varepsilon + \varepsilon a e^{at'}$$
$$\leq \langle yx^*, f(x) \rangle + aV(x)e^{at'} + \varepsilon + M\varepsilon + L\varepsilon + \varepsilon a e^{at'}$$
$$\leq [\langle x^*, f(x) \rangle + aV(x)]e^{at'} + \varepsilon + M\varepsilon + L\varepsilon + \varepsilon a e^{at'} + \frac{M}{y}\varepsilon$$
$$\leq \varepsilon \left(1 + M + L + a e^{at'} + \frac{M}{y}\right).$$

Letting $\varepsilon \to 0$ we have $\xi \leq 0$, as was to be shown.   ●

3.5.3 Commentary and Exercises

Chain rules are important tools in analyzing subdifferentials and have been discussed in many different settings (see [71, 84, 91, 198, 209, 237, 264] and the references therein). The chain rules discussed here and their applications in the stability of dynamic systems largely follows [277]. An alternative approach based on the weak invariance-viability type theorems has been discussed in [6, 169, 256, 257]. The survey paper [90] provides an excellent account of this approach and other related issues in dynamic systems. Allowing extended-valued lsc Lyapunov functions provides much flexibility. Many operations such as truncation, taking maximum or absolute value, and using indicator functions become possible. Exercises 3.5.2,3.5.3 and 3.5.4 provide some examples.

Exercise 3.5.1 Prove Theorem 3.5.2.

Exercise 3.5.2 Construct a Lyapunov function to show that for any $r > 0$, $S(r) := \{(x_1, x_2) \in \mathbb{R}^2 : x_1^2 + x_2^2 = r\}$ is a stable set of differential equations

$$x_1' = x_2, x_2' = -x_1. \tag{3.5.8}$$

Hint: Use the Lyapunov function $V(x_1, x_2) = |x_1^2 + x_2^2 - r|, r \geq 0$.

Exercise 3.5.3 The following system naturally occurs in population models.

$$x' = f(x) := -x(x+2)(x-1). \tag{3.5.9}$$

Define

$$V(x) := \begin{cases} (x+2)^2, & x \leq -1/2, \\ +\infty, & x \in (-1/2, 1/2), \\ (x-1)^2, & x \geq 1/2. \end{cases}$$

(i) Check that both $S_1 = \{1\}$ and $S_2 = \{-2\}$ are critical sets of V.

(ii) Verify that for $x < -1/2$ or $x > 1/2$, V is C^1 and it satisfies

$$V'(x)f(x) = \left\{ \begin{array}{ll} -2x(x-1)V(x), & x < -1/2 \\ -2x(x+2)V(x), & x > 1/2 \end{array} \right\} \leq -3/2V(x).$$

(iii) Show that at $x = -1/2$ and $1/2$ we have $\partial_F V(-1/2) = [2(-1/2+2), +\infty)$ and $\partial_F V(1/2) = (-\infty, 2(1/2-1)]$, respectively.

(iv) Conclude that both 1 and -2 are asymptotic equilibrium points.

Reference: [277].

*Exercise 3.5.4 Prove that $x_1^2 + x_2^2 = 1$ is a stable limit circle of the following system of differential equations.

$$\begin{aligned} x_1' &= -x_2 + x_1(1 - x_1^2 - x_2^2) \\ x_2' &= x_1 + x_2(1 - x_1^2 - x_2^2). \end{aligned} \qquad (3.5.10)$$

Hint: Construct the function

$$V(x_1, x_2) = \begin{cases} |1 - x_1^2 - x_2^2|, & x_1^2 + x_2^2 \geq 1/2, \\ +\infty, & x_1^2 + x_2^2 < 1/2. \end{cases}$$

(i)   Verify that $S = \{(x_1, x_2) : x_1^2 + x_2^2 = 1\}$ is a critical set of V.
(ii) Check that the function V is C^1 at any point $(x_1, x_2) \notin S$ and $x_1^2 + x_2^2 > 1/2$ and, for these points V' is the only element in $\partial_F V$ which satisfies

$$\langle V', f \rangle + V \leq 0.$$

Here f represent the right hand side of (3.5.10).
(iii) For $x_1^2 + x_2^2 < 1/2$ we do not need to verify (3.5.6) because

$$\partial_F V(x_1, x_2) = \emptyset.$$

(iv) Show that for points $x = (x_1, x_2)$ satisfying $x_1^2 + x_2^2 = 1/2$ we have

$$\partial_F V(x) = \{kx : k \in (-\infty, -2]\}.$$

These points also satisfy (3.5.6) with $a = 1$.

Reference: [277].

    In the above examples the sets of nonsmooth points for the V functions are all of measure zero. One may wonder then whether it is necessary to check inequality (3.5.6) for these points. The following guided exercise shows that the answer is positive. In other words, handling the nonsmooth points of lsc Lyapunov functions is a crucial part of this method that cannot be omitted.

*Exercise 3.5.5 Let C be the Cantor ternary set on $[0, 1]$ consisting of every ternary decimal involving only 0 and 2 in its expression.

    Since C is closed, $[0, 1] \backslash C$ is the union of denumerable disjoint open intervals. We write

$$[0, 1] \backslash C := \bigcup_{k=1}^{\infty} (a_k, b_k).$$

Consider the classical Cantor ternary function $V : C \to [0, 1]$ defined as follows:

$$V(x) := \sum_{i=0}^{\infty} \frac{x_i}{2^{i+1}}.$$

where x_i is the ith digit of the ternary decimal expression $x = 0.x_1 x_2 \cdots$ of x.

As for each k, a_k and b_k must have "dual" ternary expressions

$$a_k = 0.c_1 c_2 \cdots c_i 0222 \cdots \text{ and } b_k = 0.c_1 c_2 \cdots c_i 2000 \cdots ,$$

we can check that $V(a_k) = V(b_k)$. Thus, we can extend V to $[0,1]$ by defining

$$V(x) := V(a_k) = V(b_k), \quad \text{for all } x \in (a_k, b_k).$$

We further extend V to $[0, +\infty)$ by setting $V(x) := 1, x > 1$. It is well known that V is continuous and non-decreasing. Next we extend V to $\mathbb{R}$ as an even function. It is easy to check that $\{0\}$ is a critical set of V. Moreover, it follows from the calculation in Example 9.2 of [71] and the symmetry of V with respect to 0 that

$$\partial_F V(x) = \begin{cases} \emptyset & x \in [C \cup (-C)] \backslash (\{-b_k\} \cup \{b_k\} \cup \{0\}), \\ [0, \infty) & x \in (\{b_k\}), \\ (-\infty, 0] & x \in (\{-b_k\}), \\ 0 & x \in R \backslash [C \cup (-C)]. \end{cases}$$

Note that the set of nonsmooth points of V is of measure zero and V has derivative 0 at any of its differentiable points. Thus, if we use V to determine the stability of $\{0\}$ by checking only its differentiable points satisfying the inequality (3.5.6) we would arrive at the conclusion that every differential equation is stable at $\{0\}$, which is absurd.

The method used in the proof of the chain rules in this section can also be used to prove other forms of chain rules that are more closely related to the other calculus rules for subdifferentials. Guided exercises below will help the readers to develop these results.

Exercise 3.5.6 Prove the following chain rule in which the inside function may be continuous or lower semicontinuous.

Theorem 3.5.6 (Approximate Chain Rule without Lipschitz Assumption) *Let X be a Fréchet smooth Banach space. Suppose that $f_1, \ldots, f_M \colon X \to \mathbb{R} \cup \{+\infty\}$ are lsc functions, $f_{M+1}, \ldots, f_N$ are continuous functions and $f \colon \mathbb{R}^N \to \mathbb{R} \cup \{+\infty\}$ is a lsc function nondecreasing for each of its first M variables ($M \le N$). Suppose that $x^* \in \partial_F f(f_1, \ldots, f_N)(\bar{x})$. Then for any positive number $\varepsilon > 0$ and any weak-star neighborhood U of 0 in X^*, there exist $(x_n, f_n(x_n)) \in B_\varepsilon((\bar{x}, f_n(\bar{x}))), n = 0, 1, \ldots, N$, $(r, f(r)) \in B_\varepsilon(\bar{r}, f(\bar{r}))$ where $\bar{r} = (f_1(\bar{x}), \ldots, f_N(\bar{x}))$ and $\mu = (\mu_1, \ldots, \mu_N) \in \partial_F f(r) + \varepsilon B_{\mathbb{R}^N}$ such that $\mu_n > 0$ for $n = 1, \ldots, M$, and*

$$x^* \in \sum_{n=1}^{N} \partial_F (\mu_n f_n)(x_n) + U.$$

Hint: Let g be a C^1 function such that $g'(\bar{x}) = x^*$ and $f(f_1, \ldots, f_N) - g$ attains a local minimum at $\bar{x}$. Denote $r = (r_1, \ldots, r_N) \in \mathbb{R}^N$. Observing that the function

$$(x, r) \to f(r) + \sum_{n=1}^{M} \iota_{\mathrm{epi} f_n}(x, r_n) + \sum_{n=M+1}^{N} \iota_{\mathrm{graph} f_n}(x, r_n) - g(x) \quad (3.5.11)$$

attains a local minimum at $(\bar{x}, f_1(\bar{x}), \ldots, f_N(\bar{x}))$, the conclusion follows from the weak local approximate sum rule of Theorem 3.3.3.

When f is a smooth function we can sharpen the results in Theorem 3.5.6.

Exercise 3.5.7 Prove the following refined smooth chain rule.

Theorem 3.5.7 (Refined Smooth Approximate Chain Rule) *Let X be a Fréchet smooth Banach space. Suppose that $f_1, \ldots, f_N \colon X \to \mathbb{R} \cup \{+\infty\}$ are lsc functions and $f \colon \mathbb{R}^N \to \mathbb{R}$ is a C^1 function strictly increasing for each of its variables. Suppose that $x^* \in \partial_F f(f_1, \ldots, f_N)(\bar{x})$. Then, for any positive number $\varepsilon > 0$ and any weak-star neighborhood U of 0 in X^*, there exist $(x_n, f_n(x_n)) \in B_\varepsilon((\bar{x}, f_n(\bar{x}))), n = 0, 1, \ldots, N$ such that*

$$x^* \in \sum_{n=1}^{N} \partial_F(\mu_n f_n)(x_n) + U,$$

where $\mu = (\mu_1, \ldots, \mu_N) = f'(f_1(\bar{x}), \ldots, f_N(\bar{x}))$.

Exercise 3.5.8 Deduce the weak local approximate sum rule of Theorem 3.3.3 from Theorem 3.5.7.

Exercise 3.5.9 Apply the chain rule of Theorem 3.5.6 to $\max(f_1, \ldots, f_N)$ to deduce the following result.

Theorem 3.5.8 (Subdifferential of the Max Function) *Let X be a Fréchet smooth Banach space and let $f_n \colon X \to \mathbb{R} \cup \{+\infty\}, n = 1, \ldots, N$ be lsc functions. Suppose that $x^* \in \partial_F \max(f_1, \ldots, f_N)(\bar{x})$. Then, for any $\varepsilon > 0$ and any weak-star neighborhood U of 0 in X^*, there exist $(x_n, f_n(x_n)) \in B_\varepsilon((\bar{x}, f_n(\bar{x}))), x_n^* \in \partial_F f_n(x_n)$ and $\lambda_n \geq 0$ with $|\sum_{n=1}^{N} \lambda_n - 1| < \varepsilon$ such that*

$$x^* \in \sum_{n=1}^{N} \lambda_n x_n^* + U.$$

Exercise 3.5.10 Show that Theorem 3.5.1 remains valid if we assume f is w^*-lsc and F is linear. (In particular, this is valid when f is a convex continuous function.)

∗**Exercise 3.5.11** Show that when $Y = \mathbb{R}^N$ and f and f_n, $n = 1, \ldots, N$ are Lipschitz functions, Theorem 3.5.1 provides a sharper result than Theorem 3.5.6. Construct an example (in a finite dimensional space X) to show this refinement is impossible if f_n, $n = 1, \ldots, N$ are merely continuous. Reference: [274].

3.6 Multidirectional Mean Value Inequalities and Solvability

In this section we discuss a multidirectional generalization of the approximate mean value theorem of Theorem 3.4.7. We replace the endpoint b by a set S and $r < f(b)$ by the decoupled infimum $\bigwedge[f](S)$. Applications of this result to solvability and representation of superdifferentials are also discussed.

3.6.1 Multidirectional Mean Value Inequalities

Recall that for a point x and a set S in a Banach space, $[x, S] = \mathrm{conv}(\{x\} \cup S)$ represents the drop associated with x and S.

Theorem 3.6.1 (Approximate Multidirectional Mean Value Inequality) *Let X be a Fréchet smooth Banach space. Let S be a nonempty, closed and convex subset of X, let $f \colon X \to \mathbb{R} \cup \{+\infty\}$ be a lsc function and let $x \in \mathrm{dom} f$. Suppose that for some $h > 0$, f is bounded below on $B_h([x, S])$ and*

$$\bigwedge[f](S) - f(x) > r. \tag{3.6.1}$$

Then for any $\varepsilon > 0$, there exist $z \in B_\varepsilon([x, S])$ and $z^* \in \partial_F f(z)$ such that

$$f(z) < \bigwedge[f]([x, S]) + |r| + \varepsilon, \tag{3.6.2}$$

and

$$r < \langle z^*, y - x \rangle + \varepsilon \|y - x\| \text{ for all } y \in S. \tag{3.6.3}$$

Proof. We divide the proof into two steps.
(1) The special case in which $r = 0$. Let $\tilde{f} := f + \iota_{B_h([x,S])}$. Then $\tilde{f}$ is bounded below on X. It follows from (3.6.1) that we can fix an $\eta \in (0, h/2)$ such that $\inf_{y \in B_{2\eta}(S)} f(y) > f(x)$. Without loss of generality we may assume that

$$\varepsilon < \min \Big\{ \inf_{y \in B_{2\eta}(S)} f(y) - f(x), \eta \Big\}.$$

Applying the nonlocal approximate sum rule of Theorem 3.2.3 to $f_1 := f + \iota_{B_h([x,S])}$ and $f_2 := \iota_{[x,S]}$ we obtain that there exist $z \in \mathrm{dom} f \cap B_h([x, S])$ and $u \in [x, S]$ with $\|z - u\| < \varepsilon$, $z^* \in \partial_F f_1(z) = \partial_F f(z)$ and $u^* \in \partial_F \iota_{[x,S]}(u) = N_F([x, S]; u)$ satisfying

$$\max(\|z^*\|, \|u^*\|) \times \|z - u\| < \varepsilon \tag{3.6.4}$$

and

$$f(z) < \bigwedge[f]([x, S]) + \varepsilon \leq f(x) + \varepsilon \tag{3.6.5}$$

such that

$$\|z^* + u^*\| < \varepsilon. \tag{3.6.6}$$

Since $[x, S]$ is convex, $N_F([x, S]; u)$ coincides with the normal cone of $[x, S]$ at u in the sense of convex analysis. Thus, $u^* \in \partial_F \iota_{[x,S]}(u) = N_F([x, S]; u)$ implies that

$$\langle u^*, w - u \rangle \leq 0, \quad \text{for all } w \in [x, S]. \tag{3.6.7}$$

Combining (3.6.6) and (3.6.7) yields

$$\begin{aligned}
\langle z^*, w - u \rangle &= \langle z^* + u^*, w - u \rangle - \langle u^*, w - u \rangle \\
&\geq \langle z^* + u^*, w - u \rangle \\
&> -\varepsilon \|w - u\|, \quad \text{for all } w \in [x, S] \backslash \{u\}.
\end{aligned}$$

That is

$$0 < \langle z^*, w - u \rangle + \varepsilon \|w - u\|, \quad \text{for all } w \in [x, S] \backslash \{u\}. \tag{3.6.8}$$

Moreover, we must have $d(S; u) \geq \eta$ for otherwise we would have $d(S; z) \leq 2\eta$ and $f(z) \geq \inf_{y \in B_{2\eta}(S)} f(y) > f(x) + \varepsilon$ which contradicts (3.6.5). Let $u := x + \bar{t}(\bar{y} - x)$ for some $\bar{t} \in [0, 1]$ and $\bar{y} \in S$. Then $\eta \leq \|u - \bar{y}\| = (1 - \bar{t}) \|x - \bar{y}\|$ implies $1 - \bar{t} > 0$. Clearly $x \notin S$. For any $y \in S$ setting $w := y + \bar{t}(\bar{y} - y) \neq u$ in (3.6.8) yields

$$0 < \langle z^*, y - u \rangle + \varepsilon \|y - u\|, \quad \text{for all } y \in S. \tag{3.6.9}$$

(2) The general case. Consider $X \times \mathbb{R}$ with the Euclidean product norm $\|(x, \gamma)\| := \sqrt{\|x\|^2 + \gamma^2}$. Take an $\varepsilon' \in (0, \varepsilon/2)$ small enough so that

$$\bigwedge[f](S) - f(x) > r + \varepsilon'$$

and define

$$F(z, t) := f(z) - (r + \varepsilon')t. \tag{3.6.10}$$

Obviously F is lsc on $X \times \mathbb{R}$ and is bounded below on $B_h([(x, 0), S \times \{1\}])$. Moreover,

$$\bigwedge[F](S \times \{1\}) = \bigwedge[f](S) - (r + \varepsilon') > f(x) = F(x, 0).$$

Applying the special case proved above with f, x and S replaced by F, $(x, 0)$ and $S \times \{1\}$ we conclude that there exist $(z, s) \in B_\varepsilon([(x, 0), S \times \{1\}])$ (so that $z \in B_\varepsilon([x, S])$) and $(z^*, s^*) \in \partial_F F(z, s) \subset \partial_F f(z) \times \{-(r + \varepsilon')\}$ satisfying

$$f(z) - (r + \varepsilon')s = F(z, s) < \bigwedge[F]([(x, 0), S \times \{1\}]) + \varepsilon',$$

in other words,

$$f(z) < \bigwedge[F]([(x,0), S \times \{1\}]) + (r + \varepsilon')s + \varepsilon'$$
$$\leq \bigwedge[f]([x, S]) + |r| + \varepsilon$$

such that, for all $(y, 1) \in S \times \{1\}$,

$$0 < \langle (z^*, s^*), (y, 1) - (x, 0) \rangle + \varepsilon'(\|(y - x, 1)\|)$$
$$\leq \langle z^*, y - x \rangle - (r + \varepsilon') + \varepsilon'(\|y - x\| + 1)$$
$$= \langle z^*, y - x \rangle - r + \varepsilon'\|y - x\| \leq \langle z^*, y - x \rangle - r + \varepsilon\|y - x\|.$$

This completes the proof. ●

It is clear that when S is bounded the term $\varepsilon\|y - x\|$ in (3.6.3) can be eliminated. This is not the case in general. One can convince oneself by examining the simple example when $X = Y = \mathbb{R}$ and $f(y) = e^y$ (Exercise 3.6.1). When S is a compact subset of X or f is uniformly continuous in a neighborhood of S we can verify that $\bigwedge[f](S) = \inf_S f$ (Exercise 3.6.2). In general $\bigwedge[f](S)$ in Theorem 3.6.1 cannot be replaced by $\inf_S f$ (Exercises 3.6.5 and 3.6.6). In general, it is impossible to ensure the mean value z belongs to $[x, S]$ (Exercise 3.6.3). Yet, this refinement is possible under additional conditions. One such condition is to assume f to be convex. This will be discussed later.

The following corollary is often useful in various applications.

Theorem 3.6.2 (Decrease Principle) *Let X be a Fréchet smooth Banach space, let $f : X \to \mathbb{R} \cup \{+\infty\}$ be a lsc function bounded from below and let $r > 0$. Suppose that for any $x \in B_r(\bar{x})$, $\xi \in \partial_F f(x)$ implies that $\|\xi\| > \sigma > 0$. Then*

$$\inf_{x \in B_r(\bar{x})} f(x) \leq f(\bar{x}) - \sigma r.$$

Proof. Exercise 3.6.9. ●

3.6.2 Solvability

Consider a lsc function $f : X \times Y \to \mathbb{R} \cup \{+\infty\}$. As an application of the multidirectional mean value inequality, we seek infinitesimal conditions in terms of the Fréchet subdifferential for the solvability of inequality $f(x, y) \leq 0$ in terms of parameter y or the nonemptiness of $G(y) := \{x \in X \mid f(x, y) \leq 0\}$. We use $\partial_{F,x}$ to signify the Fréchet subdifferential with respect to x. We will see later that this is closely related to the more general implicit multifunction theorems.

Theorem 3.6.3 (Solvability) *Let X and Y be Fréchet smooth Banach spaces and let U be an open set in $X \times Y$. Suppose that $f : U \to \mathbb{R} \cup \{+\infty\}$ satisfies the following conditions:*

(i)   *there exists* $(\bar{x}, \bar{y}) \in U$ *such that*

$$f(\bar{x}, \bar{y}) \leq 0;$$

(ii) $y \to f(\bar{x}, y)$ *is upper semicontinuous at* $\bar{y}$;
(iii) *for any fixed* y *near* $\bar{y}$, $x \to f(x, y)$ *is lower semicontinuous;*
(iv) *there exists a* $\sigma > 0$ *such that for any* $(x, y) \in U$ *with* $f(x, y) > 0$,
       $\xi \in \partial_{F,x} f(x, y)$ *implies that* $\|\xi\| \geq \sigma$.

Then there exist open sets $W \subset X$ *and* $V \subset Y$ *containing* $\bar{x}$ *and* $\bar{y}$ *respectively such that*

(a) *for any* $y \in V$, $W \cap G(y) \neq \emptyset$;
(b) *for any* $y \in V$ *and* $x \in W$,

$$d(x, G(y)) \leq \frac{f_+(x, y)}{\sigma},$$

where $f_+(x, y) := \max\{0, f(x, y)\}$.

Proof.   Let r' be a positive number such that $B_{r'}(\bar{x}) \times B_{r'}(\bar{y}) \subset U$ and let $r = r'/3$. Since $f(\bar{x}, y)$ is upper semicontinuous at $\bar{y}$ and $f(\bar{x}, \bar{y}) = 0$, there exists an open neighborhood V of $\bar{y}$ such that $V \subset B_{r'}(\bar{y})$ and $y \in V$ implies that $f(\bar{x}, y) < r\sigma$. We will show that V and $W := \mathrm{int} B_r(\bar{x})$ satisfy the conclusion of the theorem. Let y be an arbitrary element of V. We show that $W \cap G(y) \neq \emptyset$. In fact, if this is not the case, then $f(x, y) > 0$ for any $x \in B_\tau(\bar{x}), \tau < r$. Choose τ close enough to r so that $f(\bar{x}, y) < \tau\sigma$. Invoking the decrease principle of Theorem 3.6.2 we have

$$0 \leq \inf_{x \in B_\tau(\bar{x})} f(x, y) \leq f(\bar{x}, y) - \tau\sigma < 0,$$

a contradiction.

To show the estimate (b), consider $x \in W$ and $y \in V$. If

$$B(x, f_+(x, y)/\sigma) \not\subset \mathrm{int} B(\bar{x}, r')$$

then $\|x - \bar{x}\| + f_+(x, y)/\sigma \geq r'$ or $f_+(x, y)/\sigma \geq 2r$. Since conclusion (a) implies that $d(x, G(y)) < 2r$, estimate (b) holds. Now we turn to the case when $B(x, f_+(x, y)/\sigma) \subset \mathrm{int} B(\bar{x}, r')$. Take $\tau > f_+(x, y)/\sigma$ such that $B(x, \tau) \subset \mathrm{int} B(\bar{x}, r')$. Since $f(x, y) < \tau\sigma$ an argument similar to the proof of (a) yields that there exists $z \in B(x, \tau)$ such that $f(z, y) \leq 0$. Thus, $d(x, G(y)) < \tau$. Letting $\tau \to f_+(x, y)/\sigma$ we arrive at estimate (b).   ●

3.6.3 Superdifferential and Subdifferential

Recall that if $f \colon X \to [-\infty, \infty)$ is usc then we define the Fréchet superdifferential of f at x by $\partial^F f(x) = -\partial_F(-f)(x)$. Interestingly, for a continuous function a superderivative $x^* \in \partial^F f(x)$ can be represented as a convex combination of subderivatives at points in a neighborhood of x. To prove this result we need the following nonsymmetrical minimax theorem.

Theorem 3.6.4 (Nonsymmetrical Minimax Theorem: A Banach Space Version) *Let S be a closed bounded convex subset of Banach space X^* and let T be a convex subset of X. Suppose that $f: S \times T \to \mathbb{R}$ is a function convex in s and concave in t and that for each $t \in T$, $s \to f(s,t)$ is weak-star lsc. Then*

$$\inf_{s \in S} \sup_{t \in T} f(s,t) = \sup_{t \in T} \inf_{s \in S} f(s,t).$$

Proof. Let $\alpha < \inf_{s \in S} \sup_{t \in T} f(s,t)$. Define, for each $t \in T$, the set $S_t := \{s \in S \mid f(s,t) \leq \alpha\}$. Then S_t is weak-star closed subset of S and

$$\bigcap_{t \in T} S_t = \emptyset.$$

Since S is weak-star compact, the finite intersection theorem yields a finite set $t_1, \ldots, t_N$ such that

$$\bigcap_{n=1}^{N} S_{t_n} = \emptyset.$$

Now define a set

$$K := \{r \in \mathbb{R}^N \mid \exists s \in S, r_n \geq f(s, t_n), n = 1, \ldots, N\}. \qquad (3.6.11)$$

Then $(\alpha, \ldots, \alpha) \notin K$. We can check that K is a convex set (Exercise 3.6.11). Applying the finite dimensional separation theorem (see e.g., [237, Theorem 2.39]) we have that there exists $h := (h_1, \ldots, h_N) \neq 0$ such that for any $k \in K$,

$$\langle h, k \rangle > \langle h, (\alpha, \ldots, \alpha) \rangle. \qquad (3.6.12)$$

Clearly, $h_n \geq 0$ (Exercise 3.6.12) and dividing by $\sum_{n=1}^{N} h_n$ if necessary, we may assume that $\sum_{n=1}^{N} h_n = 1$. Then for any $s \in S$,

$$\alpha < \sum_{n=1}^{N} h_n f(s, t_n) \leq f\left(s, \sum_{n=1}^{N} h_n t_n\right).$$

It follows that

$$\alpha \leq \inf_{s \in S} f\left(s, \sum_{n=1}^{N} h_n t_n\right) \leq \sup_{t \in T} \inf_{s \in S} f(s,t).$$

Letting $\alpha \to \inf_{s \in S} \sup_{t \in T} f(s,t)$ we have

$$\inf_{s \in S} \sup_{t \in T} f(s,t) \leq \sup_{t \in T} \inf_{s \in S} f(s,t). \qquad (3.6.13)$$

Since the opposite inequality

$$\inf_{s \in S} \sup_{t \in T} f(s,t) \geq \sup_{t \in T} \inf_{s \in S} f(s,t) \qquad (3.6.14)$$

always holds (Exercise 3.6.13), the proof is complete.   ●

Now we turn to subderivative representation of a superderivative. We discuss the results in reflexive Banach spaces which are Fréchet smooth.

Theorem 3.6.5 (Subderivative Representation of Superderivatives) *Let X be a reflexive Banach space, let $f\colon X \to \bar{\mathbb{R}}$ be a lsc function and let $x^* \in \partial^F f(x)$. Then for any $\varepsilon > 0$, one has*

$$x^* \in \mathrm{conv}\Big\{ \bigcup_{w \in B_\varepsilon(x)} \partial_F f(w) \Big\} + \varepsilon B_{X^*}.$$

Proof. Let $x^* \in \partial^F f(x)$. Then there exists a $\delta \in (0, \varepsilon/2)$ and a C^1 function g with $g'(x) = x^*$ such that $f - g$ attains a maximum at x over $B_\delta(x)$. Taking a smaller δ if necessary, we may assume that $y \in B_\delta(x)\backslash\{x\}$ implies that

$$f(x) - f(y) \geq g(x) - g(y) > \langle x^*, x - y \rangle - \varepsilon \|x - \dot{y}\|. \qquad (3.6.15)$$

Applying the multidirectional mean value inequality of Theorem 3.6.1 (see also Exercise 3.6.1 (i)) with $r := \langle x^*, x - y \rangle - \varepsilon \|x - y\|$ and the small positive constant δ, there exists $z \in B_\delta([x,y]) \subset B_\varepsilon(x)$ and $z^* \in \partial_F f(z)$ such that

$$\langle z^*, x - y \rangle \geq \langle x^*, x - y \rangle - \varepsilon \|x - y\|. \qquad (3.6.16)$$

Writing $y = x - \delta v$ for $v \in B$, we have

$$\langle x^* - z^*, v \rangle \leq \varepsilon. \qquad (3.6.17)$$

Denote $M := \bigcup_{w \in B_\varepsilon(x)} \partial_F f(w)$. It follows that for any $v \in B$,

$$\inf_{\xi \in \mathrm{conv}\,M} \langle x^* - \xi, v \rangle \leq \inf_{\xi \in M} \langle x^* - \xi, v \rangle \leq \varepsilon. \qquad (3.6.18)$$

Taking the supremum over $v \in B$ we get

$$\sup_{v \in B} \inf_{\xi \in \mathrm{conv}\,M} \langle x^* - \xi, v \rangle \leq \varepsilon. \qquad (3.6.19)$$

In accordance with Theorem 3.6.4, this implies

$$\inf_{\xi \in \mathrm{conv}\,M} \sup_{v \in B} \langle x^* - \xi, v \rangle = \inf_{\xi \in \mathrm{conv}\,M} \|x^* - \xi\| \leq \varepsilon, \qquad (3.6.20)$$

which completes the proof.   ●

A symmetric result of representing subderivatives in terms of superderivatives can be deduced from Theorem 3.6.5 by replacing f with $-f$.

Theorem 3.6.6 (Superderivative Representation of Subderivatives) *Let X be a reflexive Banach space, let $f: X \to \mathbb{R} \cup \{-\infty\}$ be a usc function and let $x^* \in \partial_F f(x)$. Then, for any $\varepsilon > 0$, one has*

$$x^* \in \operatorname{conv}\left\{ \bigcup_{w \in B_\varepsilon(x)} \partial^F f(w) \right\} + \varepsilon B_{X^*}.$$

Proof. Exercise 3.6.15. ●

3.6.4 Commentary and Exercises

The multidirectional mean value theorems discussed in this section were discovered by Clarke and Ledyaev [88]. A similar but less general result was derived earlier by Luc [188]. Theorem 3.6.1 generalizes the original results in [88] by allowing S to be unbounded. This theorem and its proof are taken from [272]. Other generalizations and refinements have been discussed in [10, 64, 170, 174, 178, 225]. Applications to various fields can be found in [88, 91].

The solvability theorem first appeared in [82, 83]. Theorem 3.6.3 is a Fréchet subdifferential version of [91, Theorem 3.1] taken from [174]. This result is closely related to the Graves–Lyusternik Theorem (Exercise 3.6.10) and implicit multifunction theorems to be discussed later.

Theorems 3.6.6 and 3.6.5 are reflexive Banach space versions of similar results in [89, 173] where one can also find further generalizations and applications. In finite dimensional spaces these results were known to Barron and Jensen earlier in a less general form related to applications in viscosity solutions to partial differential equations [11]. Theorem 3.6.4 is a special case of Borwein and Zhuang's nonsymmetrical minimax theorem in [72] (see Exercise 3.6.16).

Exercise 3.6.1 (i) Show that if S is bounded then the term $\varepsilon \|y - x\|$ in (3.6.3) can be eliminated. (ii) Construct an example showing that in general this term cannot be dispensed with.

Exercise 3.6.2 Let f be an extended-valued lsc function on Banach space X. Prove that $\bigwedge[f](S) = \inf_S f$ if S is a compact subset of X or if f is uniformly continuous in a neighborhood of S.

Exercise 3.6.3 Verify that in Theorem 3.6.1 one cannot ensure $z \in [x, S]$ by examining the following example: $X = \mathbb{R}$, $x = 0, S := \{1\}$ and $f(y) := -\sqrt{|y|}$ for $y \leq 0$ and $f(y) := 1$ for $y > 0$.

Exercise 3.6.4 Let f be a C^1 function on $\mathbb{R}^n$. Suppose that, for any $x \in B_r(\bar{x})$, $\|f'(x)\| \geq \sigma > 0$. Prove that $\sup_{B_r(\bar{x})} f - \inf_{B_r(\bar{x})} f \geq 2r\sigma$.

*Exercise 3.6.5 Deduce the approximate local sum rule of Theorem 3.3.1 from Theorem 3.6.1. Reference: [255].

Exercise 3.6.6 Show that the decoupled infimum $\bigwedge[f](S)$ in Theorem 3.6.1 cannot be replaced by $\inf_S f$. Hint: Use Exercise 3.6.5 and Exercise 3.3.11 or 3.3.12 to get a contradiction.

Exercise 3.6.7 Deduce the following unidirectional mean value inequality from Theorem 3.6.1.

Theorem 3.6.7 Let X be a Fréchet smooth Banach space and let $f \colon X \to \mathbb{R} \cup \{+\infty\}$ be a lsc function bounded below. Then, for any $r < f(y) - f(x)$ and any $\varepsilon > 0$, there exist $z \in B_\varepsilon([x,y])$ and $z^* \in \partial_F f(z)$ such that

$$\langle z^*, y - x \rangle > r$$

and

$$f(z) \leq \min(f(x), f(y)) + |r| + \varepsilon.$$

*Exercise 3.6.8 Prove the following refined multidirectional mean value inequality and deduce from it both Theorem 3.6.1 and Theorem 3.4.6. Reference: [272].

Theorem 3.6.8 (Refined Multidirectional Mean Value Inequality) Let X be a Fréchet smooth Banach space. Let S be a nonempty, closed and convex subset of X, let $f \colon X \to \mathbb{R} \cup \{+\infty\}$ be a lsc function and let $x \in \mathrm{dom} f$. Suppose that for some $h > 0$, f is bounded below on $B_h([x,S])$ and

$$\bigwedge[f](S) - f(x) > r.$$

Then there exists $\eta > 0$ such that for any $\varepsilon > 0$, there exist $u \in [x,S]$, $d(u,S) > \eta$, $z \in B_\varepsilon(u)$ and $z^* \in \partial_F f(z)$ such that

$$\|z^*\| \times \|u - z\| < \varepsilon,$$

$$f(z) < \bigwedge[f]([x,S]) + |r| + \varepsilon,$$

$$0 \leq \langle z^*, w - u \rangle + \varepsilon\|w - u\|, \text{ for all } w \in [x,S],$$

and

$$r < \langle z^*, y - x \rangle + \varepsilon\|y - x\| \text{ for all } y \in S.$$

Exercise 3.6.9 Prove Theorem 3.6.2. Hint: Apply Theorem 3.6.1 with $S = B_{r'}(\bar{x})$, $r' < r$, $x = \bar{x}$ and let $r' \to r$.

Exercise 3.6.10 Prove the Graves–Lyusternik theorem below.

Theorem 3.6.9 (Graves–Lyusternik) *Let* $F \colon X \times Y \to Y$ *be a* C^1 *mapping, let* U *be an open set of* $X \times Y$ *and let* $(\bar{x}, \bar{y}) \in U$ *satisfy* $F(\bar{x}, \bar{y}) = 0$. *Suppose that* $F'_x(\bar{x}, \bar{y})$ *is onto. Then there exist open sets* $W \subset X$ *and* $V \subset Y$ *containing* $\bar{x}$ *and* $\bar{y}$ *respectively, such that for some* $\sigma > 0$, *for any* $y \in V$ *and* $x \in W$,

$$d(G(y); x) \leq \frac{\|F(x,y)\|}{\sigma}.$$

Here $G(y) = \{x \in X \mid F(x,y) = 0\}$.

Hint: Apply the solvability theorem to $f(x,y) = \|F(x,y)\|$ and notice that when $f(x,y) > 0$, the Fréchet subdifferential of f with respect to x at (x,y) contains only $F'_x(x,y)^* F(x,y)/\|F(x,y)\|$, whose norm is bounded away from 0 by some $\sigma > 0$ when (x,y) is sufficiently close to $(\bar{x}, \bar{y})$.

Exercise 3.6.11 Verify that the set K defined in (3.6.11) is convex.

Exercise 3.6.12 Show that the vector h in (3.6.12) has the property that $h_n \geq 0, n = 1, \ldots, N$.

Exercise 3.6.13 Let S and T be arbitrary sets and let $f \colon S \times T \to \mathbb{R}$. Show that

$$\inf_{s \in S} \sup_{t \in T} f(s,t) \geq \sup_{t \in T} \inf_{s \in S} f(s,t).$$

Exercise 3.6.14 Let f be a lsc function. Suppose that $\partial^F f(x) \neq \emptyset$.

(i) Show that that f is continuous at x.
(ii) Give an example of such f that is not continuous in any neighborhood of x.

Exercise 3.6.15 Deduce Theorem 3.6.6 from Theorem 3.6.5

∗**Exercise 3.6.16** Prove the following general nonsymmetrical minimax theorem.

Theorem 3.6.10 (Nonsymmetrical Minimax Theorem) *Let* X *and* Y *be topological vector spaces, let* S *be a compact convex subset of* X *and let* T *be a convex subset of* Y. *Suppose that* $f \colon S \times T \to \mathbb{R}$ *is a function convex in* s *and concave in* t *and that for each* $t \in T$, $s \to f(s,t)$ *is lsc. Then*

$$\inf_{s \in S} \sup_{t \in T} f(s,t) = \sup_{t \in T} \inf_{s \in S} f(s,t).$$

3.7 Extremal Principles and Multi-Objective Optimization

An extremal principle of the form we discuss in this section may be viewed as a local version of the Hahn–Banach separation theorem for nonconvex sets. It is a powerful tool for studying various forms of problems related to optimization.

3.7.1 Extremal Systems and Examples

We start with the definition of an extremal system.

Definition 3.7.1 (Extremal System) *Let X be a Banach space and let $M_n, n = 1, 2, \ldots, N$ be finitely many metric spaces. Consider closed-valued multifunctions $S_n \colon M_n \to X$, $n = 1, 2, \ldots, N$. We say that $\bar{x}$ is an extremal point of the extremal system $(S_1, S_2, \ldots, S_N)$ at $(\bar{m}_1, \bar{m}_2, \ldots, \bar{m}_N)$ provided that $\bar{x} \in S_1(\bar{m}_1) \cap S_2(\bar{m}_2) \cap \cdots \cap S_N(\bar{m}_N)$ and there is a neighborhood U of $\bar{x}$ such that for any $\varepsilon > 0$, there exists*

$$(m_1, m_2, \ldots, m_N) \in B_\varepsilon((\bar{m}_1, \bar{m}_2, \ldots, \bar{m}_N)) \backslash \{(\bar{m}_1, \bar{m}_2, \ldots, \bar{m}_N)\},$$

with $d(S_n(m_n); \bar{x}) < \varepsilon, n = 1, 2, \ldots, N$, and

$$U \cap S_1(m_1) \cap S_2(m_2) \cap \cdots \cap S_N(m_N) = \emptyset.$$

The concept of an extremal system captures, from a geometric perspective, the essence of many different structures related to optimization. Exercises 3.7.5 and 3.7.6 provide some useful examples.

3.7.2 Extremal Principles

The following extremal principle provides a convenient tool for deriving necessary optimality conditions. A limiting form of the extremal principle in finite dimensional Banach spaces can be found in Theorem 5.2.27 in Section 5.2.

Theorem 3.7.2 (Extremal Principle) *Let X be a Fréchet smooth Banach space and let $M_n, n = 1, 2, \ldots, N$ be metric spaces. Consider closed-valued multifunctions $S_n \colon M_n \to X$, $n = 1, 2, \ldots, N$. Suppose that $\bar{x}$ is an extremal point of the extremal system $(S_1, S_2, \ldots, S_N)$ at $(\bar{m}_1, \bar{m}_2, \ldots, \bar{m}_N)$. Then for any $\varepsilon > 0$, there exist $m_n \in B_\varepsilon(\bar{m}_n), x_n \in B_\varepsilon(\bar{x}), n = 1, 2, \ldots, N$ and $x_n^* \in N_F(S_n(m_n); x_n) + \varepsilon B_{X^*}, n = 1, 2, \ldots, N$ such that $\max\{\|x_n^*\| \mid n = 1, \ldots, N\} \geq 1$ and*

$$x_1^* + x_2^* + \cdots + x_N^* = 0.$$

Proof. Let U be a neighborhood of $\bar{x}$ as in the definition of an extremal point. Without loss of generality we may assume that $U = B_r(\bar{x})$. Choose $\varepsilon' \in (0, \varepsilon/2)$ satisfying (Exercise 3.7.1)

$$4N^2[(4N^2 + 1)\varepsilon' + N(\varepsilon')^2] < \varepsilon^2/4$$

and let $m_1, m_2, \ldots, m_N$ be as in the definition of the extremal point for $\varepsilon = \varepsilon'$. Let s_1 be as in Lemma 3.2.2 and define

$$f_1(y_1, y_2, \ldots, y_N) := s_1(y_1, y_2, \ldots, y_N),$$

$$f_2(y_1, y_2, \ldots, y_N) := \sum_{n=1}^{N} \iota_{S_n(m_n)}(y_n) + r(y_1, \ldots, y_N),$$

where

$$r(y_1, \ldots, y_N) := \sum_{n=1}^{N} \|y_n - \bar{x}\|^2.$$

Choose $y'_n \in S_n(m_n), n = 1, 2, \ldots, N$ such that $\|y'_n - \bar{x}\| < d(S_n(m_n); \bar{x}) + \varepsilon' < 2\varepsilon'$. Then

$$\bigwedge [f_1, f_2](X^N) \leq (f_1 + f_2)(y'_1, y'_2, \ldots, y'_N) < 4N^2 \varepsilon' + N(\varepsilon')^2.$$

Applying the nonlocal approximate sum rule of Theorem 3.2.3 we have that there exist $x = (x_1, \ldots, x_N)$, $z = (z_1, \ldots, z_N)$, $-(x_1^*, \ldots, x_N^*) \in \partial_F f_1(z) = \partial_F s_1(z)$, and $(z_1^*, \ldots, z_N^*) \in \partial_F f_2(x)$ such that

$$\|x - z\| < \varepsilon', \tag{3.7.1}$$

$$f_1(z) + f_2(x) < \bigwedge [f_1, f_2](X^N) + \varepsilon' < (4N^2 + 1)\varepsilon' + N(\varepsilon')^2, \tag{3.7.2}$$

and

$$\|(z_1^*, \ldots, z_N^*) - (x_1^*, \ldots, x_N^*)\| < \varepsilon'. \tag{3.7.3}$$

Note that (3.7.2) implies that $x_n \in S_n(m_n)$ and

$$r(x) = \sum_{n=1}^{N} \|x_n - \bar{x}\|^2 < (4N^2 + 1)\varepsilon' + N(\varepsilon')^2. \tag{3.7.4}$$

Consequently,

$$\|r'(x)\| < 2N\sqrt{(4N^2 + 1)\varepsilon' + N(\varepsilon')^2} < \varepsilon/2. \tag{3.7.5}$$

Since

$$\partial_F f_2(x) = N_F(S_1(m_1); x_1) \times \cdots \times N_F(S_N(m_N); x_N) + r'(x),$$

combining (3.7.3) and (3.7.5) we have

$$x_n^* \in N_F(S_n(m_n); x_n) + \varepsilon B_{X^*}.$$

Finally, it follows from Lemma 3.2.2 that $x_1^* + x_2^* + \cdots + x_N^* = 0$ and $\max\{\|x_n^*\| \mid n = 1, \ldots, N\} \geq 1$, which completes the proof. ●

The following corollary is obvious yet often most convenient in applications.

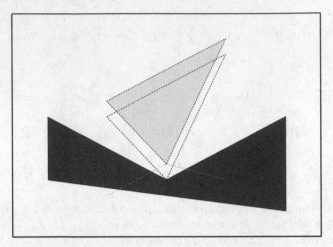

Fig. 3.3. The extremal principle for two fixed sets.

Corollary 3.7.3 *Let X be a Fréchet smooth Banach space and let $M_n, n = 1, 2, \ldots, N$ be metric spaces. Consider closed-valued multifunctions $S_n \colon M_n \to X$, $n = 1, 2, \ldots, N$. Suppose that $\bar{x}$ is an extremal point of the extremal system $(S_1, S_2, \ldots, S_N)$ at $(\bar{m}_1, \bar{m}_2, \ldots, \bar{m}_N)$. Then for any $\varepsilon > 0$, there exist $m_n \in B_\varepsilon(\bar{m}_n), x_n \in B_\varepsilon(\bar{x})$ and $x_n^* \in N_F(S_n(m_n), x_n)$, $n = 1, 2, \ldots, N$ such that $\max\{\|x_n^*\| \mid n = 1, \ldots, N\} \geq 1 - \varepsilon$ and*

$$\|x_1^* + x_2^* + \cdots + x_N^*\| < \varepsilon.$$

Proof. Exercise 3.7.4. ●

Figure 3.3 illustrates the geometry of Corollary 3.7.3 in the case of two fixed sets as described in and below Definition 3.7.5.

3.7.3 Multi-Objective Optimization

Practical decision problems often involve many factors and can be described by a vector-valued decision function whose components describe several competing objectives. The comparison between different values of the decision function is determined by a preference of the decision maker. We model such problems by using a Banach space Y with a nonreflexive *preference* $\prec$ to represent the image space of the objective function. Let X be a Banach space representing the decision variables, let S be a subset of X representing the feasible decisions and let $f \colon X \to Y$ be the objective function. Then a multi-objective optimization problem can be formulated as:

$$\mathcal{M} \qquad\qquad \text{minimize} \quad f(x)$$

$$\text{subject to} \quad x \in S.$$

Here minimization is with respect to the preference $\prec$. Namely, we say that $\bar{x}$ is a *solution* to problem $\mathcal{M}$ if there is no $x \in S$ near $\bar{x}$ such that $f(x) \prec f(\bar{x})$.

We will derive a subdifferential necessary optimality condition for a local solution to problem $\mathcal{M}$. We will denote the level set at $y \in Y$ with respect to the preference $\prec$ by $l(y) := \{z \in Y \mid z \prec y\}$.

Theorem 3.7.4 (Necessary Optimality Condition for Multi-objective Optimization) *Let X and Y be Fréchet smooth Banach spaces and let $f \colon X \to Y$ be a Lipschitz mapping. Suppose that $\bar{x}$ is a solution of the multi-objective optimization problem $\mathcal{M}$. Then for any positive number $\varepsilon > 0$, there exist $x_0, x_1 \in B_\varepsilon(\bar{x})$, $y_0, y_1 \in B_\varepsilon(f(\bar{x}))$, $y^* \in N_F(\overline{l(y_0)}, y_1)$ with $\|y^*\| = 1$ and $x^* \in N_F(S; x_1)$ such that*

$$0 \in x^* + \partial_F \langle y^*, f \rangle(x_0) + \varepsilon B_{X^*}.$$

Proof. Define $M_1 := l(f(\bar{x})) \cup \{f(\bar{x})\}$, $M_2 := \{0\}$, $S_2(0) := \{(x', f(x')) : x' \in X\}$ and $S_1(y) := S \times \overline{l(y)}$ for any $y \in M_1$. Then $(\bar{x}, f(\bar{x}))$ is an extremal point of (S_1, S_2) at $(f(\bar{x}), 0)$ (Exercise 3.7.6). Let ε be an arbitrary positive number and choose $\varepsilon' < \min(2\varepsilon L_f/(1 + L_f), \varepsilon/2, 1/4)$, where L_f is the Lipschitz constant of f. By the approximate extremal principle of Corollary 3.7.3 there exist $y_0 \in B_{\varepsilon'}(f(\bar{x}))$, $(x_i, y_i) \in B_{\varepsilon'}(\bar{x}, f(\bar{x}))$, $(x_1^*, y_1^*) \in N_F(S_1(y_0), (x_1, y_1))$ and $(x_2^*, y_2^*) \in N_F(S_2(0), (x_2, y_2))$ such that $\max\{\|(x_1^*, y_1^*)\|, \|(x_2^*, y_2^*)\|\} > 1$ and

$$\|(x_1^*, y_1^*) + (x_2^*, y_2^*)\| < \varepsilon'. \tag{3.7.6}$$

It follows that

$$\|(x_1^*, y_1^*)\|, \|(x_2^*, y_2^*)\| > 1 - 2\varepsilon' > 1/2. \tag{3.7.7}$$

By the definition of a Fréchet normal cone we have that for $(x, y) \in S_2(0)$ sufficiently close to (x_2, y_2),

$$0 \geq \langle x_2^*, x - x_2 \rangle + \langle y_2^*, y - y_2 \rangle - \varepsilon' \|(x - x_2, y - y_2)\|.$$

Observing that $y_2 = f(x_2)$ and $y = f(x)$, we have

$$x \to -\left(\langle x_2^*, x - x_2 \rangle + \langle y_2^*, f(x) - f(x_2) \rangle - \varepsilon' \|(x - x_2, f(x) - f(x_2))\| \right)$$

attains a local minimum 0 at $x = x_2$. Combining the local sum rule of Theorem 3.3.1 and the chain rule of Theorem 3.5.1 there exists $x_0 \in B_{\varepsilon'}(x_2) \subset B_{2\varepsilon'}(\bar{x})$ such that

$$0 \in -x_2^* - \partial_F \langle y_2^*, f \rangle(x_0) + (1 + L_f)\varepsilon' B_{X^*}. \tag{3.7.8}$$

It follows from (3.7.6) and (3.7.8) that

$$0 \in x_1^* + \partial_F \langle y_1^*, f \rangle(x_0) + 2(1 + L_f)\varepsilon' B_{X^*}. \tag{3.7.9}$$

It follows from (3.7.7) and (3.7.9) that $\|y_1^*\| > 1/4L_f$. Dividing (3.7.9) by $\|y_1^*\|$ and set $x^* := x_1^*/\|y_1^*\| \in N_F(S; x_1)$ and $y^* := y_1^*/\|y_1^*\| \in N_F(\overline{l(y_0)}; y_1)$ we have

$$0 \in x^* + \partial_F \langle y^*, f \rangle (x_0) + \varepsilon B_{X^*},$$

as was to be shown.   ●

So far we have discussed a multi-objective optimization problem with only set constraints. This is not a severe restriction. In fact, consider a more general problem with inequality, equality and set constraints:

$$\mathcal{M}_g \qquad \begin{aligned} &\text{minimize} \quad f(x) \\ &\text{subject to} \quad f_n(x) \leq 0, \quad n = 1, 2, \ldots, N, \\ &\qquad\qquad\quad f_n(x) = 0, \quad n = N+1, \ldots, M, \\ &\qquad\qquad\quad x \in S. \end{aligned}$$

Here, S is a closed subset of X, $f_n \colon X \to \mathbb{R} \cup \{+\infty\}, n = 1, 2, \ldots, N$ are lsc functions and $f_n \colon X \to \mathbb{R}, n = N+1, \ldots, M$ are continuous functions. Denote the feasible set by S_1. Note that by the definition of the Fréchet normal cone $x^* \in N_F(S_1; x)$ implies that $\langle x^*, y - x \rangle - (\varepsilon/2)\|y - x\| \leq 0$ for all $y \in S_1$ sufficiently close to x. Then x is a local solution to the following minimization problem

$$\begin{aligned} &\text{minimize} \quad -\langle x^*, y - x \rangle \\ &\text{subject to} \quad f_n(y) \leq 0, \; n = 1, 2, \ldots, N, \\ &\qquad\qquad\quad f_n(y) = 0, \; n = N+1, \ldots, M, \\ &\qquad\qquad\quad y \in S. \end{aligned}$$

Thus, we can deduce a necessary optimality condition for problem $\mathcal{M}_g$ by combining Theorem 3.7.4 and Theorem 3.3.7. We leave the detail as an exercise.

3.7.4 Commentary and Exercises

The term extremal principle appeared in Mordukhovich [200] while the essence of this result goes back to Mordukhovich [195] and Kruger–Mordukhovich [167]. Extremal systems were defined first for fixed sets as follows.

Definition 3.7.5 (Extremal System of Fixed Sets [167]) *Let $S_n, n = 1, \ldots, N$ be closed subsets of X. We say $(S_1, \ldots, S_N)$ is an extremal system and $\bar{x}$ is an extremal point provided that there exists a neighborhood U of $\bar{x}$ such that for any $\varepsilon > 0$, there exist $m_n \in \varepsilon B_X, n = 1, \ldots, N$ such that*

$$U \cap (S_1 + m_1) \cap \cdots \cap (S_N + m_N) = \emptyset.$$

In [275], Zhu observed that the variational proof of the extremal principle actually applied to the separation of sets formed by more general deformation and used this fact to discuss necessary optimality conditions for multi-objective optimal control problems. Later a similar approach was used to derive necessary optimality conditions for constrained multi-objective optimization problems [269]. The general definition of an extremal system and its applications to multi-objective optimization problems in this section follow [210]. This concept of a general extremal system covers a wide variety of problems related to optimization and game theory.

Exercise 3.7.1 Find an explicit estimate for ε' in the proof of Theorem 3.7.2 in terms of ε and N.

Exercise 3.7.2 Show that the extremal system of fixed sets defined in Definition 3.7.5 gives rise naturally to an extremal system.

Exercise 3.7.3 Define $S_1(s_1) := \{(x,y) \mid |x| - 2|y| \geq s_1\}$ and $S_2(s_2) := \{(x,y) \mid |y| - 2|x| \geq s_2\}$.

(i) Show that $(S_1(s_1), S_2(s_2))$ is an extremal system at $(0,0)$.

(ii) Show that $(S_1(0), S_2(0))$ is not an extremal system of fixed sets.

Exercise 3.7.4 Prove Corollary 3.7.3.

Exercise 3.7.5 (Constrained Minimization Problem) Consider the problem of minimizing $f(x)$ subject to $x \in C$ where f is a lsc function and C a closed subset of Banach space X. Let $\bar{x}$ be a solution of this problem. Define $S_1 = \mathrm{epi} f$ and $S_2 = C \times (-\infty, f(\bar{x})]$. Show that $(\bar{x}, f(\bar{x}))$ is an extremal point for the extremal system (S_1, S_2) in the sense of Definition 3.7.5.

Exercise 3.7.6 Suppose that $\bar{x}$ is a solution to $\mathcal{M}$. Define $M_1 := l(f(\bar{x})) \cup \{f(\bar{x})\}$, $M_2 := \{0\}$, $S_1(y) := S \times \overline{l(y)}$ for $y \in M_1$ and $S_2(0) := \{(x', f(x')) \mid x' \in X\}$. Show that $(\bar{x}, f(\bar{x}))$ is an extremal point of the extremal system (S_1, S_2) at $(f(\bar{x}), 0)$.

Exercise 3.7.7 (Two Person Game) Consider a two player game in which the players A and B have strategy sets C and D which are closed subsets of Banach spaces X and Y, respectively. Let $f \colon X \times Y \to \mathbb{R}$ be the payoff of the game. The objective of player A is to maximize the payoff while that of B is to minimize it. In other words, we consider the game problem of

$$\mathcal{G} \quad \max_{x \in C} f(x,y) \text{ and } \min_{y \in D} f(x,y)$$

We say $(\bar{x}, \bar{y})$ is a solution to the game or a saddle pair provided that for all $(x,y) \in C \times D$, $f(x, \bar{y}) \leq f(\bar{x}, \bar{y}) \leq f(\bar{x}, y)$.

Show that $(\bar{x}, f(\bar{x}, \bar{y}), \bar{y}, f(\bar{x}, \bar{y}))$ is an extremal point for the extremal system (S_1, S_2) at $(f(\bar{x}, \bar{y}), f(\bar{x}, \bar{y}))$. (For discussions of the relevant game theory see the classic book [261]).

Exercise 3.7.8 Derive necessary optimality conditions for problem $\mathcal{M}_g$.

4

Variational Techniques in Convex Analysis

Convex analysis is now a rich branch of modern analysis. The purpose of this chapter is merely to point out the applications of variational techniques in convex analysis. In most of the cases direct proofs in the convex case lead to sharper results.

4.1 Convex Functions and Sets

4.1.1 Definitions and Basic Properties

Let X be a Banach space. We say that a subset C of X is *convex* if, for any $x, y \in C$ and any $\lambda \in [0,1]$, $\lambda x + (1-\lambda)y \in C$. We say an extended-valued function $f\colon X \to \mathbb{R} \cup \{+\infty\}$ is *convex* if its domain is convex and for any $x, y \in \mathrm{dom}\, f$ and any $\lambda \in [0,1]$, one has

$$f(\lambda x + (1-\lambda)y) \leq \lambda f(x) + (1-\lambda)f(y).$$

We call a function $f\colon X \to [-\infty, +\infty)$ *concave* if $-f$ is convex. In some sense convex functions are the simplest functions next to linear functions. Convex functions and convex sets are intrinsically related. For example, if C is a convex set then ι_C and d_C are convex functions. On the other hand if f is a convex function then epi f and $f^{-1}((-\infty, a])$, $a \in \mathbb{R}$ are convex sets (Exercises 4.1.1, 4.1.2 and 4.1.3). Two other important functions related to a convex set C are the *gauge function* defined by

$$\gamma_C(x) := \inf\{r > 0 \mid x \in rC\},$$

and the *support function* defined on the dual space X^* by

$$\sigma_C(x^*) = \sigma(C; x^*) := \sup\{\langle x, x^* \rangle \mid x \in C\}.$$

Several useful properties of the gauge function and the support function are discussed in Exercises 4.1.6 and 4.1.10.

4.1.2 Local Lipschitz Property of Convex Functions

Lower semicontinuous convex functions are actually locally Lipschitz in the interior of their domains. This is, in fact, a combination of two facts: (a) a convex function f locally bounded above is locally Lipschitz in int dom f and (b) a lsc convex function f is locally bounded above in int dom f. Fact (a) is quite useful itself and we describe it in two propositions.

Proposition 4.1.1 *Let X be a Banach space and let $f\colon X \to \mathbb{R} \cup \{+\infty\}$ be a convex function. Suppose that f is locally bounded above at $\bar{x} \in D :=$ int(dom f). Then f is locally bounded at $\bar{x}$.*

Proof. Suppose f is bounded above by M, say, in $B_r(x) \subset$ int(dom f) for some $r > 0$, then it is bounded below in $B_r(x)$. Indeed, if $y \in B_r(x)$ then so is $2x - y$ and

$$f(x) \le \frac{1}{2}[f(y) + f(2x - y)] \le \frac{1}{2}[f(y) + M]$$

so $f(y) \ge 2f(x) - M$ for all $y \in B_r(x)$.   ●

Proposition 4.1.2 *Let X be a Banach space and let $f\colon X \to \mathbb{R} \cup \{+\infty\}$ be a convex function. Suppose that f is locally bounded at $\bar{x} \in D :=$ int(dom f). Then f is locally Lipschitz at $\bar{x}$.*

Proof. Suppose that $|f|$ is bounded by M over $B_{2r}(\bar{x}) \subset D$. Consider distinct points $x, y \in B_r(\bar{x})$. Let $a = \|y - x\|$ and let $z = y + (r/a)(y - x)$. Then $z \in B_{2r}(\bar{x})$. Since

$$y = \frac{a}{a+r}z + \frac{r}{a+r}x$$

is a convex combination lying in $B_{2r}(\bar{x})$, we have

$$f(y) \le \frac{a}{a+r}f(z) + \frac{r}{a+r}f(x).$$

Thus,

$$f(y) - f(x) \le \frac{a}{a+r}(f(z) - f(x)) \le \frac{2Ma}{r} = \frac{2M}{r}\|y - x\|.$$

Interchange x and y gives

$$|f(y) - f(x)| \le \frac{2M}{r}\|y - x\|.$$

●

Theorem 4.1.3 (Lipschitz Property of Convex Functions) *Let X be a Banach space and let $f\colon X \to \mathbb{R} \cup \{+\infty\}$ be a lsc convex function. Then f is locally Lipschitz on int(dom f).*

Proof.   By Propositions 4.1.1 and 4.1.2 we need only show f is locally bounded above. For each natural number i, define $D_i := \{x \in X : f(x) \leq i\}$. The sets D_i are closed and $D \subset \bigcup_{i=1}^{\infty} D_i$. Since D is an open set, by Baire's category theorem, we must have for some i, $\operatorname{int} D_i$ is nonempty. Suppose that $B_s(x) \subset \operatorname{int} D_i$. Then f is bounded above by i over $B_s(x)$. Also since D is open, if $y \in D$ and $y \neq x$, then there exist $\mu > 1$ such that $z := x + \mu(y-x) \in D$. Let $\lambda = 1/\mu \in (0,1)$. The set $U = \{\lambda z + (1-\lambda)b : b \in B_s(x)\}$ is a neighborhood of y in D. For any point $u = \lambda z + (1-\lambda)b \in U$ (where $b \in B_s(x)$) we have

$$f(u) \leq \lambda f(z) + (1-\lambda)i,$$

so f is bounded above in U and therefore locally Lipschitz at y.   ●

4.1.3 Convex Series Closed Sets

The condition in Theorem 4.1.3 can be weakened. To understand this deeper result we need the following concepts.

Definition 4.1.4 (Convex Series Closed and Compact) *Let X be a Banach space and let C be a subset of X. We say that C is* convex series closed (cs-closed) *if $\bar{x} = \sum_{i=1}^{\infty} \lambda_i x_i$ with $\lambda_i \geq 0$, $\sum_{i=1}^{\infty} \lambda_i = 1$ and $x_i \in C$ implies $\bar{x} \in C$. We say that C is* convex series compact (cs-compact) *if for any sequence $x_i \in C, i = 1, 2, \ldots$, and any sequence $\lambda_i \geq 0, i = 1, 2, \ldots$, with $\sum_{i=1}^{\infty} \lambda_i = 1$ we have $\sum_{i=1}^{\infty} \lambda_i x_i$ converges to a point of C.*

Some simple yet useful facts related to the cs-closed and cs-compact sets are given below.

Lemma 4.1.5 *Closed convex sets, open convex sets and G_δ convex sets in a Banach space are cs-closed.*

Proof.   We prove the lemma for open convex sets and the proofs for the other two cases are left as exercises. Let C be a convex open set in a Banach space and let $\bar{x} = \sum_{i=1}^{\infty} \lambda_i x_i$ with $\lambda_i \geq 0$, $\sum_{i=1}^{\infty} \lambda_i = 1$ and $x_i \in C$. We show that $\bar{x} \in C$. Suppose on the contrary that $\bar{x} \notin C$. Then according to the Hahn–Banach separation theorem there exists a nonzero linear functional $x^* \in X^*$ such that $\langle x^*, c \rangle > \langle x^*, \bar{x} \rangle$ for all $c \in C$. In particular, $0 > \langle x^*, \bar{x} - x_i \rangle$ for $i = 1, 2, \ldots$, and therefore for any $\lambda_i > 0$, $0 > \langle x^*, \lambda_i(\bar{x} - x_i) \rangle$. This leads to $0 > \langle x^*, \bar{x} - \sum_{i=1}^{\infty} \lambda_i x_i \rangle = 0$, a contradiction.   ●

Lemma 4.1.6 *Let X and Y be two Banach spaces and let $A: X \to Y$ be a continuous linear mapping. Suppose that C is a cs-compact subset of X. Then $A(C)$ is cs-closed.*

Proof. Exercise 4.1.13.   ●

An important fact about cs-closed sets is that they share their interior points with their closure.

Theorem 4.1.7 (Open Mapping Theorem: cs-Closed Sets) *Let S be a cs-closed subset of a Banach space X. Then*

$$\text{int } S = \text{int } \overline{S}.$$

Proof. We consider the nontrivial case when int $\overline{S} \neq \emptyset$. Let $x \in$ int $\overline{S}$. Shifting S and multiplying it by a constant if necessary we may assume (see Exercise 4.1.16) that

$$0 = x \in B_X \subset \overline{S} \subset S + \frac{1}{2}B_X. \tag{4.1.1}$$

For $i = 1, 2, \ldots$ multiplying (4.1.1) by $1/2^i$ we have

$$\frac{1}{2^i}B_X \subset \frac{1}{2^i}S + \frac{1}{2^{i+1}}B_X. \tag{4.1.2}$$

It follows from (4.1.2) that for any $i = 1, 2, \ldots,$

$$\frac{1}{2}B_X \subset \frac{1}{2}S + \frac{1}{4}S + \cdots + \frac{1}{2^i}S + \frac{1}{2^{i+1}}B_X. \tag{4.1.3}$$

That is to say, for any $u \in B_X/2$ there exist $s_1, \ldots, s_i \in S$ such that

$$u \in \frac{1}{2}s_1 + \frac{1}{4}s_2 + \cdots + \frac{1}{2^i}s_i + \frac{1}{2^{i+1}}B_X. \tag{4.1.4}$$

Taking limits as $i \to \infty$ in (4.1.4) we have $u = \sum_{i=1}^{\infty} s_i/2^i \in S$ because S is cs-closed. Thus, $0 \in 2^{-1}B_X \subset S$, and therefore int $\overline{S} \subset$ int S. Hence int $S =$ int $\overline{S}$.   ●

We now turn to the promised sharper results on the local Lipschitz property for a convex function. Let S be a subset of a Banach space X. We say s is in the *core* of S, denote $s \in \text{core}(S)$, provided that $\bigcup_{\lambda > 0} \lambda(S - s) = X$. Clearly, $\text{int}(S) \subset \text{core}(S)$ and the inclusion could be proper (Exercises 4.1.17 and 4.1.18). Our next result says that if S is the domain of a lsc convex function then the interior and the core of S coincide. The importance of this result is due to the fact that it is much easier to verify that a point belongs to the core than to the interior.

Theorem 4.1.8 *Let X be a Banach space and let $f: X \to \mathbb{R} \cup \{+\infty\}$ be a lsc convex function. Then*

$$\text{core}(\text{dom } f) = \text{int}(\text{dom } f).$$

Proof. We need only show that

$$\text{core}(\text{dom } f) \subset \text{int}(\text{dom } f).$$

Suppose that $\bar{x} \in \text{core}(\text{dom } f)$. For each natural number i, define $D_i := \{x \in X : f(x) \le i\}$. The sets D_i are closed and

$$X = \bigcup_{j=1}^{\infty} j(\text{dom } f - \bar{x}) = \bigcup_{j,i=1}^{\infty} j(D_i - \bar{x}). \qquad (4.1.5)$$

By Baire's category theorem, $\text{int}(D_i - \bar{x})$ (and therefore $\text{int } D_i$) is nonempty for some i. Suppose that $B_r(x) \subset \text{int } D_i$. Then f is bounded above by i over $B_r(x)$. Moreover, by (4.1.5) there exist integers $j, k > 0$ such that $\bar{x} - x \in j(D_k - \bar{x})$. Letting $\mu = (1 + 1/j)$, we have $z := x + \mu(\bar{x} - x) \in D_k$. Note that D_k and D_i are contained in the convex set $D_{\max(i,k)}$. Let $\lambda = 1/\mu \in (0, 1)$. The set $U = \{\lambda z + (1 - \lambda)b : b \in B_r(x)\}$ is a neighborhood of $\bar{x}$ in $D_{\max(i,k)} \subset \text{dom } f$.

●

4.1.4 Commentary and Exercises

Although there is a long history of using the convexity of both functions and sets in analysis, the systematical study of convex functions and sets starts in the 1950's associated with the names of Fenchel, Moreau, and Rockafellar. A classical reference for convex analysis is Rockafellar [229]. For a nice short introduction that provides details missed in this chapter we recommend Phelps [221]. More discussion on convex series closed and compact sets can be found in Jameson [152].

Exercise 4.1.1 Let C be a convex subset of a Banach space. Show that d_C and ι_C are convex functions.

Exercise 4.1.2 Let f be a convex function on a Banach space. Show that for any $a \in \mathbb{R}$, $f^{-1}((-\infty, a])$ is a convex set.

Exercise 4.1.3 Let X be a Banach space and let $f \colon X \to \mathbb{R} \cup \{+\infty\}$ be an extended-valued function. Show that f is convex if and only if $\text{epi } f$ is a convex subset of $X \times \mathbb{R}$.

Exercise 4.1.4 Show that the intersection of a family of arbitrary convex sets is convex. Conclude that $f(x) := \sup\{f_\alpha(x) : \alpha \in A\}$ is convex (and lsc) when $\{f_\alpha\}_{\alpha \in A}$ is a collection of convex (and lsc) functions.

Exercise 4.1.5 Calculate the gauge function for $C := \text{epi } 1/x \cap \mathbb{R}^2_+$ and conclude that a gauge function is not necessarily lsc.

Exercise 4.1.6 Let C be a convex subset of a Banach space X and let γ_C be the gauge function of C.

(i)   Show that γ_C is convex and when $0 \in C$ it is subadditive.

(ii) Show that if $x \in \text{core}\,C$ then $\text{dom}\,\gamma_{C-x} = X$.

(iii) Suppose $0 \in \text{core}\,C$. Prove that $\text{cl}\,C \subset \{x \in X \mid \gamma_C(x) \leq 1\}$.

Exercise 4.1.7 Let X be a Banach space and let C be a cs-closed subset of X. Prove that $\text{int}\,C = \text{core}\,C$.

Exercise 4.1.8 Let X be a Banach space and let C be a convex subset of X. Suppose that C is cs-closed and $0 \in \text{core}\,C$.

(i) Show that $\text{int}\,C = \{x \in X \mid \gamma_C(x) < 1\}$.

(ii) Deduce that γ_C is defined on X and is continuous.

*∗**Exercise 4.1.9** Construct an example showing that the conclusion in Exercise 4.1.8 fails when C is not cs-closed. Hint: Use the existence of a Hamel basis in a vector space to show that in every infinite dimensional Banach space there is a finite linear functional, ϕ which is (everywhere) discontinuous. Deduce that $C := \phi^{-1}[-1,1]$ is a symmetric convex set with a nonempty core that contains 0 but an empty interior. Yet $\gamma_C(0) = 0 < 1$.

Exercise 4.1.10 Let C_1 and C_2 be closed convex subsets of a Banach space X. Then $C_1 \subset C_2$ if and only if, for any $x^* \in X^*$, $\sigma(C_1; x^*) \leq \sigma(C_2; x^*)$. Thus, a closed convex set is characterized by its support function.

Exercise 4.1.11 Prove that if f is a convex lsc function then $\partial f(x) = \partial_F f(x)$.

Exercise 4.1.12 Prove Lemma 4.1.5 for the cases of closed convex sets and convex G_δ sets.

Exercise 4.1.13 Prove Lemma 4.1.6.

Exercise 4.1.14 Let X be a Banach space and let C be a subset of X. Show that C is cs-compact if and only if C is cs-closed and bounded. In particular, both the open and closed unit balls in a Banach space are cs-compact.

Exercise 4.1.15 Let X be a Banach space and let A and B be subsets of X. Suppose that A is cs-compact and B is cs-closed. Then $A+B$ and $\text{conv}(A \cup B)$ are cs-closed.

Exercise 4.1.16 Suppose that S is cs-closed and $\bar{x} \in S$. Show that for any $\delta > 0$ $(S - \bar{x})/\delta$ is also cs-closed.

Exercise 4.1.17 Let S be a subset of a Banach space. Show that $\text{int}(S) \subset \text{core}(S)$.

Exercise 4.1.18 (Core Versus Interior) Consider the set in $\mathbb{R}^2$

$$S = \{(x,y) \mid y = 0 \text{ or } |y| \geq x^2\}.$$

Prove $0 \in \text{core}(S) \setminus \text{int}(S)$.

Exercise 4.1.19 Show that in the proof of Theorem 4.1.8 the set U can be expressed explicitly as $U = B_{r(1-\lambda)}(\bar{x})$.

4.2 Subdifferential

4.2.1 The Subdifferential and the Normal Cone

Let X be a Banach space. We define the *convex subdifferential* of a convex function $f: X \to \mathbb{R} \cup \{+\infty\}$ at $x \in \operatorname{dom} f$ by

$$\partial f(x) := \{x^* \in X^* : f(y) - f(x) \geq \langle x^*, y - x \rangle, \quad \text{for all } y \in X\}, \quad (4.2.1)$$

and we define its *domain*

$$\operatorname{dom} \partial f = \{x \in X \mid \partial f(x) \neq \emptyset\}.$$

An element of $\partial f(x)$ is called a subgradient of f at x. Although the domain of a convex function is always convex it is not necessarily so for $\operatorname{dom} \partial f$ (see Exercise 4.2.6).

For a closed convex set $C \subset X$, we define the *normal cone* of C at $\bar{x} \in C$ by $N(C; \bar{x}) = \partial \iota_C(\bar{x})$. Sometimes we will also use the notation $N_C(\bar{x}) = N(C; \bar{x})$. A useful characterization of the normal cone is $x^* \in N(C; x)$ if and only if, for all $y \in C$, $\langle x^*, y - x \rangle \leq 0$ (Exercise 4.2.7). The following easy observation suggests the fundamental significance of subdifferential in optimization.

Proposition 4.2.1 (Subdifferential at Optimality) *Let X be a Banach space and let $f: X \to \mathbb{R} \cup \{+\infty\}$ be a proper convex function. Then the point $\bar{x} \in X$ is a (global) minimizer of f if and only if the condition $0 \in \partial f(x)$ holds.*

Proof. Exercise 4.2.9. ●

Alternatively put, minimizers of f correspond exactly to "zeroes" of ∂f. It is obvious that $\partial f(x) \subset \partial_F f(x)$ and we actually have equality (Exercise 4.2.10).

We have seen that a general lsc function in a Fréchet smooth Banach space is densely subdifferentiable in its domain. For convex functions we have a similar but much stronger result: the subdifferential of a lsc convex function is nonempty at every point in $\operatorname{core}(\operatorname{dom} f)$. This will be the focus of the next two subsections.

4.2.2 Directional Derivatives of Convex Functions

One useful tool in analyzing the convex subdifferential is the directional derivative. Let $f: X \to \mathbb{R} \cup \{+\infty\}$ and let $x \in \operatorname{dom} f$ and $d \in X$. The directional derivative of f at x in the direction of d is defined by

$$f'(x; d) := \lim_{t \to 0+} \frac{f(x + td) - f(x)}{t}$$

when this limit exists. It turns out that the directional derivative of a convex function is again convex. Furthermore, the directional derivative has an even stronger property of characterizing subgradients as seen in Proposition 4.2.5.

If a convex function f satisfies the stronger condition

$$f(\lambda x + \mu y) \le \lambda f(x) + \mu f(y) \quad \text{for all } x, y \in X, \ \lambda, \mu \ge 0$$

we say f is *sublinear*. If $f(\lambda x) = \lambda f(x)$ for all x in X and $\lambda \ge 0$ then f is *positively homogeneous*; in particular this implies $f(0) = 0$. (Here we use the convention $0 \times (+\infty) = 0$.) If $f(x + y) \le f(x) + f(y)$ for all x and y in X then we say f is *subadditive*. It is an easy exercise to show that these two properties characterize a sublinear function.

Proposition 4.2.2 (Sublinearity) *Let X be a Banach space and let $f \colon X \to \mathbb{R} \cup \{+\infty\}$ be an extended-valued function. Then f is sublinear if and only if it is positively homogeneous and subadditive.*

Proof. Exercise 4.2.11.   ●

It is immediate that if the function f is sublinear then $-f(x) \le f(-x)$ for all x in X. The *linearity space* of a sublinear function f is the set

$$\mathrm{lin}\, f = \{x \in X \mid -f(x) = f(-x)\}.$$

The following result shows this set is a subspace.

Proposition 4.2.3 (Linearity Space) *Let X be a Banach space and let $f \colon X \to \mathbb{R} \cup \{+\infty\}$ be a sublinear function. Then, the linearity space $\mathrm{lin}\, f$ of f is the largest subspace of X on which f is linear.*

Proof. It is clear that if Y is a subspace on which f is linear then $Y \subset \mathrm{lin}\, f$. We need only show that $\mathrm{lin}\, f$ is a subspace. Let $x \in \mathrm{lin}\, f$ and $a \in \mathbb{R}$. Since f is homogeneous we have

$$f(ax) = |a| f\left(\frac{a}{|a|} x\right) = -|a| f\left(-\frac{a}{|a|} x\right) = -f\left(|a|\left(-\frac{a}{|a|} x\right)\right) = -f(-ax),$$

so that $ax \in \mathrm{lin}\, f$. Let $x, y \in \mathrm{lin}\, f$. Since f is subadditive we have

$$f(x + y) \le f(x) + f(y) = -f(-x) - f(-y) \le -f(-x - y) = -f(-(x + y)),$$

so that $x + y \in \mathrm{lin}\, f$. Thus, $\mathrm{lin}\, f$ is a subspace.   ●

It is easy to check that if the point $\bar{x}$ lies in the core of the domain of a convex function f then the directional derivative $f'(\bar{x}; \cdot)$ is well-defined and positively homogeneous.

Proposition 4.2.4 (Sublinearity of the Directional Derivative) *Let X be a Banach space and let $f \colon X \to \mathbb{R} \cup \{+\infty\}$ be a convex function. Suppose that $\bar{x} \in \mathrm{core}(\mathrm{dom}\, f)$. Then the directional derivative $f'(\bar{x}; \cdot)$ is everywhere finite and sublinear.*

Proof. For d in X and nonzero t in $\mathbb{R}$, define

$$g(d;t) = \frac{f(\bar{x}+td) - f(\bar{x})}{t}.$$

By convexity we deduce (Exercise 4.2.1) for $0 < t \leq s \in \mathbb{R}$, the inequality

$$g(d;-s) \leq g(d;-t) \leq g(d;t) \leq g(d;s).$$

Since $\bar{x}$ lies in core(dom f), for small $s > 0$ both $g(d;-s)$ and $g(d;s)$ are finite, so as $t \downarrow 0$ we have

$$+\infty > g(d;s) \geq g(d;t) \downarrow f'(\bar{x};d) \geq g(d;-s) > -\infty. \tag{4.2.2}$$

Again by convexity we have for any directions d and e in X and real $t > 0$,

$$g(d+e;t) \leq g(d;2t) + g(e;2t).$$

Now letting $t \downarrow 0$ we see that $f'(\bar{x};\cdot)$ is subadditive. The positive homogeneity is easy to check.   ●

Next we show that the directional derivative characterizes subgradients. That explains why it is useful in analyzing the subdifferential.

Proposition 4.2.5 (Subgradients and Directional Derivatives) *Let X be a Banach space, let $f\colon X \to \mathbb{R}\cup\{+\infty\}$ be a convex function and let $\bar{x} \in$ dom f. Then $x^* \in X^*$ is a subgradient of f at $\bar{x}$ if and only if it satisfies $x^* \leq f'(\bar{x};\cdot)$.*

Proof. For the "only if" part, let $x^* \in \partial f(\bar{x})$. Then, for any $h \in X$ and $t > 0$,

$$\langle x^*, th \rangle \leq f(\bar{x}+th) - f(\bar{x}).$$

Dividing by t and taking limits as $t \to 0$ we have $\langle x^*, h \rangle \leq f'(\bar{x};h)$.

For the reverse direction, it follows from the proof of Proposition 4.2.4 that for any $h \in X$ and $t > 0$,

$$\langle x^*, h \rangle \leq f'(\bar{x};h) \leq \frac{f(\bar{x}+th) - f(\bar{x})}{t}.$$

Let x be an arbitrary element of X. Setting $h = x - \bar{x}$ and $t = 1$ in the above inequality we have

$$\langle x^*, x - \bar{x} \rangle \leq f(x) - f(\bar{x}),$$

that is $x^* \in \partial f(\bar{x})$.   ●

4.2.3 Nonemptiness of the Subdifferential

The main result of this section is that the set of subgradients of a convex function is usually nonempty. We prove this by actually constructing a subgradient. The idea is rather simple. We recursively construct a decreasing sequence of sublinear functions which, after translation, minorize f. At each step we guarantee one extra direction of linearity. The basic step is summarized in the following lemma.

Lemma 4.2.6 *Let X be a Banach space and let $p\colon X \to \mathbb{R} \cup \{+\infty\}$ be a sublinear function. Suppose that $d \in \text{core}(\text{dom}\, p)$. Then the function $q(\cdot) = p'(d; \cdot)$ satisfies the conditions*

(i) $q(\lambda d) = \lambda p(d)$ *for all real λ,*
(ii) $q \le p$,
(iii) $\lim q \supset \lim p + \text{span}\{d\}$, *and*
(iv) $p = q$ *on* $\lim p$.

Proof. Exercise 4.2.12.   ●

With these tools we are now ready for the main result, which gives conditions guaranteeing the existence of a subgradient of a convex function. Proposition 4.2.5 showed how to identify subgradients from directional derivatives; this next result shows how to move in the reverse direction. For an extended-valued function f we use cont f to denote the set of all points where f is finite and continuous.

Theorem 4.2.7 (Max Formula) *Let X be a Banach space, $d \in X$ and let $f\colon X \to \mathbb{R} \cup \{+\infty\}$ be a convex function. Suppose that either*

Q1 $\bar{x} \in \text{core}(\text{dom}\, f)$ *and f is lsc or*
Q2 $\bar{x} \in \text{cont}\, f$.

Then,

$$f'(\bar{x}; d) = \max\{\langle x^*, d\rangle : x^* \in \partial f(\bar{x})\}. \tag{4.2.3}$$

Proof. In view of Proposition 4.2.5, we simply have to show that for any fixed d in X there is a subgradient x^* satisfying $\langle x^*, d\rangle = f'(\bar{x}; d)$.

Let $p(\cdot) := f'(\bar{x}; \cdot)$. Then p is a sublinear function defined on X. Consider the family $\mathcal{S}$ of all sublinear functions dominated by p and coinciding with p at d, with a partial order defined by: $p_2 \prec p_1$ if and only if $\lim p_2 \subset \lim p_1$ and $p_1 \le p_2$ with equality holding on $\lim p_2$. We can check that any chain $\{p_a\}_{a \in A} \subset \mathcal{S}$ has an upper bound $\bar{p} := \inf_{a \in A} p_a$ defined on $\bigcup_{a \in A} \text{dom}(p_a)$ (Exercise 4.2.13). Thus, by Zorn's lemma $\mathcal{S}$ has a maximum element x^*. By Lemma 4.2.6 we must have $\lim x^* = X$. Under the constraint qualification condition Q1 or Q2 f is locally Lipschitz at $\bar{x}$. When Q1 holds this follows from Theorems 4.1.3 and 4.1.8 and when Q2 holds this follows directly from

Proposition 4.1.2. Let L be a Lipschitz constant of f in a neighborhood of $\bar{x}$. Then $|f'(\bar{x}; h)| \leq L\|h\|$ for all $h \in X$. Since x^* is dominated by $p(\cdot) = f'(\bar{x}; \cdot)$, we must have $x^* \in X^*$, and therefore $x^* \in \partial f(\bar{x})$. The max formula follows from $\langle x^*, d \rangle = p(d) = f'(\bar{x}; d)$.   ●

As an easy corollary of the max formula we have the following key result in subdifferential theory due to Fenchel and Rockafellar.

Theorem 4.2.8 (Nonemptiness of Subdifferential) *Let X be a Banach space and let $f: X \to \mathbb{R} \cup \{+\infty\}$ be a convex function. Suppose that either*

Q1 $\bar{x} \in \operatorname{core}(\operatorname{dom} f)$ *and f is lsc or*
Q2 $\bar{x} \in \operatorname{cont} f$.

Then the subdifferential $\partial f(\bar{x})$ is nonempty.

Proof. Follows directly from Theorem 4.2.7.   ●

The constraint qualification conditions in Theorems 4.2.7 and 4.2.8 are indispensible in any infinite dimensional spaces because in any infinite dimensional normed space there exists a discontinuous linear functional defined on the whole space (Exercise 4.2.14).

A differentiability result for convex functions follows immediately. Recall that a function $f: X \to \mathbb{R}$ is *Gâteaux differentiable* at x provided that there exists $x^* \in X^*$ such that for any $v \in X$, the directional derivative $f'(x; v)$ exists and $f'(x; v) = \langle x^*, v \rangle$.

Corollary 4.2.9 (Differentiability of Convex Functions) *Let X be a Banach space, let $f: X \to \mathbb{R} \cup \{+\infty\}$ be a convex function and let $\bar{x} \in \operatorname{core}(\operatorname{dom} f)$. Then f is Gâteaux differentiable at $\bar{x}$ exactly when f has a unique subgradient at $\bar{x}$ (in which case this subgradient is the derivative).*

Proof. Exercise 4.2.15   ●

The conclusion $\partial f(\bar{x}) \neq \emptyset$ can be stated alternatively as there exists a linear functional x^* such that $f - x^*$ attains its minimum at $\bar{x}$. Thus, in a certain sense Theorem 4.2.8 plays a role in the analysis of convex functions similar to that of a variational principle in the analysis of general lower semicontinuous functions. Note that due to the nice properties of convex functions the conclusion here is stronger in comparison to the variational principles. One can view the variational principles as approximate versions of Theorem 4.2.8 for (nonconvex) lsc functions.

The condition $\bar{x} \in \operatorname{core}(\operatorname{dom} f)$ is crucial in ensuring $\partial f(\bar{x}) \neq \emptyset$. Without this condition the subdifferential may be an emptyset. As a simple example one can check that $\partial f(0) = \emptyset$ for function $f: \mathbb{R} \to \mathbb{R} \cup \{+\infty\}$ defined by $f(x) = -\sqrt{x}, x \geq 0$ and $+\infty$ otherwise. The following is a systematical scheme for generating such functions in infinite dimensional spaces.

Example 4.2.10 Let X be an infinite dimensional separable Banach space and let C be a symmetric compact convex set whose core is empty but whose span is dense. (The Hilbert cube in ℓ_2 is a typical example of such a set, see Exercise 4.2.18.) Let $\bar{x} \notin \text{span}(C)$. Define $f\colon X \to \mathbb{R} \cup \{+\infty\}$ by $f(x) :=$ $\min\{\lambda \in \mathbb{R} \mid x + \lambda\bar{x} \in C\}$, where we use the convention that $\min(\emptyset) = +\infty$. It is easy to check that f is a convex function and for any $s \in \mathbb{R}$ and $c \in C$, $f(c + s\bar{x}) = -s$ (Exercise 4.2.19). It follows that

$$f'(0; y) = \begin{cases} -s & \text{if } y = rc + s\bar{x} \text{ for some } c \in C \text{ and } r, s \in \mathbb{R}, \\ +\infty & \text{otherwise.} \end{cases}$$

Now we show that $\partial f(0) = \emptyset$. Suppose on the contrary that $x^* \in \partial f(0)$. Since $\text{span}(C)$ is dense in X, for any $s \in \mathbb{R}$ we can find $r \in \mathbb{R}$ and $c \in C$ such that $rc + s\bar{x}$ is close to a unit vector so that

$$-\|x^*\| - 1 \le \langle x^*, rc + s\bar{x} \rangle \le f'(0, rc + s\bar{x}) = -s,$$

which is a contradiction.

4.2.4 Commentary and Exercises

The nonemptiness of subdifferentials and the more delicate max formula are core results of the convex analysis. Fenchel, Moreau, Rockafellar, Valadier and many others contributed to the current form of these results. Besides providing a convex version of the variational principle, they also characterize convexity which leads to a number of important ways of recognizing convex functions (see Exercise 4.2.16). The algebraic proof of the max formula we follow here is due to [29]. The convexity of a function is also characterized by the monotonicity of its subdifferential. Exercise 4.2.21 provides a taste of the more general results along this line. Examples of emptiness of the subdifferential in the absence of the qualification conditions are discussed in [60, 221].

Exercise 4.2.1 Let X be a Banach space and let $f\colon X \to \mathbb{R} \cup \{+\infty\}$ be a convex function. Suppose that $\bar{x} \in \text{core}(\text{dom } f)$. Show that for any $d \in X$, $t \to g(d; t) := (f(\bar{x} + td) - f(\bar{x}))/t$ is a nondecreasing function.

Exercise 4.2.2 Prove the subdifferential of a convex function at a given point is a closed convex set.

Exercise 4.2.3 Prove the following functions $x \in \mathbb{R} \mapsto f(x)$ are convex and calculate ∂f:

(i) $|x|$;

(ii) $\iota_{\mathbb{R}_+}$;

(iii) $\begin{cases} -\sqrt{x} & \text{if } x \ge 0, \\ +\infty & \text{otherwise;} \end{cases}$

(iv)
$$\begin{cases} 0 & \text{if } x < 0, \\ 1 & \text{if } x = 0, \\ +\infty & \text{otherwise.} \end{cases}$$

Exercise 4.2.4 (Subgradients of Norm) Calculate $\partial \|\cdot\|$. Generalize your result to an arbitrary sublinear function.

Exercise 4.2.5 (Subgradients of Maximum Eigenvalue) Denote the largest eigenvalue of an N by N symmetric matrix by λ_1. Prove that $\partial \lambda_1(0)$ is the set of all N by N symmetric matrices with trace 1.

Exercise 4.2.6 (Domain of Subdifferential) If the function $f \colon \mathbb{R}^2 \to \mathbb{R} \cup \{+\infty\}$ is defined by

$$f(x_1, x_2) = \begin{cases} \max\{1 - \sqrt{x_1}, |x_2|\} & \text{if } x_1 \geq 0, \\ +\infty & \text{otherwise,} \end{cases}$$

prove that f is convex but that $\operatorname{dom} \partial f$ is not convex.

Exercise 4.2.7 (Normal Cone's Characterization) Let C be a closed convex subset of X. Prove that $x^* \in N_C(x)$ if and only if, for all $y \in C$,

$$\langle x^*, y - x \rangle \leq 0.$$

Exercise 4.2.8 Let $K \subset X$ be a closed convex cone. Show that both d_K and ι_K are convex functions and, for any $x \in X$,

$$\partial d_K(x) \subset \partial \iota_K(0) \cap B_{X^*},$$

and

$$\partial \iota_K(x) \subset \partial \iota_K(0).$$

Exercise 4.2.9 Prove Proposition 4.2.1.

Exercise 4.2.10 (The Fréchet Subdifferential of Convex Functions) Prove that for a lsc convex function $f \colon X \to \mathbb{R} \cup \{+\infty\}$ and $x \in X$,

$$\partial f(x) = \partial_F f(x).$$

Exercise 4.2.11 Prove Proposition 4.2.2 (Sublinearity).

Exercise 4.2.12 Prove Lemma 4.2.6.

Exercise 4.2.13 Show that the chain $\{p_a\}_{a \in A} \subset \mathcal{S}$ defined in the proof of Theorem 4.2.7 has an upper bound $\bar{p} := \inf_{a \in A} p_a$ defined on $\bigcup_{a \in A} \operatorname{dom}(p_a)$.

Exercise 4.2.14 Let X be a normed space. Show that the following are equivalent

(i)   X is finite dimensional.
(ii) Every linear function f is continuous.

(iii) Every absorbing convex set has zero in its interior.

Hint: (iii) $\Rightarrow$ (ii): $f^{-1}(-1,1)$ is absorbing and convex and symmetric. (ii) $\Rightarrow$ (i): use the existence of an infinite linearly independent set $\{e_i\}$ to define a discontinuous everywhere finite linear functional satisfying $f(e_i/\|e_i\|) = i$. (i) $\Rightarrow$ (iii) is obvious.

Exercise 4.2.15 Prove Corollary 4.2.9.

*Exercise 4.2.16** (Recognizing Convex Functions) Suppose the set $C \subset \mathbb{R}^N$ is open and convex, and consider a function $f: C \to \mathbb{R}$. For points $x \notin C$, define $f(x) = +\infty$.

(i)   Prove $\partial f(x)$ is nonempty for all x in C if and only if f is convex. Hint: For points u and v in C and real λ in $[0,1]$, use the subgradient inequality (4.2.1) at the points $\bar{x} = \lambda u + (1-\lambda)v$ and $x = u, v$ to check the definition of convexity.

(ii) Prove that if $T \subset \mathbb{R}$ is an open interval and $g: T \to \mathbb{R}$ is differentiable then g is convex if and only if g' is nondecreasing on T, and g is strictly convex if and only if g' is strictly increasing on T. Deduce that if g is twice differentiable then g is convex if and only if g'' is nonnegative on T and g is strictly convex if g'' is strictly positive on T.

(iii) Deduce that if f is twice continuously differentiable on C then f is convex if and only if its Hessian matrix is positive semidefinite everywhere on C, and f is strictly convex if its Hessian matrix is positive definite everywhere on C. Hint: Apply part (ii) to the function g defined by $g(t) = f(x + td)$ for small real t, points x in C, and directions d in X.

(iv) Find a strictly convex function $f: (-1,1) \to \mathbb{R}$ with $f''(0) = 0$.

(v)   Prove that a continuous function $h:\ \text{cl}\, C \to \mathbb{R}$ is convex if and only if its restriction to C is convex. What about strictly convex functions?

Exercise 4.2.17 (DC function) We say that $f: X \to \mathbb{R} \cup \{+\infty\}$ is a DC function if it can be written as the difference of two real valued lsc convex functions. Prove that a DC function is locally Lipschitz and directional differentiable at any $x \in \text{core dom}\, f$.

Exercise 4.2.18 The Hilbert cube in ℓ_2 is defined by

$$H := \left\{ x = (x_1, x_2, \dots) \in \ell^2 \mid |x_i| \le 1/2^i, i = 1, 2, \dots \right\}.$$

Show that the Hilbert cube is a symmetric compact convex set of ℓ_2 satisfying $\text{core}\, H = \emptyset$ and $\overline{\text{span}(H)} = \ell_2$.

Exercise 4.2.19 Prove that the function f defined in Example 4.2.10 is convex and has the property that for any $s \in \mathbb{R}$ and $c \in C$, $f(c + s\bar{x}) = -s$.

*Exercise 4.2.20** With some additional work we can also construct a convex function whose subdifferential is empty on a dense subset of its domain. Let

$X = \ell_2$ and H be the Hilbert cube defined in Exercise 4.2.18 and define $f \colon X \to \mathbb{R} \cup \{+\infty\}$ by

$$f(x) = \begin{cases} -\sum_{i=1}^{\infty} \sqrt{2^{-i} + x_i} & \text{if } x \in H, \\ +\infty & \text{otherwise.} \end{cases}$$

Show that f is lsc and $\partial f(x) = \emptyset$ for any $x \in H$ such that $x_i > -2^{-i}$ for infinitely many i. Reference: [221, Example 3.8].

*Exercise 4.2.21 (Monotonicity of Gradients) Suppose that the set $C \subset \mathbb{R}^N$ is open and convex and that the function $f \colon C \to \mathbb{R}$ is differentiable. Prove f is convex if and only if

$$\langle f'(x) - f'(y), x - y \rangle \geq 0 \quad \text{for all } x, y \in S,$$

and f is strictly convex if and only if the above inequality holds strictly whenever $x \neq y$. (You may use Exercise 4.2.16.)

*Exercise 4.2.22 We consider an *objective function* p_N involved in the coupon collection problem given by

$$p_N(q) = \sum_{\sigma \in S_N} \left(\prod_{i=1}^{N} \frac{q_{\sigma(i)}}{\sum_{j=i}^{N} q_{\sigma(j)}} \right) \left(\sum_{i=1}^{N} \frac{1}{\sum_{j=i}^{N} q_{\sigma(j)}} \right),$$

summed over *all* $N!$ permutations; so a typical term is

$$\left(\prod_{i=1}^{N} \frac{q_i}{\sum_{j=i}^{N} q_j} \right) \left(\sum_{i=1}^{N} \frac{1}{\sum_{j=i}^{n} q_j} \right).$$

For example, with $N = 3$ this is

$$q_1 q_2 q_3 \left(\frac{1}{q_1 + q_2 + q_3} \right) \left(\frac{1}{q_2 + q_3} \right) \left(\frac{1}{q_3} \right) \left(\frac{1}{q_1 + q_2 + q_3} + \frac{1}{q_2 + q_3} + \frac{1}{q_3} \right).$$

Show that p_N is *convex* on the positive orthant. Further more show that $1/p_N$ is concave.

Hint:

(i)   Establish

$$p_N(x_1, \ldots, x_N) = \int_0^1 \left(1 - \prod_{n=1}^{N} (1 - t^{x_n}) \right) \frac{dt}{t}. \tag{4.2.4}$$

(ii) Use

$$1 - e^{-t x_n} = x_n \int_0^t e^{-x_n y_n} \, dy_n,$$

to establish

$$1 - \prod_{n=1}^{N}(1 - e^{-tx_n}) = \left(\prod_{n=1}^{N} x_n\right)\left(\int_{\mathbb{R}_+^N} e^{-\langle x,y\rangle}\, dy - \int_{S_t^N} e^{-\langle x,y\rangle}\, dy\right),$$

where
$$S_t^N = \{y \in \mathbb{R}_+^N \mid 0 < y_n \leq t \text{ for } n = 1, \ldots, N\}.$$

(iii) Derive

$$\int_0^\infty \left(1 - \prod_{n=1}^{N}(1 - e^{-tx_n})\right) dt = \left(\prod_{n=1}^{N} x_n\right) \int_0^\infty dt \int_{\mathbb{R}_+^N \setminus S_t^N} e^{-\langle x,y\rangle}\, dy$$

$$= \left(\prod_{n=1}^{N} x_n\right) \int_0^\infty dt \int_{\mathbb{R}_+^N} e^{-\langle x,y\rangle} \chi_t(y)\, dy,$$

where

$$\chi_t(y) = \begin{cases} 1 & \text{if } \max(y_1, \ldots, y_N) > t, \\ 0 & \text{otherwise.} \end{cases}$$

(iv) Show that the integral in (iii) can be expressed as the *joint expectation* of Poisson distributions. Explicitly, if $x = (x_1, \ldots, x_N)$ is a point in the positive orthant $\mathbb{R}_+^N$, then

$$\int_0^\infty \left(1 - \prod_{n=1}^{N}(1 - e^{-tx_n})\right) dt = \left(\prod_{n=1}^{N} x_i\right) \int_{\mathbb{R}_+^N} e^{-\langle x,y\rangle} \max(y_1, \ldots, y_N)\, dy.$$

(v) Deduce that

$$p_N(x_1, \ldots, x_N) = \int_{\mathbb{R}_+^N} e^{-(y_1 + \cdots + y_N)} \max\left(\frac{y_1}{x_1}, \ldots, \frac{y_N}{x_N}\right) dy,$$

and hence that p_N is positive, decreasing and convex, as is the integrand.

(vi) To derive the stronger result that $1/p_N$ is concave. Let

$$h(a, b) = \frac{2ab}{a + b}.$$

Then h is concave and show that the concavity of $1/p_N$ is equivalent to

$$p_N\left(\frac{x + x'}{2}\right) \leq h(p_N(x), p_N(x')) \text{ for all } x, x' \in \mathbb{R}_+^N. \qquad (4.2.5)$$

Reference: The history of this problem and additional details can be found in Borwein, Bailey and Girgensohn [35, p. 36]. This book and its sister volume by Borwein and Bailey [34] also discuss how to use methods of experimental mathematics to gain insights on this and other related problems.

4.3 Sandwich Theorems and Calculus

4.3.1 A Decoupling Lemma

As in the case of the Fréchet subdifferential, to apply the convex subdifferential we need a convenient calculus for it. It turns out the key for developing such a calculus is again to combine a decoupling mechanism with the existence of subgradient that plays a role similar to that of the variational principles. We summarize this idea in the following lemma.

Lemma 4.3.1 (Decoupling Lemma) *Let X and Y be Banach spaces, let the functions $f \colon X \to \mathbb{R}$ and $g \colon Y \to \mathbb{R}$ be convex and the map $A \colon X \to Y$ be linear and bounded. Suppose that f, g and A satisfy either the condition*

$$0 \in \operatorname{core}(\operatorname{dom} g - A \operatorname{dom} f) \qquad (4.3.1)$$

and both f and g are lsc, or the condition

$$A \operatorname{dom} f \cap \operatorname{cont} g \neq \emptyset. \qquad (4.3.2)$$

Then there is a $y^* \in Y^*$ such that for any $x \in X$ and $y \in Y$,

$$p \leq [f(x) - \langle y^*, Ax \rangle] + [g(y) + \langle y^*, y \rangle], \qquad (4.3.3)$$

where $p = \inf_X \{f(x) + g(Ax)\}$.

Proof. Define an optimal value function $h \colon Y \to [-\infty, +\infty]$ by

$$h(u) = \inf_{x \in X} \{f(x) + g(Ax + u)\}.$$

It is easy to check h is convex and $\operatorname{dom} h = \operatorname{dom} g - A \operatorname{dom} f$. We will show that $0 \in \operatorname{int} h(0)$ under the constraint qualification condition (4.3.1) or (4.3.2).

First assume condition (4.3.1) is satisfied so that f and g are lsc functions. We may assume $f(0) = g(0) = 0$, and define $S := \bigcup_{x \in B_X} \{u \in Y \mid f(x) + g(Ax + u) \leq 1\}$. Clearly S is a convex set. We check that S is absorbing and cs-closed.

Let $y \in Y$ be an arbitrary element. Since $0 \in \operatorname{core}(\operatorname{dom} g - A \operatorname{dom} f)$ there exists $t > 0$ such that $ty \in \operatorname{dom} g - A \operatorname{dom} f$. Choose an element $x \in \operatorname{dom} f$ such that $Ax + ty \in \operatorname{dom} g$. Then,

$$f(x) + g(Ax + ty) = k < \infty. \qquad (4.3.4)$$

Choose $m \geq \max\{\|x\|, |k|, 1\}$. Dividing (4.3.4) by m and observing f and g are convex and $f(0) = g(0) = 0$ we have

$$f\left(\frac{x}{m}\right) + g\left(A\frac{x}{m} + \frac{ty}{m}\right) \leq 1.$$

Thus, $ty/m \in S$ and S is absorbing.

To show S is cs-closed let $y = \sum_{i=1}^{\infty} \lambda_i y_i$ where $\lambda_i \geq 0$, $\sum_{i=1}^{\infty} \lambda_i = 1$ and $y_i \in S$. By the definition of S for each i there exists $x_i \in B_X$ such that

$$f(x_i) + g(Ax_i + y_i) \leq 1. \qquad (4.3.5)$$

Since B_X is cs-compact (see Exercise 4.1.14), $\sum_{i=1}^{\infty} \lambda_i x_i$ converges to a point $x \in B_X$. Multiplying (4.3.5) by λ_i and sum over all $i = 1, 2, \ldots$ we have

$$\sum_{i=1}^{\infty} f(x_i) + \sum_{i=1}^{\infty} g(Ax_i + y_i) \leq 1.$$

Since f and g are convex and lsc and A is continuous we have

$$f(x) + g(Ax + y) \leq 1$$

or $y \in S$, proving S is cs-closed. It follows from Exercise 4.1.7 that $0 \in \operatorname{core} S = \operatorname{int} S$. Note that h is bounded above by 1 on S and therefore continuous (actually locally Lipschitz) in a neighborhood of 0 by Propositions 4.1.1 and 4.1.2.

Next assume condition (4.3.2) holds. Choose $y \in A \operatorname{dom} f \cap \operatorname{cont} g$. Then there exists $r > 0$ such that for any $u \in rB_Y$, $g(y + u) \leq g(y) + 1$. Let $x \in \operatorname{dom} f$ be an element satisfying $y = Ax$. It follows that for all $u \in rB_Y$, $h(u) \leq f(x) + g(Ax + u) \leq f(x) + g(y) + 1$. Again by Propositions 4.1.1 and 4.1.2, $0 \in \operatorname{cont} h$.

Now, Theorem 4.2.8 implies that $\partial h(0) \neq \emptyset$. Suppose that $-y^* \in \partial h(0)$. Then for all u in Y and x in X,

$$\begin{aligned} h(0) = p &\leq h(u) + \langle y^*, u \rangle \\ &\leq f(x) + g(Ax + u) + \langle y^*, u \rangle. \end{aligned} \qquad (4.3.6)$$

For arbitrary $y \in Y$, set $u = y - Ax$ in (4.3.6) we arrive at (4.3.3).   ●

We can see the two basic variational techniques at work here. The attainment of a minimum of the perturbed function takes the form of $\partial h(0) \neq \emptyset$ and the decoupling of variables is achieved through the perturbation u. In this aspect this lemma is rather similar to the nonlocal approximate sum rule of Theorem 3.2.3. Again, due to the nice property of convex functions the conclusion here is more precise.

4.3.2 Sandwich Theorems

We apply the decoupling lemma of Lemma 4.3.1 to establish a sandwich theorem.

Theorem 4.3.2 (Sandwich Theorem) *Let X and Y be Banach spaces, let $f: X \to \mathbb{R} \cup \{+\infty\}$ and $g: Y \to \mathbb{R} \cup \{+\infty\}$ be convex functions and let*

$A: X \to Y$ *be a bounded linear map. Suppose that* $f \geq -g \circ A$ *and* f, g *and* A *satisfy either condition* (4.3.1) *or condition* (4.3.2). *Then there is an affine function* $\alpha: X \to R$ *of the form* $\alpha(x) = \langle A^* y^*, x \rangle + r$ *satisfying* $f \geq \alpha \geq -g \circ A$. *Moreover, for any* $\bar{x}$ *satisfying* $f(\bar{x}) = -g \circ A(\bar{x})$, *we have* $-y^* \in \partial g(A\bar{x})$.

Proof. By Lemma 4.3.1 there exists $y^* \in Y^*$ such that for any $x \in X$ and $y \in Y$,

$$0 \leq p \leq [f(x) - \langle y^*, Ax \rangle] + [g(y) + \langle y^*, y \rangle]. \tag{4.3.7}$$

For any $z \in X$ setting $y = Az$ in (4.3.7) we have

$$f(x) - \langle A^* y^*, x \rangle \geq -g(Az) - \langle A^* y^*, z \rangle. \tag{4.3.8}$$

Thus,

$$a := \inf_{x \in X} [f(x) - \langle A^* y^*, x \rangle] \geq b := \sup_{z \in X} [-g(Az) - \langle A^* y^*, z \rangle].$$

Picking any $r \in [a, b]$, $\alpha(x) := \langle A^* y^*, x \rangle + r$ is an affine function that separates f and $-g \circ A$. Finally, when $f(\bar{x}) = -g \circ A(\bar{x})$, it follows from (4.3.7) that $-x^* \in \partial g(A\bar{x})$. ●

4.3.3 Calculus for the Subdifferential

We now use the tools established above to deduce calculus rules for the convex functions. We start with a sum rule that can be viewed as a convex function version of the local approximate sum rule of Theorem 3.3.1.

Theorem 4.3.3 (Convex Subdifferential Sum Rule) *Let* X *and* Y *be Banach spaces, let* $f: X \to \mathbb{R} \cup \{+\infty\}$ *and* $g: Y \to \mathbb{R} \cup \{+\infty\}$ *be convex functions and let* $A: X \to Y$ *be a bounded linear map. Then at any point* x *in* X, *we have the sum rule*

$$\partial(f + g \circ A)(x) \supset \partial f(x) + A^* \partial g(Ax), \tag{4.3.9}$$

with equality if either condition (4.3.1) *or* (4.3.2) *holds.*

Proof. Inclusion (4.3.9) is easy and left as an exercise. We prove the reverse inclusion under condition (4.3.1) or (4.3.2). Suppose $x^* \in \partial(f + g \circ A)(\bar{x})$. Since shifting by a constant does not change the subdifferential of a convex function, we may assume without loss of generality that

$$x \to f(x) + g(Ax) - \langle x^*, x \rangle$$

attains its minimum 0 at $x = \bar{x}$. By the sandwich theorem there exists an affine function $\alpha(x) := \langle A^* y^*, x \rangle + r$ with $-y^* \in \partial g(A\bar{x})$ such that

$$f(x) - \langle x^*, x \rangle \geq \alpha(x) \geq -g(Ax).$$

Clearly equality is attained at $x = \bar{x}$. It is now an easy matter to check that $x^* + A^* y^* \in \partial f(\bar{x})$. $\bullet$

Note that when A is the identity mapping, Theorem 4.3.3 sharpens the Fréchet subdifferential sum rules discussed in Section 3.3.1. The geometrical interpretation of this is that one can find a hyperplane in $X \times \mathbb{R}$ that separates the epigraph of f and hypograph of $-g$. Also, by applying the subdifferential sum rule to the indicator functions of two convex sets we have parallel results for the normal cones to the intersection of convex sets.

Theorem 4.3.4 (Normals to an Intersection) *Let C_1 and C_2 be two convex subsets of X and let $x \in C_1 \cap C_2$. Suppose that C_1 and C_2 are closed $0 \in \text{core}(C_1 - C_2)$ or $C_1 \cap \text{int}\, C_2 \neq \emptyset$. Then*

$$N(C_1 \cap C_2; x) = N(C_1; x) + N(C_2; x).$$

Proof. Exercise 4.3.6. $\bullet$

The condition (4.3.1) or (4.3.2) is often referred to as a constraint qualification. Without it the equality in the convex subdifferential sum rule may not hold (Exercise 4.3.12).

Using the convex subdifferential sum rule we can also get a version of the multidirectional mean value inequality for convex functions that refines Theorem 3.6.1.

Theorem 4.3.5 (Convex Multidirectional Mean Value Inequality) *Let X be a Banach space, let C be a nonempty, closed and convex subset of X and $x \in X$ and let $f \colon X \to \mathbb{R}$ be a continuous convex function. Suppose that f is bounded below on $[x, C]$ and*

$$\inf_{y \in C} f(y) - f(x) > r.$$

Then for any $\varepsilon > 0$, there exist $z \in [x, C]$ and $z^* \in \partial f(z)$, the convex subdifferential of f at z, such that

$$f(z) < \inf_{[x,C]} f + |r| + \varepsilon,$$

and

$$r < \langle z^*, y - x \rangle + \varepsilon \|y - x\| \text{ for all } y \in C.$$

Proof. As in the proof of Theorem 3.6.1 we can convert the general case to the special case when $r = 0$. So we will only prove this special case. Let $\tilde{f} := f + \iota_{[x,C]}$. Then $\tilde{f}$ is bounded below on X. By taking a smaller $\varepsilon > 0$ if necessary, we may assume that

$$\varepsilon < \inf_{y \in C} f(y) - f(x).$$

Applying Ekeland's variational principle of Theorem 2.1.2 we conclude that there exists z such that

$$\tilde{f}(z) < \inf \tilde{f} + \varepsilon \qquad (4.3.10)$$

and

$$\tilde{f}(z) \leq \tilde{f}(u) + \varepsilon \|u - z\|, \text{ for all } u \in X. \qquad (4.3.11)$$

That is to say

$$u \to f(u) + \iota_{[x,C]}(u) + \varepsilon \|u - z\|$$

attains a minimum at z. By (4.3.10) $\tilde{f}(z) < +\infty$ hence $z \in [x, C]$. The sum rule for convex subdifferential of Theorem 4.3.3 (with A being the identity mapping) implies that there exists $z^* \in \partial f(z)$ such that $0 \leq \langle z^*, w - z \rangle + \varepsilon \|w - z\|$, for all $w \in [x, C]$. Using a smaller ε to begin with if necessary, we have for $w \neq z$,

$$0 < \langle z^*, w - z \rangle + \varepsilon \|w - z\|, \text{ for all } w \in [x, C] \backslash \{z\}. \qquad (4.3.12)$$

Moreover by inequality (4.3.10) we have $f(z) = \tilde{f}(z) \leq f(x) + \varepsilon < \inf_C f$, so $z \notin C$. Thus we can write $z = x + \bar{t}(\bar{y} - x)$ where $\bar{t} \in [0, 1)$. For any $y \in C$ set $w = y + \bar{t}(\bar{y} - y) \neq z$ in (4.3.12) yields

$$0 < \langle z^*, y - x \rangle + \varepsilon \|y - x\|, \text{ for all } y \in C. \qquad (4.3.13)$$

•

4.3.4 The Pshenichnii–Rockafellar Conditions

We turn to discuss the simple convex programming problem of

$$\mathcal{CP} \qquad \text{minimize} \quad f(x) \qquad (4.3.14)$$

$$\text{subject to} \quad x \in C \subset X,$$

where X is a Banach space, C is a closed convex subset of X and $f \colon X \to \mathbb{R} \cup \{+\infty\}$ is a convex lsc function. The convex subdifferential calculus developed in this section enables us to derive sharp necessary optimality conditions for $\mathcal{CP}$.

Theorem 4.3.6 (Pshenichnii–Rockafellar Conditions) *Let X be a Banach space, let C be a closed convex subset of X and let $f \colon X \to \mathbb{R} \cup \{+\infty\}$ be a convex function. Suppose that $C \cap \mathrm{cont}\, f \neq \emptyset$ or $\mathrm{int}\, C \cap \mathrm{dom}\, f \neq \emptyset$ and f is bounded below on C. Then there is an affine function $\alpha \leq f$ with $\inf_C f = \inf_C \alpha$. Moreover, $\bar{x}$ is a solution of $\mathcal{CP}$ if and only if it satisfies*

$$0 \in \partial f(\bar{x}) + N(C; \bar{x}).$$

Proof. Apply the convex subdifferential sum rule of Theorem 4.3.3 to $f + \iota_C$ at $\bar{x}$.   ●

4.3.5 The Extension and Separation Theorems

We have seen that the sandwich theorem and the convex subdifferential calculus are intimately related to the separation theorem. We now explicitly deduce the separation theorem and the Hahn–Banach theorem from these results.

Theorem 4.3.7 (Hahn–Banach Extension) *Let X be a Banach space and let $f\colon X \to \mathbb{R}$ be a continuous sublinear function with $\operatorname{dom} f = X$. Suppose that L is a linear subspace of X and the function $h\colon L \to \mathbb{R}$ is linear and dominated by f, that is, $f \geq h$ on L. Then there exists $x^* \in X^*$, dominated by f, such that*
$$h(x) = \langle x^*, x \rangle, \ \text{for all } x \in L.$$

Proof. Apply the sandwich theorem of Theorem 4.3.2 to f and $g := -h + \iota_L$.   ●

Theorem 4.3.8 (Separation Theorem) *Let X be a Banach space and let C_1 and C_2 be two convex subsets of X. Suppose that $\operatorname{int} C_1 \neq \emptyset$ and $C_2 \cap \operatorname{int} C_1 = \emptyset$. Then there exists an affine function α on X such that*
$$\sup_{c_1 \in C_1} \alpha(c_1) \leq \inf_{c_2 \in C_2} \alpha(c_2).$$

Proof. Without loss of generality we may assume that $0 \in \operatorname{int} C_1$ and then apply the sandwich theorem with $f = \iota_{\operatorname{cl} C_2}$, A the identity mapping of X and $g = \gamma_{C_1} - 1$.   ●

4.3.6 Commentary and Exercises

We can view convex analysis as a natural next step from linear functional analysis. Thus, it is not surprising to see fundamental results of linear functional analysis follow from those of convex analysis. There are different ways of developing basic results in convex analysis. Here we follow our short notes [67] using the Decoupling Lemma as a starting point. This development actually works in a more general setting (for details see [67]). In particular, the space Y need not be complete (Exercise 4.3.3). The proof of this result is a typical variational argument similar to those of the basic results in the calculus of Fréchet subdifferentials. We illustrate the potential of this theorem by deducing the sandwich theorem, the sum rule for convex subdifferential and a convex version of the multidirectional mean value theorem due to Ledyaev and Zhu [174]. The Pshenichnii–Rockafellar condition [223, 229] provides a prototype for necessary conditions for nonsmooth constrained optimization problems. Subsection 4.3.5 and several exercises below further highlight how to use the sandwich theorem to deduce other important results in convex and linear functional analysis.

Exercise 4.3.1 Define

$$h(u) = \inf_{x \in X} \{f(x) + g(Ax + u)\}.$$

Prove that

$$\operatorname{dom} h = \operatorname{dom} g - A \operatorname{dom} f.$$

Exercise 4.3.2 Show that condition (4.3.2) implies the inclusion in (4.3.1).

*Exercise 4.3.3** Show in the Decoupling Lemma that it suffices to assume only that X is a topological vector space that has a cs-compact absorbing set, A is linear and closed and that Y is a *barrelled* topological vector space. Hint: When condition (4.3.2) is satisfied the proof is the same. When condition (4.3.1) is satisfied, let C be a cs-compact absorbing set in X and define $S :=$ $\bigcup_{x \in C} \{u \in Y \mid f(x) + g(Ax + u) \le 1\}$. Prove $T = S \cap (-S)$ is cs-closed and the closure of T is a barrel in Y. Reference: [67].

Exercise 4.3.4 Supply details for the proof of Theorem 4.3.3 by

(i) Proving (4.3.9).
(ii) Verifying $x^* + A^* y^* \in \partial f(\bar{x})$.

Exercise 4.3.5 Interpret the sandwich theorem geometrically in the case when A is the identity map.

Exercise 4.3.6 Prove Theorem 4.3.4.

Exercise 4.3.7 Give the details of the proof of Theorem 4.3.6.

Exercise 4.3.8 Apply the Pshenichnii–Rockafellar conditions to the following two cases:

(i) C a single point $\{x^0\} \subset X$,
(ii) C a polyhedron $\{x \mid Ax \le b\}$, where $b \in \mathbb{R}^N = Y$.

Exercise 4.3.9 Provide details for the proof of Theorem 4.3.7.

Exercise 4.3.10 Provide details for the proof of Theorem 4.3.8.

Exercise 4.3.11 (Subdifferential of a Max-Function) Suppose that I is a finite set of integers, and $g_i \colon X \to \mathbb{R} \cup \{+\infty\}, i \in I$ are lower semicontinuous convex functions with

$$\operatorname{dom} g_j \cap \bigcap_{i \in I \setminus \{j\}} \operatorname{cont} g_i \ne \emptyset$$

for some index j in I. Prove

$$\partial(\max_i g_i)(\bar{x}) = \text{conv} \bigcup_{i \in I} \partial g_i(\bar{x}).$$

*Exercise 4.3.12 (Failure of Convex Calculus)

(i) Find convex functions $f, g : \mathbb{R} \to \mathbb{R} \cup \{+\infty\}$ with

$$\partial f(0) + \partial g(0) \neq \partial(f + g)(0).$$

(ii) Find a convex function $g : \mathbb{R}^2 \to \mathbb{R} \cup \{+\infty\}$ and a linear map $A : \mathbb{R} \to \mathbb{R}^2$ with $A^* \partial g(0) \neq \partial(g \circ A)(0)$.

Exercise 4.3.13 Let $K(x^*, \varepsilon) = \{x \in X \mid \varepsilon \|x^*\| \|x\| \leq \langle x^*, x \rangle\}$ be a Bishop–Phelps cone. Show that

$$N(K(x^*, \varepsilon); 0) = \partial \iota_{K(x^*, \varepsilon)}(0) \subset \bigcup_{r \geq 0} r B_{\varepsilon \|x^*\|}(-x^*).$$

4.4 Fenchel Conjugate

In this section we give a concise sketch of the Fenchel conjugation theory. One can regard it as a natural generalization of the linear programming duality, or as a form of the Legendre transform in the convex setting. More relevant to the context of this book is to think of it as an elegant primal-dual space representation of the basic variational techniques for convex functions.

4.4.1 The Fenchel Conjugate

Let X be a Banach space. The *Fenchel conjugate* of a function $f : X \to [-\infty, +\infty]$ is the function $f^* : X^* \to [-\infty, +\infty]$ defined by

$$f^*(x^*) = \sup_{x \in X} \{\langle x^*, x \rangle - f(x)\}.$$

The function f^* is convex and if the domain of f is nonempty then f^* never takes the value $-\infty$. Clearly the conjugacy operation is *order-reversing*: for functions $f, g : X \to [-\infty, +\infty]$, the inequality $f \geq g$ implies $f^* \leq g^*$. We can consider the conjugate of f^* called the *biconjugate* of f and denoted f^{**}. This is a function on X^{**}.

4.4.2 The Fenchel–Young Inequality

This is an elementary but important result that relates conjugation with the subgradient.

Proposition 4.4.1 (Fenchel–Young Inequality) *Let X be a Banach space and let $f: X \to \mathbb{R} \cup \{+\infty\}$ be a convex function. Suppose that $x^* \in X^*$ and $x \in \operatorname{dom} f$. Then satisfy the inequality*

$$f(x) + f^*(x^*) \geq \langle x^*, x \rangle. \tag{4.4.1}$$

Equality holds if and only if $x^* \in \partial f(x)$.

Proof. The inequality (4.4.1) follows directly from the definition. Now we have the equality

$$f(x) + f^*(x^*) = \langle x^*, x \rangle,$$

if and only if, for any $y \in X$,

$$f(x) + \langle x^*, y \rangle - f(y) \leq \langle x^*, x \rangle.$$

That is

$$f(y) - f(x) \geq \langle x^*, y - x \rangle,$$

or $x^* \in \partial f(x)$. ●

4.4.3 Weak Duality

Conjugate functions are ubiquitous in optimization. Our next result is phrased in terms of convex programming problems. The formulation is in many aspects similar to the duality theory in linear programming.

Theorem 4.4.2 (Fenchel Weak Duality) *Let X and Y be Banach spaces, let $f: X \to \mathbb{R} \cup \{+\infty\}$ and $g: Y \to \mathbb{R} \cup \{+\infty\}$ be convex functions and let $A: X \to Y$ be a bounded linear map. Define the primal and dual values $p, d \in [-\infty, +\infty]$ by the Fenchel problems*

$$p = \inf_{x \in X} \{f(x) + g(Ax)\}$$
$$d = \sup_{x^* \in Y^*} \{-f^*(A^*x^*) - g^*(-x^*)\}. \tag{4.4.2}$$

Then these values satisfy the weak duality inequality $p \geq d$.

Proof. Exercise 4.4.1. ●

4.4.4 Strong Duality

The Fenchel duality theorem can be viewed as a dual representation of the sandwich theorem.

Theorem 4.4.3 (Fenchel Duality) *Let X and Y be Banach spaces, $f \colon X \to \mathbb{R} \cup \{+\infty\}$ and $g \colon Y \to \mathbb{R} \cup \{+\infty\}$ be convex functions and $A \colon X \to Y$ be a bounded linear map. Suppose that f, g and A satisfy either condition (4.3.1) or condition (4.3.2). Then $p = d$, and the supremum in the dual problem (4.4.2) is attained if finite. Here $p, d \in [-\infty, +\infty]$ are defined as in Theorem 4.4.2.*

Proof. If p is $-\infty$ there is nothing to prove, while if condition (4.3.1) or (4.3.2) holds and p is finite then by Lemma 4.3.1 there is a $x^* \in X^*$ such that (4.3.3) holds. For any $u \in Y$, setting $y = Ax + u$ in (4.3.3), we have

$$p \leq f(x) + g(Ax + u) + \langle x^*, u \rangle$$
$$= \{ f(x) - \langle A^* x^*, x \rangle \} + \{ g(Ax + u) - \langle -x^*, Ax + u \rangle \}.$$

Taking the infimum over all points u, and then over all points x, gives the inequalities

$$p \leq -f^*(A^* x^*) - g^*(-x^*) \leq d \leq p.$$

Thus x^* attains the supremum in problem (4.4.2), and $p = d$.   ●

To relate Fenchel duality and convex programming with linear constraints, we let g be the indicator function of a point, which gives the following particularly elegant and useful corollary.

Corollary 4.4.4 (Fenchel Duality for Linear Constraints) *Given any function $f \colon X \to \mathbb{R} \cup \{+\infty\}$, any bounded linear map $A \colon X \to Y$, and any element b of Y, the weak duality inequality*

$$\inf_{x \in X} \{ f(x) \mid Ax = b \} \geq \sup_{x^* \in Y} \{ \langle b, x^* \rangle - f^*(A^* x^*) \}$$

holds. If f is lsc and convex and b belongs to $\mathrm{core}(A \operatorname{dom} f)$ then equality holds, and the supremum is attained when finite.

Proof. Exercise 4.4.8.   ●

Fenchel duality can be used to conveniently calculate polar cones. Recall that for a set K in a Banach space X, the (*negative*) *polar cone* of K is the convex cone

$$K^\circ = \{ x^* \in X^* \mid \langle x^*, x \rangle \leq 0, \ \text{for all } x \in K \}.$$

The cone $K^{\circ\circ}$ is called the *bipolar* – sometimes in the second dual and sometimes in the predual, X. Here, we take it in X. An important example of the polar cone is the normal cone to a convex set $C \subset X$ at a point $x \in C$, since $N(C; x) = (C - x)^\circ$.

The following calculus for polar cones is a direct consequence of the Fenchel duality theorem.

Corollary 4.4.5 *Let X and Y be Banach spaces, let $K \subset X$ and $H \subset Y$ be cones and let $A: X \to Y$ be a bounded linear map. Then*

$$K^\circ + A^* H^\circ \subset (K \cap A^{-1}H)^\circ.$$

Equality holds if H and K are closed and convex and satisfy $H - AK = Y$.

Proof. Observe that for any cone K, we have $K^\circ = \partial \iota_K(0)$. The result follows directly from Theorem 4.4.3. ●

4.4.5 Commentary and Exercises

Fenchel's original work is [117]. Much of the exposition here follows the concise book [56] which is also a good source for additional examples and applications.

Exercise 4.4.1 Prove Theorem 4.4.3. Hint: This follows immediately from the Fenchel–Young inequality (4.4.1).

Exercise 4.4.2 Many important convex functions f on a reflexive Banach space equal their biconjugate f^{**}. Such functions thus occur as natural pairs, f and f^*. Table 4.1 shows some elegant examples on $\mathbb{R}$, and Table 4.2 describes some simple transformations of these examples. Check the calculation of f^* and check $f = f^{**}$ for functions in Table 4.1. Verify the formulas in Table 4.2.

Exercise 4.4.3 Calculate the conjugate and biconjugate of the function

$$f(x_1, x_2) = \begin{cases} \dfrac{x_1^2}{2x_2} + x_2 \log x_2 - x_2 & \text{if } x_2 > 0, \\ 0 & \text{if } x_1 = x_2 = 0, \\ +\infty & \text{otherwise.} \end{cases}$$

∗Exercise 4.4.4 (Maximum Entropy Example)

(i) Let $a^0, a^1, \ldots, a^N \in X$. Prove the function

$$g(z) := \inf_{x \in \mathbb{R}^{N+1}} \left\{ \sum_{n=0}^N \exp^*(x_n) \,\Big|\, \sum_{n=0}^N x_n = 1, \sum_{n=0}^N x_n a^n = z \right\}$$

is convex.

(ii) For any point y in $\mathbb{R}^{N+1}$, prove

$$g^*(y) = \sup_{x \in \mathbb{R}^{N+1}} \left\{ \sum_{n=0}^N (x_n \langle a^n, y \rangle - \exp^*(x_n)) \,\Big|\, \sum_{n=0}^N x_n = 1 \right\}.$$

$f(x) = g^*(x)$	dom f	$g(y) = f^*(y)$	dom g
0	$\mathbb{R}$	0	$\{0\}$
0	$\mathbb{R}_+$	0	$-\mathbb{R}_+$
0	$[-1,1]$	$|y|$	$\mathbb{R}$
0	$[0,1]$	y^+	$\mathbb{R}$
$|x|^p/p, \ p>1$	$\mathbb{R}$	$|y|^q/q \ \ (\frac{1}{p}+\frac{1}{q}=1)$	$\mathbb{R}$
$|x|^p/p, \ p>1$	$\mathbb{R}_+$	$|y^+|^q/q \ \ (\frac{1}{p}+\frac{1}{q}=1)$	$\mathbb{R}$
$-x^p/p, \ 0<p<1$	$\mathbb{R}_+$	$-(-y)^q/q \ \ (\frac{1}{p}+\frac{1}{q}=1)$	$-\operatorname{int}\mathbb{R}_+$
$-\log x$	$\operatorname{int}\mathbb{R}_+$	$-1-\log(-y)$	$-\operatorname{int}\mathbb{R}_+$
e^x	$\mathbb{R}$	$\begin{cases} y\log y - y & (y>0) \\ 0 & (y=0) \end{cases}$	$\mathbb{R}_+$

Table 4.1. Conjugate pairs of convex functions on $\mathbb{R}$.

$f = g^*$	$g = f^*$
$f(x)$	$g(y)$
$h(ax) \ \ (a \neq 0)$	$h^*(y/a)$
$h(x+b)$	$h^*(y) - by$
$ah(x) \ \ (a > 0)$	$ah^*(y/a)$

Table 4.2. Transformed conjugates.

(iii) Deduce the conjugacy formula

$$g^*(y) = 1 + \ln\Big(\sum_{n=0}^{N} \exp\langle a^n, y\rangle\Big).$$

(iv) Compute the conjugate of the function of $x \in \mathbb{R}^{N+1}$,

$$\begin{cases} \sum_{n=0}^{N} \exp^*(x_n) & \text{if } \sum_{n=0}^{N} x_n = 1, \\ +\infty & \text{otherwise.} \end{cases}$$

Exercise 4.4.5 Give the details for the proof of Theorem 4.4.2 (Fenchel Weak Duality).

Exercise 4.4.6 (Conjugate of Indicator Function) Let X be a reflexive Banach space and let C be a closed convex subset of X. Show that $\iota_C^* = \sigma_C$ and $\iota_C^{**} = \iota_C$.

Exercise 4.4.7 Let X be a reflexive Banach space. Suppose that $A: X \to X^*$ is a bounded linear operator, C a convex subset of X and D a nonempty closed bounded convex subset of X^*. Show that

$$\inf_{x \in C} \sup_{y \in D} \langle y, Ax \rangle = \max_{y \in D} \inf_{x \in C} \langle y, Ax \rangle.$$

Hint: Apply the Fenchel duality theorem to $f = \iota_C$ and $g = \iota_D^*$.

Exercise 4.4.8 Prove Corollary 4.4.4 (Fenchel Duality for Linear Constraints). Deduce duality theorems for the following separable problems.

$$\inf \left\{ \sum_{n=1}^{N} p(x_n) \,\Big|\, Ax = b \right\},$$

where the map $A: \mathbb{R}^N \to \mathbb{R}^M$ is linear, $b \in \mathbb{R}^M$, and the function $p: \mathbb{R} \to \mathbb{R} \cup \{+\infty\}$ is convex, defined as follows:

(i)   (Nearest Points in Polyhedra) $p(t) = t^2/2$ with domain $\mathbb{R}_+$.
(ii)   (Analytic Center) $p(t) = -\log t$ with domain $\operatorname{int} \mathbb{R}_+$.
(iii)   (Maximum Entropy) $p = \exp^*$.

What happens if the objective function is replaced by $\sum_{n=1}^{N} p_n(x_n)$?

Exercise 4.4.9 (Symmetric Fenchel Duality) Let X be a Banch space. For functions $f, g: X \to [-\infty, +\infty]$, define the *concave conjugate* $g_*: X \to [-\infty, +\infty]$ by

$$g_*(x^*) = \inf_{x \in X} \{ \langle x^*, x \rangle - g(x) \}.$$

Prove

$$\inf(f - g) \geq \sup(g_* - f^*),$$

with equality if f is lower semicontinuous and convex, g is upper semicontinuous and concave, and

$$0 \in \operatorname{core}(\operatorname{dom} f - \operatorname{dom}(-g)),$$

or f is convex, g is concave and

$$\operatorname{dom} f \cap \operatorname{cont} g \neq \emptyset.$$

Exercise 4.4.10 Let X be a Banach space and let $K \subset X$ be a cone. Show that $\iota_{K^\circ} = \iota_K^*$, and therefore $\iota_{K^{\circ\circ}} = \iota_K^{**}$.

Exercise 4.4.11 (Sum of Closed Cones) Let X be a finite dimensional Banach space.

(i)   Prove that any cones $H, K \subset X$ satisfy $(H + K)^o = H^o \cap K^o$.
(ii) Deduce that if H and K are closed convex cones then they satisfy $(H \cap K)^o = \text{cl}\,(H^o + K^o)$.

In $\mathbb{R}^3$, define sets

$$H = \{x \mid x_1^2 + x_2^2 \le x_3^2, x_3 \le 0\} \text{ and}$$
$$K = \{x \mid x_2 = -x_3\}.$$

(iii) Prove H and K are closed convex cones.
(iv) Calculate the polar cones H^o, K^o and $(H \cap K)^o$.
(v)   Prove $(1, 1, 1) \in (H \cap K)^o \backslash (H^o + K^o)$, and deduce that the sum of two closed convex cones is not necessarily closed.

4.5 Convex Feasibility Problems

Let X be a Hilbert space and let $C_n, n = 1, \ldots, N$ be convex closed subsets of X. The convex feasibility problem is to find some points $x \in \bigcap_{n=1}^N C_n$ when this intersection is nonempty. In this section we discuss projection algorithms for finding such a feasible point. These kinds of algorithms have wide ranging applications in many different problems, such as solution of convex inequalities, minimization of convex nonsmooth functions, medical imaging, computerized tomography and electron microscopy. Following the theme of this book we approach this problem by converting it to a convex optimization problem.

4.5.1 Projection

We start by defining projection to a closed convex set and its basic properties. This is based on the following theorem.

Theorem 4.5.1 (Existence and Uniqueness of Nearest Point) *Let X be a Hilbert space and let C be a closed convex subset of X. Then for any $x \in X$, there exists a unique element $\bar{x} \in C$ such that*

$$\|x - \bar{x}\| = d(C; x).$$

Proof. If $x \in C$ then $\bar{x} = x$ satisfies the conclusion. Suppose that $x \notin C$. Then there exists a sequence $x_i \in C$ such that $d(C; x) = \lim_{i \to \infty} \|x - x_i\|$. Clearly, x_i is bounded and therefore has a subsequence weakly converging to some $\bar{x} \in X$. Since a closed convex set is weakly closed (Mazur's Theorem), we have $\bar{x} \in C$ and $d(C; x) = \|x - \bar{x}\|$. We show such $\bar{x}$ is unique. Suppose that

$z \in C$ also has the property that $d(C; x) = \|x - z\|$. Then for any $t \in [0, 1]$ we have $t\bar{x} + (1 - t)z \in C$. It follows that

$$d(C; x) \leq \|x - (t\bar{x} + (1 - t)z)\| = \|t(x - \bar{x}) + (1 - t)(x - z)\|$$
$$\leq t\|x - \bar{x}\| + (1 - t)\|x - z\| = d(C; x).$$

That is to say

$$t \to \|x - z - t(\bar{x} - z)\|^2 = \|x - z\|^2 - 2t\langle x - z, \bar{x} - z \rangle + t^2\|\bar{x} - z\|^2$$

is a constant mapping, which implies $\bar{x} = z$. ●

The nearest point can be characterized by the normal cone as follows.

Theorem 4.5.2 (Normal Cone Characterization of Nearest Point) *Let X be a Hilbert space and let C be a closed convex subset of X. Then for any $x \in X$, $\bar{x} \in C$ is a nearest point to x if and only if*

$$x - \bar{x} \in N(C; \bar{x}).$$

Proof. Noting that the convex function $f(y) = \|y - x\|^2/2$ attains a minimum at $\bar{x}$ over set C, this directly follows from the Pshenichnii–Rockafellar conditions in Theorem 4.3.6. ●

Definition 4.5.3 (Projection) *Let X be a Hilbert space and let C be a closed convex subset of X. For any $x \in X$ the unique nearest point $y \in C$ is called the projection of x on C and we define the projection mapping P_C by $P_C x = y$.*

We summarize some useful properties of the projection mapping in the next proposition whose elementary proof is left as an exercise.

Proposition 4.5.4 (Properties of Projection) *Let X be a Hilbert space and let C be a closed convex subset of X. Then the projection mapping P_C has the following properties.*

(i)   *for any $x \in C$, $P_C x = x$;*
(ii) *$P_C^2 = P_C$;*
(iii) *for any $x, y \in X$, $\|P_C y - P_C x\| \leq \|y - x\|$.*

Proof. Exercise 4.5.3. ●

Projection to a convex set can also be represented as the Fréchet derivative of a convex function and therefore is a monotone operator.

Theorem 4.5.5 (Potential Function of Projection) *Let X be a Hilbert space and let C be a closed convex subset of X. Define*

$$f(x) = \sup\left\{ \langle x, y \rangle - \frac{\|y\|^2}{2} \mid y \in C \right\}.$$

Then f is convex, $P_C(x) = f'(x)$, and therefore P_C is a monotone operator.

Proof. It is easy to check (Exercise 4.5.6) that f is convex and

$$f(x) = \frac{1}{2}(\|x\|^2 - \|x - P_C(x)\|^2).$$

We need only show $P_C(x) = f'(x)$. Fix $x \in X$. For any $y \in X$ we have

$$\|(x + y) - P_C(x + y)\| \leq \|(x + y) - P_C(x)\|,$$

so

$$\|(x + y) - P_C(x + y)\|^2 \leq \|x + y\|^2 - 2\langle x + y, P_C(x) \rangle + \|P_C(x)\|^2$$
$$= \|x + y\|^2 + \|x - P_C(x)\|^2 - \|x\|^2 - 2\langle y, P_C(x) \rangle,$$

hence $f(x+y) - f(x) - \langle P_C(x), y \rangle \geq 0$. On the other hand, since $\|x - P_C(x)\| \leq \|x - P_C(x + y)\|$ we get

$$f(x + y) - f(x) - \langle P_C(x), y \rangle \leq \langle y, P_C(x + y) - P_C(x) \rangle$$
$$\leq \|y\| \times \|P_C(x + y) - P_C(x)\| \leq \|y\|^2,$$

which implies $P_C(x) = f'(x)$. ●

4.5.2 Projection Algorithms as Minimization Problems

We start with the simple case of the intersection of two convex sets. Let X be a Hilbert space and let C and D be two closed convex subsets of X. Suppose that $C \cap D \neq \emptyset$. Define a function

$$f(c, d) := \frac{1}{2}\|c - d\|^2 + \iota_C(c) + \iota_D(d).$$

We see that f attains a minimum at $(\bar{c}, \bar{d})$ if and only if $\bar{c} = \bar{d} \in C \cap D$. Thus, the problem of finding a point in $C \cap D$ becomes one of minimizing function f.

We consider a natural descending process for f by alternately minimizing f with respect to its two variables. More concretely, start with any $x_0 \in D$. Let x_1 be the solution of minimizing

$$x \to f(x, x_0).$$

It follows from Theorem 4.5.2 that

$$x_0 - x_1 \in N(C; x_1).$$

That is to say $x_1 = P_C x_0$. We then let x_2 be the solution of minimizing

$$x \to f(x_1, x).$$

Similarly, $x_2 = P_D x_1$. In general, we define

$$x_{i+1} = \begin{cases} P_C x_i & i \text{ is even,} \\ P_D x_i & i \text{ is odd.} \end{cases} \tag{4.5.1}$$

This algorithm is a generalization of the classical von Neumann projection algorithm for finding points in the intersection of two half spaces. We will show that in general x_i weakly converge to a point in $C \cap D$ and when $\operatorname{int}(C \cap D) \neq \emptyset$ we have norm convergence.

4.5.3 Attracting Mappings and Fejér Sequences

We discuss two useful tools for proving the convergence of the projection algorithm.

Definition 4.5.6 (Nonexpansive Mapping) *Let X be a Hilbert space, let C be a closed convex nonempty subset of X and let $T\colon C \to X$. We say that T is* nonexpansive *provided that $\|Tx - Ty\| \le \|x - y\|$.*

Definition 4.5.7 (Attracting Mapping) *Let X be a Hilbert space, let C be a closed convex nonempty subset of X and let $T\colon C \to C$ be a nonexpansive mapping. Suppose that D is a closed nonempty subset of C. We say that T is attracting with respect to D if for every $x \in C \backslash D$ and $y \in D$,*

$$\|Tx - y\| \le \|x - y\|.$$

We say that T is k-attracting with respect to D if for every $x \in C \backslash D$ and $y \in D$,

$$k\|x - Tx\|^2 \le \|x - y\|^2 - \|Tx - y\|^2.$$

Lemma 4.5.8 (Attractive Property of Projection) *Let X be a Hilbert space and let C be a convex closed subset of X. Then $P_C\colon X \to X$ is 1-attracting with respect to C.*

Proof. Let $y \in C$. We have

$$\begin{aligned} \|x - y\|^2 - \|P_C x - y\|^2 &= \langle x - P_C x, x + P_C x - 2y \rangle \\ &= \langle x - P_C x, x - P_C x + 2(P_C x - y) \rangle \\ &= \|x - P_C x\|^2 + 2\langle x - P_C x, P_C x - y \rangle \\ &\ge \|x - P_C x\|^2. \end{aligned}$$

Note that if T is attracting (k-attracting) with respect to a set D, then it is attracting (k-attracting) with respect to any subsets of D.

We now turn to Fejér monotonicity.

Definition 4.5.9 (Fejér Monotone Sequence) *Let X be a Hilbert space, let C be closed convex set and let (x_i) be a sequence in X. We say that (x_i) is* Fejér monotone *with respect to C if*

$$\|x_{i+1} - c\| \le \|x_i - c\|, \quad \text{for all } c \in C \text{ and } i = 1, 2, \dots$$

Our next theorem summarizes important properties of Fejér monotone sequences.

Theorem 4.5.10 (Properties of Fejér Monotone Sequences) *Let X be a Hilbert space, let C be a closed convex set and let (x_i) be a Fejér monotone sequence with respect to C. Then*

(i)   *(x_i) is bounded and $d(C; x_{i+1}) \le d(C; x_i)$.*
(ii)   *(x_i) has at most one weak cluster point in C.*
(iii) *If the interior of C is nonempty then (x_i) converges in norm.*
(iv)   *$(P_C x_i)$ converges in norm.*

Proof. (i) is obvious.

Observe that, for any $c \in C$ the sequence $(\|x_i - c\|^2)$ converges and so does

$$(\|x_i\|^2 - 2\langle x_i, c\rangle). \tag{4.5.2}$$

Now suppose $c_1, c_2 \in C$ are two weak cluster points of (x_i). Letting c in (4.5.2) be c_1 and c_2, respectively, and taking limits of the difference, yields $\langle c_1, c_1 - c_2\rangle = \langle c_2, c_1 - c_2\rangle$ so that $c_1 = c_2$, which proves (ii).

To prove (iii) suppose that $B_r(c) \subset C$. For any $x_{i+1} \ne x_i$, simplifying

$$\left\| x_{i+1} - \left(c - h\frac{x_{i+1} - x_i}{\|x_{i+1} - x_i\|}\right)\right\|^2 \le \left\| x_i - \left(c - h\frac{x_{i+1} - x_i}{\|x_{i+1} - x_i\|}\right)\right\|^2$$

we have

$$2h\|x_{i+1} - x_i\| \le \|x_i - c\|^2 - \|x_{i+1} - c\|^2.$$

For any $j > i$, adding the above inequality from i to $j - 1$ yields

$$2h\|x_j - x_i\| \le \|x_i - c\|^2 - \|x_j - c\|^2.$$

Since $(\|x_i - c\|^2)$ is a convergent sequence we conclude that (x_i) is a Cauchy sequence.

Finally, for natural numbers i, j with $j > i$, apply the parallelogram law $\|a - b\|^2 = 2\|a\|^2 + 2\|b\|^2 - \|a + b\|^2$ to $a := P_C x_j - x_j$ and $b := P_C x_i - x_j$ we obtain

$$\begin{aligned}
\|P_C x_j - P_C x_i\|^2 &= 2\|P_C x_j - x_j\|^2 + 2\|P_C x_i - x_j\|^2 \\
&\quad - 4\left\|\frac{P_C x_j + P_C x_i}{2} - x_j\right\|^2 \\
&\le 2\|P_C x_j - x_j\|^2 + 2\|P_C x_i - x_j\|^2 - 4\|P_C x_j - x_j\|^2 \\
&\le 2\|P_C x_i - x_j\|^2 - 2\|P_C x_j - x_j\|^2 \\
&\le 2\|P_C x_i - x_i\|^2 - 2\|P_C x_j - x_j\|^2.
\end{aligned}$$

We identify $(P_C x_i)$ as a Cauchy sequence, because $(\|x_i - P_C x_i\|)$ converges by (i).   ●

4.5.4 Convergence of Projection Algorithms

Let X be a Hilbert space. We say a sequence (x_i) in X is *asymptotically regular* if

$$\lim_{i \to \infty} \|x_i - x_{i+1}\| = 0.$$

Lemma 4.5.11 (Asymptotical Regularity of Projection Algorithm) *Let X be a Hilbert space and let C and D be closed convex subsets of X. Suppose $C \cap D \neq \emptyset$. Then the sequence (x_i) defined by the projection algorithm*

$$x_{i+1} = \begin{cases} P_C x_i & i \text{ is even,} \\ P_D x_i & i \text{ is odd.} \end{cases}$$

is asymptotically regular.

Proof. By Lemma 4.5.8 both P_C and P_D are 1-attracting with respect to $C \cap D$. Let $y \in C \cap D$. Since x_{i+1} is either $P_C x_i$ or $P_D x_i$ it follows that

$$\|x_{i+1} - x_i\|^2 \leq \|x_i - y\|^2 \quad \|x_{i+1} - y\|^2.$$

Since $(\|x_i - y\|^2)$ is a monotone decreasing sequence, therefore the right-hand side of the inequality converges to 0 and the result follows.   ●

Now, we are ready to prove the convergence of the projection algorithm.

Theorem 4.5.12 (Convergence of Projection Algorithm for the Intersection of Two Sets) *Let X be a Hilbert space and let C and D be closed convex subsets of X. Suppose $C \cap D \neq \emptyset$ ($\text{int}(C \cap D) \neq \emptyset$). Then the projection algorithm*

$$x_{i+1} = \begin{cases} P_C x_i & i \text{ is even,} \\ P_D x_i & i \text{ is odd.} \end{cases}$$

converges weakly (in norm) to a point in $C \cap D$.

Proof. Let $y \in C \cap D$. Then, for any $x \in X$, we have $\|P_C x - y\| = \|P_C x - P_C y\| \leq \|x - y\|$ and $\|P_D x - y\| = \|P_D x - P_D y\| \leq \|x - y\|$. Since x_{i+1} is either $P_C x_i$ or $P_D x_i$ we have that

$$\|x_{i+1} - y\| \leq \|x_i - y\|.$$

That is to say (x_i) is a Fejér monotone sequence with respect to $C \cap D$. By item (i) of Theorem 4.5.10 the sequence (x_i) is bounded, and therefore has a weakly convergent subsequence. We show that all weak cluster points of (x_i)

belong to $C \cap D$. In fact, let (x_{i_k}) be a subsequence of (x_i) converging to x weakly. Taking a subsequence again if necessary we may assume that (x_{i_k}) is a subset of either C or D. For the sake of argument let us assume that it is a subset of C and, thus, the weak limit x is also in C. On the other hand by the asymptotical regularity of (x_i) in Lemma 4.5.11 $(P_D x_{i_k}) = (x_{i_k+1})$ also weakly converges to x. Since $(P_D x_{i_k})$ is a subset of D we conclude that $x \in D$, and therefore $x \in C \cap D$. By item (ii) of Theorem 4.5.10 (x_i) has at most one weak cluster point in $C \cap D$, and we conclude that (x_i) weakly converges to a point in $C \cap D$. When $\mathrm{int}(C \cap D) \neq \emptyset$ it follows from item (iii) of Theorem 4.5.10 that (x_i) converges in norm.   ●

Whether the alternating projection algorithm converged in norm without the assumption that $\mathrm{int}(C \cap D) \neq \emptyset$, or more generally of metric regularity, was a long-standing open problem. Recently Hundal constructed an example showing that the answer is negative [140]. The proof of Hundal's example is self-contained and elementary. However, it is quite long and delicate, therefore we will be satisfied in stating the example.

Example 4.5.13 (Hundal) Let $X = \ell_2$ and let $\{e_i \mid i = 1, 2, \dots\}$ be the standard basis of X. Define $v \colon [0, +\infty) \to X$ by

$$v(r) = \exp(-100r^3)e_1 + \cos\left((r - [r])\pi/2\right)e_{[r]+2} + \sin\left((r - [r])\pi/2\right)e_{[r]+3},$$

where $[r]$ signifies the integer part of r and further define

$$C = \{e_1\}^\perp \text{ and } D = \mathrm{conv}\{v(r) \mid r \geq 0\}.$$

Then the hyperplane C and cone D satisfies $C \cap D = \{0\}$. However, Hundal's sequence of alternating projections x_i given by

$$x_{i+1} = P_D P_C x_i$$

starting from $x_0 = v(1)$ (necessarily) converges weakly to 0, but not in norm.

A related useful example is the moment problem.

Example 4.5.14 (Moment Problem) Let X be a Hilbert lattice – a Banach lattice in the Hilbert norm – with lattice cone $D = X^+$. All Hilbert lattices are realized as $L_2(\Omega, \mu)$ in the natural ordering for some measure space.

Consider a linear continuous mapping A from X onto $\mathbb{R}^N$. The moment problem seeks the solution of $A(x) = y \in \mathbb{R}^N, x \in D$ (see [53] for a recent survey). Define $C = A^{-1}(y)$. Then the moment problem is feasible if and only if $C \cap D \neq \emptyset$. A natural question is whether the projection algorithm converges in norm. This problem is answered affirmatively in [13] for $N = 1$ yet remains open in general when $N > 1$.

4.5.5 Projection Algorithms for Multiple Sets

We now turn to the general problem of finding some points in

$$\bigcap_{n=1}^{N} C_n,$$

where $C_n, n = 1, \ldots, N$ are closed convex sets in a Hilbert space X.

Let $a_n, n = 1, \ldots, N$ be positive numbers. Denote

$$X^N := \{x = (x_1, x_2, \ldots, x_N) \mid x_n \in X, n = 1, \ldots, N\}$$

the product space of N copies of X with inner product

$$\langle x, y \rangle = \sum_{n=1}^{N} a_n \langle x_n, y_n \rangle.$$

Then X^N is a Hilbert space. Define $C := C_1 \times C_2 \times \cdots \times C_N$ and $D := \{(x_1, \ldots, x_N) \in X^N : x_1 = x_2 = \cdots = x_N\}$. Then C and D are closed convex sets in X^N and $x \in \bigcap_{n=1}^{N} C_n$ if and only if $(x, x, \ldots, x) \in C \cap D$ (Exercise 4.5.2).

Applying the projection algorithm (4.5.1) to the convex sets C and D defined above we have the following generalized projection algorithm for finding some points in

$$\bigcap_{n=1}^{N} C_n.$$

Denote $P_n = P_{C_n}$. The algorithm can be expressed by

$$x_{i+1} = \Big(\sum_{n=1}^{N} \lambda_n P_n \Big) x_i, \qquad (4.5.3)$$

where $\lambda_n = a_n / \sum_{m=1}^{N} a_m$. In other words, each new approximation is the convex combination of the projections of the previous step to all the sets $C_n, n = 1, \ldots, N$. It follows from the convergence theorem in the previous subsection that the algorithm (4.5.3) converges weakly to some point in $\bigcap_{n=1}^{N} C_n$ when this intersection is nonempty.

Theorem 4.5.15 (Weak Convergence of Projection Algorithm for the Intersection of N Sets) *Let X be a Hilbert space and let $C_n, n = 1, \ldots, N$ be closed convex subsets of X. Suppose that $\bigcap_{n=1}^{N} C_n \neq \emptyset$ and $\lambda_n \geq 0$ satisfies $\sum_{n=1}^{N} \lambda_n = 1$. Then the projection algorithm*

$$x_{i+1} = \Big(\sum_{n=1}^{N} \lambda_n P_n \Big) x_i,$$

converges weakly to a point in $\bigcap_{n=1}^{N} C_n$.

Proof. This follows directly from Theorem 4.5.12.   ●

When the interior of $\bigcap_{n=1}^{N} C_n$ is nonempty we also have that the algorithm (4.5.3) converges in norm. However, since D does not have interior this conclusion cannot be derived from Theorem 4.5.12. Rather it has to be proved by directly showing that the approximation sequence is Fejér monotone with respect to $\bigcap_{n=1}^{N} C_n$.

Theorem 4.5.16 (Strong Convergence of Projection Algorithm for the Intersection of N Sets) *Let X be a Hilbert space and let $C_n, n = 1, \ldots, N$ be closed convex subsets of X. Suppose that* int $\bigcap_{n=1}^{N} C_n \neq \emptyset$ *and* $\lambda_n \geq 0$ *satisfies* $\sum_{n=1}^{N} \lambda_n = 1$. *Then the projection algorithm*

$$x_{i+1} = \Big(\sum_{n=1}^{N} \lambda_n P_n \Big) x_i,$$

converges to a point in $\bigcap_{n=1}^{N} C_n$ in norm.

Proof. Let $y \in \bigcap_{n=1}^{N} C_n$. Then

$$\|x_{i+1} - y\| = \Big\| \Big(\sum_{n=1}^{N} \lambda_n P_n \Big) x_i - y \Big\| = \Big\| \sum_{n=1}^{N} \lambda_n (P_n x_i - P_n y) \Big\|$$

$$\leq \sum_{n=1}^{N} \lambda_n \|P_n x_i - P_n y\| \leq \sum_{n=1}^{N} \lambda_n \|x_i - y\| = \|x_i - y\|.$$

That is to say (x_i) is a Fejér monotone sequence with respect to $\bigcap_{n=1}^{N} C_n$. The norm convergence of (x_i) then follows directly from Theorems 4.5.10 and 4.5.15.   ●

4.5.6 Commentary and Exercises

The projection algorithm can be traced back to von Neumann [260] and has been studied extensively. Here we emphasize the relationship between the projection algorithm and variational methods in Hilbert spaces. While projection operators can be defined outside the setting of a Hilbert space, they are not necessarily nonexpansive (Exercises 4.5.4 and 4.5.5). Indeed, the nonexpansive property of the projection operator characterizes Hilbert spaces in two or more dimensions. Thus, Hilbert space is the natural setting for the analysis of projection algorithms.

The survey paper [14] and the book [97] discuss many possible generalizations, provide historical perspective and are a rich source for additional literature. Many interesting applications are presented in [242]. Hundal's example is constructed in [140]. A simplification can be found in [191]. This example also clarifies many other related problems about convergence such as convergence of averages of projectors, the classical proximal point algorithm and the string-averaging projection method. Details of these applications can be found in [19]. Bregman distance (see [74] and Exercise 4.5.1) provides an alternative perspective into many of the generalizations of the projection algorithm. We refer the readers to [17, 18] for details and additional references.

*Exercise 4.5.1 Let X be a Hilbert space and let $f\colon X \to \mathbb{R} \cup \{+\infty\}$ be strictly convex and differentiable on $\mathrm{int}(\mathrm{dom}\, f)$. Define the Bregman distance $d_f\colon \mathrm{dom}\, f \times \mathrm{int}(\mathrm{dom}\, f) \to \mathbb{R}$ by

$$d_f(x,y) = f(x) - f(y) - \langle f'(y), x - y \rangle.$$

(i)   Prove $d_f(x,y) \geq 0$ with equality if and only if $x = y$.

(ii) Compute d_f when $f(t) = t^2/2$ and when f is the Boltzmann–Shannon entropy defined in 4.7.3.

(iii) Suppose f is three times differentiable. Prove d_f is convex if and only if $-1/f''$ is convex on $\mathrm{int}(\mathrm{dom}\, f)$.

(iv) Extend the results in (ii) and (iii) to the function $D_f\colon (\mathrm{dom}\, f)^N \times (\mathrm{int}(\mathrm{dom}\, f))^N \to \mathbb{R}$ defined by $D_f(x,y) = \sum_{n=1}^{N} d_f(x_n, y_n)$. (See [16] for the more general case when D_f is not separately defined.)

Exercise 4.5.2 Show that $x \leftarrow \bigcap_{n=1}^{N} C_n$ if and only if $(x, x, \ldots, x) \in C \cap D$.

Exercise 4.5.3 Prove Proposition 4.5.4.

Exercise 4.5.4 Show that Theorem 4.5.1, and therefore Definition 4.5.3, can be extended to reflexive Banach spaces.

Exercise 4.5.5 Show that in more than one dimension, with respect to the $\|\cdot\|_p$, for $1 < p < \infty$ and $p \neq 2$, the projection operator is not nonexpansive.

Exercise 4.5.6 Let X be a Hilbert space and let C be a closed convex subset of X. Show that

$$f(x) = \sup\{\langle x, y \rangle - \frac{\|y\|^2}{2} \mid y \in C\}$$

is convex and

$$f(x) = \frac{1}{2}(\|x\|^2 - \|x - P_C(x)\|^2).$$

*Exercise 4.5.7 (Infimal Convolution) If the functions $f, g\colon X \to (-\infty, +\infty]$ are convex, we define the *infimal convolution* $f \square g\colon X \to [-\infty, +\infty]$ by

$$(f \square g)(y) = \inf_x \{f(x) + g(y - x)\}.$$

(i)   Prove $f \square g$ is convex. (On the other hand, if g is concave prove so is $f \square g$.)

(ii) Prove $(f \square g)^* = f^* + g^*$.

(iii) If $\mathrm{dom}\, f \cap \mathrm{cont}\, g \neq \emptyset$, prove $(f + g)^* = f^* \square g^*$.

(iv) Define the *Lambert W-function* $W : \mathbb{R}_+ \to \mathbb{R}_+$ as the inverse of $y \in \mathbb{R}_+ \mapsto ye^y$. Prove the conjugate of the function

$$x \in \mathbb{R} \mapsto \exp^*(x) + \frac{x^2}{2}$$

is the function

$$y \in \mathbb{R} \mapsto W(e^y) + \frac{(W(e^y))^2}{2}.$$

∗**Exercise 4.5.8** Given a nonempty set $C \subset X$, consider the *distance function*

$$d_C(x) = \inf_{y \in C} \|x - y\|.$$

(i) Prove d_C^2 is a difference of convex functions, by observing

$$(d_C(x))^2 = \frac{\|x\|^2}{2} - \left(\frac{\|\cdot\|^2}{2} + \iota_C \right)^*(x).$$

Now suppose C is convex.

(ii) Prove d_C is convex and $d_C^* = \iota_{B_{X^*}} + \iota_C^* = \iota_{B_{X^*}} + \sigma_C$.
(iii) If C is closed and $x \notin C$, prove

$$d_C'(x) = d_C(x)^{-1}(x - P_C(x)),$$

where $P_C(x)$ is the nearest point to x in C.
(iv) If C is closed, prove

$$\left(\frac{d_C^2}{2} \right)'(x) = x - P_C(x)$$

for all points x.

4.6 Duality Inequalities for Sandwiched Functions

We derive duality inequalities that involve a mix of convex and nonconvex functions. These inequalities can be used to derive the striking Clarke–Ledyaev two-set multidirectional mean value theorem [87] and its elegant reformulation by Lewis and Ralph [180].

4.6.1 Sandwiched Functions

To avoid technical distraction we establish our fundamental inequality in $\mathbb{R}^N$ and we consider the simple case when the nonconvex function is smooth.

Theorem 4.6.1 (Sandwiched Functions) *Let C be a nonempty compact convex subset of $\mathbb{R}^N$. Suppose that f and h are proper convex lsc functions with f^* and h^* continuously differentiable and $\operatorname{dom}(f) \cap \operatorname{dom}(h) \subset C$. Then, for any continuously differentiable function $g\colon C \to \mathbb{R}$, there exists $z \in C$ such that*

$$\max_C(g - f) + \max_C(-g - h) \geq f^*(g'(z)) + h^*(-g'(z)).$$

Proof. Let $M := 2\sup\{\|c\| \mid c \in C\}$ and $W := \{x : [0,1] \to C \mid x$ is Lipschitz with a Lipschitz constant no more than $M\}$. Then W is compact in the uniform norm topology, by the Arzela–Ascoli Theorem [102]. For $x \in W$ define

$$Tx(t) := \int_0^t (f^*)' \circ g' \circ x(s)\, ds + \int_t^1 (h^*)' \circ (-g') \circ x(s)\, ds. \quad (4.6.1)$$

Then $T\colon W \to W$ is continuous (Exercise 4.6.1). Since W is compact and convex, the Schauder fixed point theorem [96, p. 60] shows that there is $x \in W$ such that $x = Tx$. That is

$$x(t) = \int_0^t (f^*)' \circ g' \circ x(s)\, ds + \int_t^1 (h^*)' \circ (-g') \circ x(s)\, ds.$$

Thus

$$g(x(1)) - g(x(0)) = \int_0^1 \langle g' \circ x(s), x'(s) \rangle\, ds$$

$$= \int_0^1 \langle g' \circ x(s), (f^*)' \circ g' \circ x(s) - (h^*)' \circ (-g') \circ x(s) \rangle\, ds.$$

$$= \int_0^1 \langle g' \circ x(s), (f^*)' \circ g' \circ x(s) \rangle\, ds$$

$$+ \int_0^1 \langle -g' \circ x(s), (h^*)' \circ (-g') \circ x(s) \rangle\, ds.$$

By Fenchel's equality of Proposition 4.4.1 we have

$$g(x(1)) - g(x(0)) = \int_0^1 (f^* \circ g' \circ x(s) + f \circ (f^*)' \circ g' \circ x(s))\, ds$$

$$+ \int_0^1 (h^* \circ (-g') \circ x(s) + h \circ (h^*)' \circ (-g') \circ x(s))\, ds.$$

By the integral form of Jensen's inequality we have

$$g(x(1)) - g(x(0)) \geq \int_0^1 (f^* \circ g' \circ x(s) + h^* \circ (-g') \circ x(s))\, ds$$

$$+ f\left(\int_0^1 (f^*)' \circ g' \circ x(s)\, ds \right)$$

$$+ h\left(\int_0^1 (h^*)' \circ (-g') \circ x(s)\, ds \right)$$

$$= \int_0^1 (f^* \circ g' \circ x(s) + h^* \circ (-g') \circ x(s))\, ds$$

$$+ f(x(1)) + h(x(0)).$$

Then, for some $z = x(t) \in C$ we have

$$g(x(1)) - g(x(0)) - f(x(1)) - h(x(0)) \geq f^*(g'(z)) + h^*(-g'(z)).$$

Thus, $x(1) \in C$ and $x(0) \in C$ will give the required inequality. ●

4.6.2 Two-Set Mean Value Inequalities

We will deduce the Clarke–Ledyaev two-set multidirectional mean value inequality and its reformulation due to Lewis and Ralph from the fundamental inequality in the previous subsection. The idea is to let f and h be the indicator functions of the sets involved. Since such f and h may not have smooth duals we need to generalize Theorem 4.6.1. For this purpose we need the following lemma which is a useful tool to smooth the dual of a convex function.

Lemma 4.6.2 *Let* $f\colon \mathbb{R}^N \to \mathbb{R} \cup \{+\infty\}$ *be a proper convex lsc function. Suppose that* $\mathrm{dom}(f)$ *is a bounded subset of* $\mathbb{R}^N$. *Then, for any* $\varepsilon > 0$, $(f + \varepsilon\|\cdot\|^2)^*$ *is continuously differentiable.*

Now we have the following generalization of Theorem 4.6.1 whose proof is left as a guided exercise.

Theorem 4.6.3 *Let C be a nonempty compact convex subset of* $\mathbb{R}^N$. *Suppose that f and h are proper convex lsc functions with* $\mathrm{dom}(f) \cap \mathrm{dom}(h) \subset C$. *Then, for any continuously differentiable function $g\colon C \to \mathbb{R}$, there exists $z \in C$ such that*

$$\max_C(g - f) + \max_C(-g - h) \geq f^*(g'(z)) + h^*(-g'(z)).$$

Proof. Exercise 4.6.3. ●

This result remains valid with g assumed only Lipschitz and with gradients replaced by Clarke subdifferentials (see [42] for this and other extensions). We illustrate what Theorem 4.6.3, so generalized, does and does not say in Figure 4.1. Therein, one has a subdifferential which does not separate the convex and

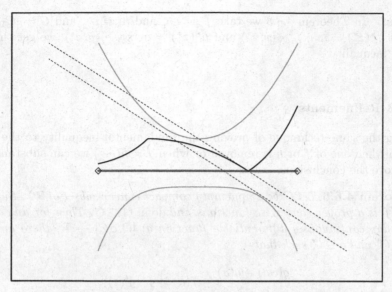

Fig. 4.1. The duality sandwich on a compact interval.

concave functions but a lower translate with the same slope does. However, there are other subgradients that will work. It is not known whether this is always true.

The following corollary is immediate.

Corollary 4.6.4 (The Lewis–Ralph Sandwich Theorem) *Let C be a non-empty compact convex subset of $\mathbb{R}^N$. Suppose that f and h are proper convex lsc functions with $\mathrm{dom}(f) \cap \mathrm{dom}(h) \subset C$. Then, for any continuously differentiable function $g\colon C \to \mathbb{R}$ such that $f \geq g \geq -h$, there exists $z \in C$ such that*

$$0 \geq f^*(g'(z)) + h^*(-g'(z)).$$

Proof. Exercise 4.6.4.   ●

Now we can deduce a two-set multidirectional mean value inequality. For two-sets $C_1, C_2 \subset X$ we denote $[C_1, C_2] := \mathrm{conv}(C_1 \cup C_2)$.

Corollary 4.6.5 (Two-Set Multidirectional Mean Value Inequality) *Let C_1 and C_2 be nonempty compact convex subsets of $\mathbb{R}^N$. Suppose that*

$$g\colon [C_1, C_2] \to \mathbb{R}$$

is a continuously differentiable function. Then there exists $z \in [C_1, C_2]$ such that for any $x \in C_1$ and $y \in C_2$,

$$\langle x - y, g'(z)\rangle \leq \max_{C_1} g - \min_{C_2} g.$$

Proof. In Theorem 4.6.3 we take $f := \iota_{C_1}$ and $h := \iota_{C_2}$ and $C = [C_1, C_2]$. Since $f^*(z^*) = \max_{x \in C_1} \langle x, z^* \rangle$ and $h^*(z^*) = \max_{y \in C_2} \langle y, z^* \rangle$, we get the desired inequality.   ●

4.6.3 Refinements

Using the same technique of proving the fundamental inequality to the situation when one of f or h is removed or when $f = h(-\cdot)$ we can substantially improve the conclusions.

Theorem 4.6.6 *Let C be a nonempty compact convex subset of $\mathbb{R}^N$. Suppose that f is a proper convex lsc functions and $\mathrm{dom}(f) \subset C$. Then for any $\alpha \neq 1$ and any continuously differentiable function $g \colon [C, \alpha C] \to \mathbb{R}$, there are $z \in [C, \alpha C]$ and $a \in C$ such that*

$$\frac{g(\alpha a) - g(a)}{\alpha - 1} - f(a) \geq f^*(g'(z)).$$

Proof. The method used to deduce Theorem 4.6.3 from Theorem 4.6.1 allows us to assume without loss of generality that f^* is continuously differentiable.

Let $M := (1 + |\alpha|) \sup\{\|c\| \mid c \in C\}$ and $W := \{x : [0,1] \to [C, \alpha C] \mid x$ is Lipschitz with a Lipschitz constant less than $M\}$. Then W is compact in the uniform norm topology, again by the Arzela–Ascoli Theorem. For $x \in W$ define

$$Tx(t) := \alpha \int_0^t (f^*)' \circ g' \circ x(s)\, ds + \int_t^1 (f^*)' \circ g' \circ x(s)\, ds. \qquad (4.6.2)$$

Then $T \colon W \to W$ is continuous (Exercise 4.6.2). Since W is compact and convex, once more the Schauder fixed point theorem shows that there is $x \in W$ such that $x = Tx$. That is

$$x(t) = \alpha \int_0^t (f^*)' \circ g' \circ x(s)\, ds + \int_t^1 (f^*)' \circ g' \circ x(s)\, ds.$$

Thus

$$g(x(1)) - g(x(0)) = (\alpha - 1) \int_0^1 \langle g' \circ x(s), (f^*)' \circ g' \circ x(s) \rangle\, ds.$$

By Fenchel's inequality (Proposition 4.4.1) we have

$$g(x(1)) - g(x(0)) = (\alpha - 1) \int_0^1 (f^* \circ g' \circ x(s) + f \circ (f^*)' \circ g' \circ x(s))\, ds.$$

By the integral form of Jensen's inequality we have

$$\frac{g(x(1)) - g(x(0))}{\alpha - 1} \geq \int_0^1 f^* \circ g' \circ x(s)\, ds + f\left(\int_0^1 (f^*)' \circ g' \circ x(s)\, ds\right)$$

$$= \int_0^1 f^* \circ g' \circ x(s)\, ds + f(x(0)).$$

Letting $a = x(0)$ we see that there is some $z = x(t) \in [C, \alpha C]$ such that

$$\frac{g(\alpha a) - g(a)}{\alpha - 1} - f(a) \geq f^*(g'(z))$$

since $x(1) = \alpha x(0) = \alpha a$.   ●

Setting $\alpha = -1$ and 0 we have two useful corollaries below.

Corollary 4.6.7 *Let C be a nonempty compact convex subset of $\mathbb{R}^N$. Suppose that f is a proper convex lsc function and $\mathrm{dom}(f) \subset C$. Then for any $\alpha \neq 1$ and any continuously differentiable function $g\colon [C, -C] \to \mathbb{R}$, there are $z \in [C, -C]$ and $a \in C$ such that*

$$\frac{g(a) - g(-a)}{2} - f(a) \geq f^*(g'(z)).$$

In particular, if f dominates the odd part of g on C then $f^*(g'(z)) \leq 0$.

Proof. Let $\alpha = -1$ in Theorem 4.6.6.   ●

Corollary 4.6.8 *Let C be a nonempty compact convex subset of $\mathbb{R}^N$. Suppose that f is a proper convex lsc functions and $\mathrm{dom}(f) \subset C$. Then for any $\alpha \neq 1$ and any continuously differentiable function $g\colon [0, C] \to \mathbb{R}$, there are $z \in [0, C]$ and $a \in C$ such that*

$$g(a) - g(0) \geq f(a) + f^*(g'(z)).$$

Proof. Let $\alpha = 0$ in Theorem 4.6.6.   ●

4.6.4 Commentary and Exercises

Clarke and Ledyaev proved two interesting multidirectional mean value inequalities: the one-set version presented in Theorem 3.6.1 and the two-set version discussed in Corollary 4.6.5. The one-set version is proved by a variational method. The original proof for the two-set version in [87] involves flows and a fixed point theorem. Whether one can find a "pure" variational proof for the two-set version remains an open problem. The general framework of this section in terms of sandwiched functions following [42] provides, in particular, a proof for the two-set multidirectional mean value inequality that combines a duality inequality (variational in nature) and topological fixed point theory. The comparison of f to the odd part of g in Corollary 4.6.7 reinforces the suggestion that fixed point theory is central to these results. The main result on sandwiched functions in Theorem 4.6.1 also enables us to further refine the two-set multidirectional mean value inequality (see Theorem 4.6.6 and its corollaries). One can find interesting special cases when the convex set C is the unit ball (Exercises 4.6.5). The equivalent sandwich theorem form of the two-set multidirectional mean value inequality in Corollary 4.6.4 is derived in [180].

Exercise 4.6.1 Show that the operator T defined in (4.6.1) is a continuous mapping that maps W to itself.

Exercise 4.6.2 Show that the operator T defined in (4.6.2) is a continuous mapping that maps W to itself.

Exercise 4.6.3 Prove Theorem 4.6.3. Hint: First apply Theorem 4.6.1 to $f + \varepsilon \| \cdot \|^2$, $h + \varepsilon \| \cdot \|^2$ and g and then take limits when $\varepsilon \to 0$.

Exercise 4.6.4 Prove Corollary 4.6.4.

Exercise 4.6.5 Setting C to be the unit ball in Theorem 4.6.6, use Corollaries 4.6.7 and 4.6.8 to deduce the following.

Corollary 4.6.9 *Let B be the closed unit ball of $\mathbb{R}^N$. Then for any $\alpha \neq 1$ and any continuously differentiable function $g \colon B \to \mathbb{R}$, there exists $z \in B$ such that*

$$\max_{a \in B} \frac{g(\alpha a) - g(a)}{\alpha - 1} \geq \|g'(z)\|.$$

Corollary 4.6.10 *Let B be the closed unit ball of $\mathbb{R}^N$. Then for any continuously differentiable function $g \colon B \to \mathbb{R}$, there exists $z \in B$ such that*

$$\max_{a \in B} \frac{g(a) - g(-a)}{2} \geq \|g'(z)\|.$$

Corollary 4.6.11 *Let B be the closed unit ball of $\mathbb{R}^N$. Then for any continuously differentiable function $g \colon B \to \mathbb{R}$, there exists $z \in B$ such that*

$$\max_{a \in B} g(a) - g(0) \geq \|g'(z)\|.$$

Note that Corollary 4.6.11 can also be deduced from the one-set multidirectional mean value inequality Theorem 3.6.1, and therefore has a variational proof.

Exercise 4.6.6 Deduce Corollary 4.6.11 from Theorem 3.6.1. Hint: Apply Theorem 3.6.1 in the form for bounded set $S = B$ pointed out in Exercise 3.6.1 (i) to function $f = -g$. Reference: See [42] for a direct variational proof.

4.7 Entropy Maximization

Entropy maximization is a special kind of convex programming problem with a finite number of linear constraints representing the condition on moments and a convex cost function emulating the negative of an entropy. A wide variety of application problems can be covered by this model due to its physical background. In this section we discuss the general duality theory for entropy maximization problems and illustrate its applications with a number of examples. The special structure of finitely many linear constraints in this problem makes the dual problem easy to solve. This is the key feature that we are going to explore.

4.7.1 Duality for Entropy Maximizations

We consider the following general form of the entropy maximization problem.

$$\mathcal{E} \qquad \begin{array}{c} \text{minimize} \quad f(x) \\ \text{subject to} \quad Ax = b, \end{array}$$

where $f\colon X \to \mathbb{R} \cup \{+\infty\}$ is a lsc convex function on a Banach space X representing the negative of an entropy like function and $A\colon X \to \mathbb{R}^N$ is a linear operator. Our solution to this entropy maximization problem in different concrete forms will be based on the following duality theorem.

Theorem 4.7.1 (Duality for Entropy Maximization) *Let X be a Banach space, let $f\colon X \to \mathbb{R} \cup \{+\infty\}$ be a lsc convex function and let $A\colon X \to \mathbb{R}^N$ be a continuous linear operator. Suppose that $b \in \mathrm{core}(A\,\mathrm{dom}\,f)$. Then*

$$\inf_{x \in X} \{f(x) \mid Ax = b\} = \max_{\phi \in \mathbb{R}^N} \{\langle \phi, b \rangle - f^*(A^*\phi)\}. \qquad (4.7.1)$$

Proof. Let $g = \iota_{\{b\}}$. Then $g^*(\phi) = \langle \phi, b \rangle$ (Exercise 4.7.1). Thus, it follows directly from the Fenchel duality equality of Theorem 4.4.3 that

$$\begin{aligned} \inf_{x \in X} \{f(x) \mid Ax = b\} &= \inf_{x \in X} \{f(x) + g(Ax)\} \\ &= \sup_{\phi \in \mathbb{R}^N} \{-g^*(-\phi) - f^*(A^*\phi)\} \\ &= \sup_{\phi \in \mathbb{R}^N} \{\langle \phi, b \rangle - f^*(A^*\phi)\}. \qquad (4.7.2) \end{aligned}$$

The condition $b \in \mathrm{core}(A\,\mathrm{dom}\,f)$ ensures that the infimum is finite. We leave the fact that $\sup_{\phi \in \mathbb{R}^N} \{\langle \phi, b \rangle - f^*(A^*\phi)\}$ must be attained as Exercise 4.7.2. ●

In general the infimum in the equality (4.7.1) may not be attained. An example is given in Exercise 4.7.3.

4.7.2 Finite Dimensional Problems

Entropy maximization problems in finite dimensional spaces for the classical Boltzmann–Shannon entropy illustrate well the characteristics of such problems. We start with the definition of the Boltzmann–Shannon entropy function. Define

$$p(t) := \begin{cases} t\ln t - t & \text{if } t > 0, \\ 0 & \text{if } t = 0, \\ +\infty & \text{if } t < 0, \end{cases} \tag{4.7.3}$$

and $f: \mathbb{R}^N \to \mathbb{R} \cup \{+\infty\}$ by

$$f(x) := \sum_{n=1}^{N} p(x_n). \tag{4.7.4}$$

The p and f defined above are Boltzmann–Shannon entropy functions on $\mathbb{R}$ and $\mathbb{R}^N$, respectively. We note that the original Boltzmann–Shannon entropy functions are the negatives of p and f and they are maximized. The following proposition summarizes some basic properties of the Boltzmann–Shannon entropy function whose elementary proof is left as an exercise.

Proposition 4.7.2 *Let f be the Boltzmann–Shannon entropy function defined in (4.7.4). Then*

(i) *for any $c \in \mathbb{R}^N$, $f(x) + \langle c, x \rangle$ is strictly convex on $\mathbb{R}^N_+$ and has compact sublevel sets;*
(ii) *for any $\bar{x} \in \text{int}(\mathbb{R}^N_+)$ and $x \in \text{bd}(\mathbb{R}^N_+)$, $f'(x; \bar{x} - x) = -\infty$.*

Proof.   Exercise 4.7.4.   ●

We now consider the finite dimensional entropy maximization problem.

$$\mathcal{FE} \qquad\qquad \text{minimize} \quad f(x) + \langle c, x \rangle$$

$$\text{subject to} \quad Ax = b,$$

where f is the Boltzmann–Shannon entropy function defined in (4.7.4), $c \in \mathbb{R}^N$, $b \in \mathbb{R}^M$ and $A: \mathbb{R}^N \to \mathbb{R}^M$ is a linear mapping.

The general duality theorem in this case can help us conveniently derive an explicit formula for the unique solution of problem $\mathcal{FE}$ in terms of the solution to its dual problem.

Theorem 4.7.3 *Suppose that there exists a $z \in \text{int}(\mathbb{R}^N_+)$ such that $Az = b$. Then problem $\mathcal{FE}$ has a unique solution $\bar{x} = (\bar{x}_1, \ldots, \bar{x}_N)$ determined by*

$$\bar{x}_n = \exp(A^\top \bar{\phi} - c)_n, n = 1, \ldots, N,$$

where $\bar{\phi}$ is a solution to the dual problem

$$\max_{\phi \in \mathbb{R}^M} \{ \langle \phi, b \rangle - (f + c)^*(A^\top \phi) \}.$$

Proof.   The compactness of the sublevel sets of the objective function as asserted in Proposition 4.7.2 ensures the existence of solutions to problem $\mathcal{FE}$. By (ii) of Proposition 4.7.2 the directional derivative of the cost function is $-\infty$ on any boundary point x of $\mathbb{R}_+^N$, the domain of the cost function, in the direction of $z - x$. Thus, any solution of $\mathcal{FE}$ must be in the interior of $\mathbb{R}_+^N$. By (i) of Proposition 4.7.2 the cost function is strictly convex on $\mathrm{int}(\mathbb{R}_+^N)$, and therefore the solution is unique. Let us denote this unique solution of $\mathcal{FE}$ by $\bar{x}$. Then the duality result of Theorem 4.7.1 implies that

$$f(\bar{x}) + \langle c, \bar{x}\rangle = \inf_{x \in \mathbb{R}^N}\{f(x) + \langle c, x\rangle \mid Ax = b\}$$

$$= \max_{\phi \in \mathbb{R}^M}\{\langle \phi, b\rangle - (f + c)^*(A^\top \phi)\}.$$

Now let $\bar{\phi}$ be a solution to the dual problem. We have

$$f(\bar{x}) + \langle c, \bar{x}\rangle + (f + c)^*(A^\top \bar{\phi}) = \langle \bar{\phi}, b\rangle = \langle \bar{\phi}, A\bar{x}\rangle = \langle A^\top \bar{\phi}, \bar{x}\rangle.$$

It follows from Proposition 4.4.1 that $A^\top \bar{\phi} \in \partial(f + c)(\bar{x})$. Since $\bar{x} \in \mathrm{int}(\mathbb{R}_+^N)$ where f is differentiable, we have $A^\top \bar{\phi} = f'(\bar{x}) + c$. The formula for $\bar{x}$ now follows from explicit computation (Exercise 4.7.5).   ●

We note that $\bar{\phi}$ is a Lagrange multiplier for the constrained minimization problem $\mathcal{FE}$ (Exercise 4.7.6).

4.7.3 The DAD Problem

We now turn to an interesting application in matrix theory. Let $A = (a_{nm})$ be an N by N matrix. We say that A is *doubly stochastic* if each entry is nonnegative and $\sum_{n=1}^N a_{nm} = 1$ for $m = 1, \ldots, N$ and $\sum_{m=1}^N a_{nm} = 1$ for $n = 1, \ldots, N$. We say that A has a *double stochastic pattern* if there is a doubly stochastic matrix with exactly the same zero entries as A. Using the result in the previous subsection we can prove the following characterization of matrices that have a doubly stochastic pattern.

Theorem 4.7.4 (Matrices with Doubly Stochastic Pattern) *Let A be a square matrix. Then A has a double stochastic pattern if and only if there are diagonal matrices D_1 and D_2 with strictly positive diagonal entries such that $D_1 A D_2$ is doubly stochastic.*

Proof.   The sufficiency is easy and is left as an exercise (Exercise 4.7.7). We prove the necessity. Let A have a doubly stochastic pattern. Define a set $Z = \{(n, m) \mid a_{nm} > 0\}$, and let $\mathbb{R}^Z$ denote the set of vectors with components indexed by Z and $\mathbb{R}_+^Z$ denote those vectors in $\mathbb{R}^Z$ with all nonnegative components. The key is to realize that the desired doubly stochastic matrix $D_1 A D_2 = B$ is a solution of the following entropy maximization problem.

$$\text{minimize} \quad \sum_{(n,m)\in Z} (p(x_{nm}) - x_{nm} \log a_{nm}) \quad x \in \mathbb{R}^Z$$

$$\text{subject to} \quad \sum_{n:(n,m)\in Z} x_{nm} = 1 \quad \text{for } m = 1,\dots,N$$

$$\sum_{m:(n,m)\in Z} x_{nm} = 1 \quad \text{for } n = 1,\dots,N.$$

Here p is the Boltzmann–Shannon entropy function defined in (4.7.3). That matrix A has a doubly stochastic pattern implies that the constraint is satisfied at an interior point of $\mathbb{R}^Z_+$ (Exercise 4.7.8). Thus, by Theorem 4.7.3, the above entropy maximization problem has a unique solution $\bar{x} = (\bar{x}_{nm})$ whose components are given by $\bar{x}_{nm} = \exp[(G^\top \bar{\phi})_{nm} + \ln(a_{nm})]$, where G is the matrix in the linear equality constraints and $\bar{\phi}$ is a solution of the dual problem. Since $G\colon \mathbb{R}^Z \to \mathbb{R}^{2N}$, we can write $\bar{\phi} = (\lambda_1,\dots,\lambda_N,\mu_1,\dots,\mu_N)$. It follows that, for any $x = (x_{nm}) \in \mathbb{R}^Z$,

$$\langle x, G^\top \bar{\phi} \rangle = \langle Gx, \bar{\phi} \rangle$$

$$= \sum_{n=1}^N \lambda_n \sum_{m:(n,m)\in Z} x_{nm} + \sum_{m=1}^N \mu_m \sum_{n:(n,m)\in Z} x_{nm}$$

$$= \sum_{n:(n,m)\in Z} x_{nm}(\lambda_n + \mu_m).$$

Thus,

$$(G^\top \bar{\phi})_{nm} = \lambda_n + \mu_m.$$

Now we have $\bar{x}_{nm} = a_{nm}\exp(\lambda_n + \mu_m) = a_{nm}\exp(\lambda_n) \times \exp(\mu_m)$. Define $B = (b_{nm})$ by $b_{nm} = \bar{x}_{nm}$ for $(n,m) \in Z$ and $b_{nm} = 0$ otherwise. Then B is a doubly stochastic matrix and $B = D_1 A D_2$ where D_1 and D_2 are diagonal matrices with diagonal entries $\exp(\lambda_1),\dots,\exp(\lambda_N)$ and $\exp(\mu_1),\dots,\exp(\mu_N)$, respectively. ●

4.7.4 Infinite Dimensional Problems

Maximum entropy methods in image reconstruction and option pricing problems lead to the following entropy maximization problem in infinite dimensional space.

$$\mathcal{IE} \qquad \text{minimize} \quad f(x) := \int_I p(x(t))\, dt \quad x \in L^1(I)$$

$$\text{subject to} \quad \int_I a_n(t)x(t)\, dt = b_n, n = 1,\dots,N.$$

Here again p is the Bolzmann–Shannon entropy defined in (4.7.3), I is a (possibly infinite) interval, $a_n \in L^\infty(I)$ and the integral $\int_I p(x(t))\,dt$ is understood in the following sense: If there exists a function $\alpha \in L^1(I)$ such that $\alpha(t) \geq p(x(t))$ almost everywhere on I, the integral has an unambiguous classical value (finite or $-\infty$). Otherwise, we set $\int_I p(x(t))\,dt = +\infty$.

Although Theorem 4.7.1 holds for problem $\mathcal{IE}$, the constraint qualification $b \in \mathrm{core}(A\,\mathrm{dom}\,f)$ is too strong for most of the applications. Fortunately, this condition has been weakened in [51] to: b belongs to the relative interior of $A\,\mathrm{dom}\,f$, the interior relative to $\mathrm{span}(A\,\mathrm{dom}\,f)$ denoted by $\mathrm{ri}(A\,\mathrm{dom}\,f)$. We will not get into the full technical details and will merely state this duality theorem.

Theorem 4.7.5 (Duality for Entropy Maximization in Infinite Dimensional Spaces) *Let $f\colon L^1(I) \to \mathbb{R} \cup \{+\infty\}$ be the lsc convex function defined in $\mathcal{IE}$, let $A\colon L^1(I) \to \mathbb{R}^N$ be the continuous linear operator defined by*

$$Ax = \left(\int_I a_n(t)x(t)\,dt, \ldots, \int_I a_n(t)x(t)\,dt \right)$$

and let $b = (b_1, \ldots, b_N)$. Suppose that $b \in \mathrm{ri}(A\,\mathrm{dom}\,f)$. Then

$$\inf_{x \in L^1(I)} \{ f(x) \mid Ax = b \} = \max_{\phi \in \mathbb{R}^N} \{ \langle \phi, b \rangle - f^*(A^*\phi) \}. \qquad (4.7.5)$$

Proof. See [51, 36]. ●

For problem $\mathcal{IE}$ we can actually get an explicit representation of the solution. For this we need a representation of the dual of the f.

Proposition 4.7.6 (Dual of the Integral of Entropy) *For any $x^* \in L^\infty(I)$,*

$$f^*(x^*) = \int_I p^*(x^*(t))\,dt = \int_I \exp(x^*(t))\,dt.$$

Proof. Let us begin with the case when I is finite. We can compute that $p^*(s) = e^s$ (Exercise 4.7.10). Thus, for any $x^* \in L^\infty(I)$

$$f^*(x^*) = \sup\left\{ \int_I [\langle x^*(t), x(t) \rangle - p(x(t))]\,dt \;\middle|\; x \in L^1(I) \right\}$$

$$\leq \int_I \sup[\langle x^*(t), x \rangle - p(x)]\,dt$$

$$= \int_I p^*(x^*(t))\,dt = \int_I \exp(x^*(t))\,dt.$$

On the other hand, clearly $\bar{x}(t) := \exp(x^*(t)) \in L^1(I)$ and

$$\sup\left\{ \int_I [\langle x^*(t), x(t) \rangle - p(x(t))]\,dt \;\middle|\; x \in L^1(I) \right\}$$

is attained at $\bar{x}$.

When I is an infinite interval and $\exp(x^*(\cdot)) \in L^1(I)$ the result follows from the finite interval case by limiting process. If $\exp(x^*(\cdot)) \notin L^1(I)$ then both sides of the equality are $+\infty$ according to our convention. We leave the detail as an exercise (Exercise 4.7.11). ●

Combining Theorem 4.7.5 and Proposition 4.7.6 we have the following corollary that gives an explicit solution to problem $\mathcal{IE}$.

Corollary 4.7.7 *Let $f\colon L^1(I) \to \mathbb{R} \cup \{+\infty\}$ be the lsc convex function defined in $\mathcal{IE}$, let $A\colon L^1(I) \to \mathbb{R}^N$ be the continuous linear operator defined by*

$$Ax = \left(\int_I a_n(t)x(t)\, dt, \ldots, \int_I a_n(t)x(t)\, dt \right)$$

and let $b = (b_1, \ldots, b_N)$. Suppose that $b \in \mathrm{ri}(A \operatorname{dom} f)$. Then $\mathcal{IE}$ has a unique solution given by

$$\bar{x}(t) = \exp\Big(\sum_{n=1}^{N} \bar{\phi}_n a_n(t) \Big),$$

where $\bar{\phi} \in \mathbb{R}^N$ is the solution of

$$\max_{\phi \in \mathbb{R}^N} \{ \langle \phi, b \rangle - f^*(A^*\phi) \}. \tag{4.7.6}$$

Proof. Exercise 4.7.12. ●

4.7.5 Commentary and Exercises

The DAD problem and its infinite dimensional extensions to probability theory were discussed in [54]. A comprehensive foundation for infinite dimensional entropy maximization was established in [51] and other papers by Borwein and Lewis, especially [52]. The value of such formalism is that it naturally allows one to handle non-negativity and similar constraints with a natural barrier function, and that it captures much of the power of convex duality theory as shown in Section 4.7.4. A fleshed out application of the results in that subsection to option pricing can be found in [36].

Exercise 4.7.1 Let X be a Banach space and let C be a closed convex subset of X. Show that $\iota_C^*(\phi) = \sigma_C(\phi) := \sup_{c \in C}\langle \phi, c \rangle$. In particular, $\iota_{\{b\}}^* = \langle \phi, b \rangle$.

Exercise 4.7.2 Show that the last supremum in (4.7.2) must be attained.

Exercise 4.7.3 Show that the infimum in the equality (4.7.1) may not be attained.

Exercise 4.7.4 Prove Proposition 4.7.2.

Exercise 4.7.5 Let f be the Boltzmann–Shannon entropy function defined in (4.7.4). Show that $f'(\bar{x}) = A^*\bar{\phi} - c$ implies that $\bar{x} = (\bar{x}_1, \ldots, \bar{x}_N)$ is determined by

$$\bar{x}_n = \exp(A^*\bar{\phi} - c)_n, n = 1, \ldots, N.$$

Exercise 4.7.6 Verify that $\bar{\phi}$ in Theorem 4.7.3 is a Lagrange multiplier for the constrained minimization problem $\mathcal{FE}$.

Exercise 4.7.7 Let A be a square matrix. Prove that if there are diagonal matrices D_1 and D_2 with strictly positive diagonal entries such that D_1AD_2 is doubly stochastic then A has a double stochastic pattern.

Exercise 4.7.8 Show that the constraint in the optimization problem in the proof of Theorem 4.7.4 is satisfied at an interior point of $\mathbb{R}_+^Z$.

Exercise 4.7.9 Construct a square matrix with nonnegative entries that does not have a doubly stochastic pattern.

Exercise 4.7.10 Verify that for the Bolzmann–Shannon entropy p defined in (4.7.3) we have $p^*(s) = e^s$.

Exercise 4.7.11 Supplement the detail for the proof of Proposition 4.7.6 in the case when I is an infinite interval.

Exercise 4.7.12 Prove Corollary 4.7.7.

5

Variational Techniques and Multifunctions

Multifunctions arise naturally in many situations. Some frequently encountered examples are: the level sets and sublevel sets of a function, various subdifferentials of nonsmooth functions, the solution sets of an optimization problem depending on some parameters and the vector field of a control system. Here we give a concise discussion on how to apply the technique of variational analysis to problems involving multifunctions. We also discuss subdifferentials as multifunctions.

5.1 Multifunctions

5.1.1 Multifunctions and Related Functions

Let X and Y be two sets. A multifunction from X to Y is a mapping $F: X \to 2^Y$, where 2^Y represents the collection of all subsets of Y. We define the domain, range and graph of F by $\operatorname{dom} F := \{x \in X \mid F(x) \neq \emptyset\}$, $\operatorname{range} F := \{y \in Y \mid y \in F(x) \text{ for some } x \in X\}$ and $\operatorname{graph} F := \{(x,y) \in X \times Y \mid y \in F(x)\}$, respectively. The inverse of a multifunction $F: X \to 2^Y$ is a multifunction $F^{-1}: Y \to 2^X$ defined by $F^{-1}(y) := \{x \in X \mid y \in F(x)\}$. Clearly the domain of F is the range of F^{-1} and the range of F is the domain of F^{-1}. A multifunction is completely characterized by its graph. Moreover, we have the following symmetric relationship between F, F^{-1} and the graph of F: $F(x) = \{y \in Y \mid (x,y) \in \operatorname{graph} F\}$ and $F^{-1}(y) = \{x \in X \mid (x,y) \in \operatorname{graph} F\}$. The following are some examples of multifunctions.

Example 5.1.1 Let X be a Fréchet smooth Banach space and let $f: X \to \mathbb{R} \cup \{+\infty\}$ be a lsc function. Then $\partial_F f$ is a multifunction from X to X^*.

Example 5.1.2 Let X and Y be metric spaces and let $f: X \times Y \to \mathbb{R} \cup \{+\infty\}$ be a lsc function. Then the solution set to the parametric minimization problem of minimizing $x \to f(x,y)$,

$$\mathrm{argmin}(y) := \{x \in X \mid f(x,y) = \inf\{f(x',y) \mid x' \in X\}\},$$

is a multifunction from Y to X.

Example 5.1.3 Let X be a metric space and let $f\colon X \to \mathbb{R} \cup \{+\infty\}$ be a lsc function. Then the sublevel set

$$f^{-1}((-\infty, r]) = \{x \in X \mid f(x) \leq r\}$$

and the level set

$$f^{-1}(r) = \{x \in X \mid f(x) = r\}$$

are multifunctions from $\mathbb{R} \to X$.

Example 5.1.4 Let X be a metric space and let $f\colon X \to \mathbb{R}$ be a lsc function. Then the epigraphical profile mapping

$$E_f(x) = \{r \in \mathbb{R} \mid f(x) \leq r\}$$

is a multifunction from $X \to \mathbb{R}$. We can see that graph $E_f = \mathrm{epi}\, f$ (Exercise 5.1.3).

One can often study a multifunction $F\colon X \to 2^Y$ through related functions. Clearly, $\iota_{\mathrm{graph}\, F}$ completely characterizes F. When both X and Y are topological spaces, $\iota_{\mathrm{graph}\, F}$ is a lsc function on $X \times Y$ if and only if graph F is a closed subset of $X \times Y$. This is an important condition when we analyze a multifunction with variational techniques. Thus, we define a multifunction to be closed if its graph is closed. We say that multifunction F is closed (open, compact, convex) valued if, for every $x \in \mathrm{dom}\, F$, the set $F(x)$ is closed (open, compact, convex). Note that a closed multifunction is always closed valued yet the converse is not true (Exercise 5.1.2). When Y has additional structure other functions can be used to study a multifunction $F\colon X \to 2^Y$. For example when Y is a metric space we can use $(x,y) \to d(F(x);y)$ and when Y is a Banach space we can use $(x,x^*) \to \sigma(F(x);x^*)$. These functions are in general nonsmooth. We will emphasize the use of variational tools in studying multifunctions by their related nonsmooth functions.

5.1.2 An Example: The Convex Subdifferential

Subdifferentials are multifunctions from X to X^*. In Section 3.4 we have seen the interplay of properties of a function and its (Fréchet) subdifferential. Here we further discuss the subdifferential of a convex function to illustrate various nice properties of the subdifferential as a multifunction inherited from the convexity of the underlying function.

We say a multifunction $F\colon X \to 2^{X^*}$ is monotone provided that for any $x, y \in X$, $x^* \in F(x)$ and $y^* \in F(y)$,

$$\langle y^* - x^*, y - x \rangle \geq 0.$$

The convex subdifferential of a convex lsc function is a typical example of a monotone multifunction.

Theorem 5.1.5 *Let X be a Banach space and let $f\colon X \to \mathbb{R}\cup\{+\infty\}$ be a lsc convex function. Then ∂f is a monotone multifunction.*

Proof. Let $x^* \in \partial f(x)$ and $y^* \in \partial f(y)$. It follows from the definition of the convex subdifferential that

$$f(y) - f(x) \geq \langle x^*, y - x \rangle \tag{5.1.1}$$

and

$$f(x) - f(y) \geq \langle y^*, x - y \rangle \tag{5.1.2}$$

Adding (5.1.1) and (5.1.2) we have

$$\langle y^* - x^*, y - x \rangle \geq 0.$$

•

In fact the monotonicity of the subdifferential characterizes the convexity of the underlying function.

Theorem 5.1.6 (Convexity) *Let X be a Fréchet smooth Banach space and let $f\colon X \to \mathbb{R}\cup\{+\infty\}$ be a lsc function. Suppose that $\partial_F f$ is monotone. Then f is convex.*

Proof. If $\partial_F f$ is monotone then for each $x^* \in X^*$ the operator $x \to \partial_F f(x) + x^* - \partial_F(f + x^*)(x)$ is monotone, hence quasi-monotone. By Theorem 3.4.12, for each $x^* \in X^*$, the function $f + x^*$ is quasi-convex. This implies the convexity of f (Exercise 5.1.4). •

Recall that a monotone multifunction $F\colon X \to 2^{X^*}$ is said to be maximal monotone if graph F is not properly contained in the graph of any monotone multifunction. It is not hard to check that a maximal monotone multifunction is convex valued and closed (Exercise 5.1.5). We can further prove the maximal monotonicity of a monotone Fréchet subdifferential of a lsc function (which must be convex by Theorem 5.1.6).

Theorem 5.1.7 (Maximal Monotonicity) *Let X be a Fréchet smooth Banach space and let $f\colon X \to \mathbb{R}\cup\{+\infty\}$ be a proper lsc function. Suppose that dom $f \neq \emptyset$ and $\partial_F f$ is monotone. Then $\partial_F f$ is maximal monotone.*

Proof. Let $b \in X$ and $b^* \in X^*$ be such that $b^* \notin \partial_F f(b)$. We need to show that there exists $x \in X$ and $x^* \in \partial_F f(x)$ such that $\langle x^* - b^*, x - b \rangle < 0$. Observing that $0 \notin \partial_F(f - b^*)(b)$, and therefore b is not a minimum of $f - b^*$, there exists $a \in X$ such that $(f - b^*)(a) < (f - b^*)(b)$. Then it follows from the approximate mean value theorem of Theorem 3.4.6 that there exists a sequence (x_i) converging to $c \subset [a, b)$ and $x_i^* \in \partial_F f(x_i)$ such that

$y_i^* := x_i^* - b^* \in \partial_F(f - b^*)(x_i)$ satisfying $\liminf_{i \to \infty} \langle y_i^*, c - x_i \rangle \geq 0$ and $\liminf_{i \to \infty} \langle y_i^*, b - a \rangle > 0$. It follows that

$$\liminf_{i \to \infty} \langle x_i^* - b^*, b - x_i \rangle \geq \liminf_{i \to \infty} \langle y_i^*, b - c \rangle + \liminf_{i \to \infty} \langle y_i^*, c - x_i \rangle$$

$$\geq \frac{\|b - c\|}{\|b - a\|} \liminf_{i \to \infty} \langle y_i^*, b - a \rangle + \liminf_{i \to \infty} \langle y_i^*, c - x_i \rangle > 0$$

It remains to set $x := x_i$ and $x^* := x_i^*$ for i sufficiently large.   ●

We have seen in Proposition 4.1.2 and Theorem 4.1.8 that a lsc convex function is locally Lipschitz in the core of its domain. Consequently the subdifferential of a lsc convex function is locally bounded in the core of its domain. We will show this holds true in general for a maximal monotone multifunction. The proof actually reduces this more general situation to the continuity of a convex function in the core of its domain.

Theorem 5.1.8 (Boundedness of Monotone Multifunctions) *Let $F \colon X \to 2^{X^*}$ be a monotone multifunction. Suppose that $x \in$ core (dom F). Then F is locally bounded at x.*

Proof. By choosing any $x^* \in F(x)$ and replacing F by the monotone multifunction $y \to F(y + x) - x^*$, we lose no generality in assuming that $x = 0$ and that $0 \in F(0)$. Define, for $x \in X$,

$$f(x) := \sup\{\langle y^*, x - y \rangle : y \in \text{dom } F, \|y\| \leq 1 \text{ and } y^* \in F(y)\}.$$

As the supremum of affine continuous functions, f is convex and lower semicontinuous. We show that dom f is an absorbing set. First, since $0 \in F(0)$, we must have $f \geq 0$. Second, whenever $y \in$ dom F and $y^* \in F(y)$, monotonicity implies that $0 \leq \langle y^* - 0, y - 0 \rangle$, so $f(0) \leq 0$. Thus, $f(0) = 0$. Suppose $x \in X$. By hypothesis, dom F is absorbing so there exists $t > 0$ such that $F(tx) \neq \emptyset$. Choose any element $u^* \in F(tx)$. If $y \in$ dom F and $y^* \in F(y)$, then by monotonicity

$$\langle y^*, tx - y \rangle \leq \langle u^*, tx - y \rangle.$$

Consequently,

$$f(tx) \leq \sup\{\langle u^*, tx - y \rangle : y \in \text{dom } F, \|y\| \leq 1\} < \langle u^*, tx \rangle + \|u^*\| < +\infty.$$

By virtue of Proposition 4.1.2 and Theorem 4.1.8, f is continuous at 0 and hence there exists $\eta > 0$ such that $f(x) < 1$ for all $x \in 2\eta B_X$. Equivalently, if $x \in 2\eta B_X$, then $\langle y^*, x \rangle \leq \langle y^*, y \rangle + 1$ whenever $y \in$ dom F, $\|y\| \leq 1$ and $y^* \in F(y)$. Thus, if $y \in \eta B_X \cap$ dom F and $y^* \in F(y)$, then

$$2\eta\|y^*\| = \sup\{\langle y^*, x \rangle : x \in 2\eta B_X\} \leq \|y^*\|\|y\| + 1 \leq \eta\|y^*\| + 1,$$

so $\|y^*\| \leq 1/\eta$.   ●

Note that Theorem 5.1.8 does not require that the domain of F be convex.

5.1.3 Limits of Sequences of Sets

Having defined multifunctions we turn to their limits and continuity. We will take a sequential approach, and therefore need to study the limits of sequences of sets.

Definition 5.1.9 *Let Y be a Hausdorff topological space and let (F_i) be a sequence of subsets of Y. The sequential lower and upper limits of F_i are defined by*

$$\liminf_{i \to \infty} F_i = \{\, \lim_{i \to \infty} y_i \mid y_i \in F_i \text{ for all } i = 1, 2, \dots \}$$

and

$$\limsup_{i \to \infty} F_i = \{\, \lim_{k \to \infty} y_{i_k} \mid y_{i_k} \in F_{i_k} \text{ for some } i_k \to \infty \}.$$

Clearly $\liminf_{i \to \infty} F_i \subset \limsup_{i \to \infty} F_i$. When they are equal we define the common set to be the Painlevé–Kuratowski limit of the sequence (F_i) and denote it by $\lim_{i \to \infty} F_i$. In a metric space both the sequential lower and upper limits are closed. However, this is not true in general (Exercise 5.1.8).

When Y is a metric space the lower and upper limits can be represented alternatively as

$$\liminf_{i \to \infty} F_i = \bigcap_{k=1}^{\infty} \bigcup_{j=1}^{\infty} \bigcap_{i=j}^{\infty} B_{\frac{1}{k}}(F_i) \tag{5.1.3}$$

and

$$\limsup_{i \to \infty} F_i = \bigcap_{k=1}^{\infty} \bigcap_{j=1}^{\infty} \bigcup_{i=j}^{\infty} B_{\frac{1}{k}}(F_i). \tag{5.1.4}$$

We leave the proofs of these alternative representations as Exercise 5.1.6.

These lower and upper limits can also be described by using the distance between a set and a point.

Lemma 5.1.10 *Let Y be a metric space and let (F_i) be a sequence of subsets in Y. Then*

$$\liminf_{i \to \infty} F_i = \{ y \in Y \mid \limsup_{i \to \infty} d(F_i; y) = 0 \}$$

and

$$\limsup_{i \to \infty} F_i = \{ y \in Y \mid \liminf_{i \to \infty} d(F_i; y) = 0 \}.$$

Proof. Exercise 5.1.7. ●

Lemma 5.1.10 is a special case of the following more general characterization of the upper and lower limits of a sequence of sets.

Lemma 5.1.11 *Let Y be a metric space, let F be a closed subset of X and let (F_i) be a sequence of subsets in Y. Then*

$$F \subset \liminf_{i \to \infty} F_i \qquad\qquad (5.1.5)$$

if and only if for any $y \in Y$,

$$\limsup_{i \to \infty} d(F_i; y) \le d(F; y); \qquad\qquad (5.1.6)$$

and

$$\limsup_{i \to \infty} F_i \subset F \qquad\qquad (5.1.7)$$

if and only if for any $y \in Y$,

$$\liminf_{i \to \infty} d(F_i; y) \ge d(F; y). \qquad\qquad (5.1.8)$$

Consequently,

$$\lim_{i \to \infty} F_i = F$$

if and only if for any $y \in Y$,

$$\lim_{i \to \infty} d(F_i; y) = d(F; y).$$

Proof. We prove the equivalence of (5.1.5) and (5.1.6). It follows from Lemma 5.1.10 that (5.1.6) implies (5.1.5). Now suppose (5.1.5) holds and let $y \in Y$ be an arbitrary element. For any $\varepsilon > 0$ choose $x \in F$ such that $d(F; y) + \varepsilon \ge d(x, y)$ and let (x_i) be a sequence converges to x with $x_i \in F_i$. Then

$$d(F_i; y) \le d(x_i, y).$$

Taking $\limsup$ as $i \to \infty$ we have

$$\limsup_{i \to \infty} d(F_i; y) \le d(x, y) \le d(F; y) + \varepsilon.$$

Since ε is arbitrary we obtain (5.1.6).

The proof of the equivalence of (5.1.7) and (5.1.8) is similar and left as Exercise 5.1.9.   ●

Applying the Painlevé–Kuratowski limit to the epigraph of a sequence of functions leads to the concept of epi-convergence. This is particularly useful in analyzing approximations of functions when minimizing the function is a primary concern.

Definition 5.1.12 (Epi-convergence) *Let X be a metric space and let $f_i \colon X \to \mathbb{R}$ be a sequence of lsc functions. The lower epi-limit e-$\liminf_{i \to \infty} f_i$ is the function with*

$$\mathrm{epi}(\text{e-}\varliminf_{i\to\infty} f_i) = \limsup_{i\to\infty} \mathrm{epi}\, f_i,$$

and the upper epi-limit e-$\limsup_{i\to\infty} f_i$ is the function with

$$\mathrm{epi}(\text{e-}\limsup_{i\to\infty} f_i) = \liminf_{i\to\infty} \mathrm{epi}\, f_i.$$

When these two functions coincide we say that f_i epi-converges to its epi-limit

$$\text{e-}\lim_{i\to\infty} f_i = \text{e-}\liminf_{i\to\infty} f_i = \text{e-}\limsup_{i\to\infty} f_i.$$

Note that both lower and upper epi-limits are lsc functions, and so is the epi-limit when it exists (Exercise 5.1.11). Epi-limits have the following easy yet useful characterization, whose proof is left as an exercise.

Lemma 5.1.13 Let X be a metric space and let $f_i\colon X\to\mathbb{R}$ be a sequence of lsc functions. Then $f = \text{e-}\lim_{i\to\infty} f_i$ if and only if at each point $x\in X$ one has

$$\liminf_{i\to\infty} f_i(x_i) \geq f(x) \text{ for every sequence } x_i \to x \qquad (5.1.9)$$

and

$$\limsup_{i\to\infty} f_i(x_i) \leq f(x) \text{ for some sequence } x_i \to x \qquad (5.1.10)$$

Proof. Exercise 5.1.12. $\bullet$

We end this subsection with a result that illuminates the usefulness of epi-convergence in minimization problems.

Theorem 5.1.14 Let X be a metric space and let $f_i\colon X\to\mathbb{R}$ be a sequence of lsc functions. Suppose that $f = \text{e-}\lim_{i\to\infty} f_i$ and that $\mathrm{dom}\, f, \mathrm{dom}\, f_i \subset E$, $i=1,2,\ldots$ for some compact subset E of X. Then

$$\lim_{i\to\infty} \inf f_i = \inf f, \qquad (5.1.11)$$

and

$$\limsup_{i\to\infty} \mathrm{argmin}\, f_i \subset \mathrm{argmin}\, f. \qquad (5.1.12)$$

Proof. Let $\bar{x}\in\mathrm{argmin}\, f$. Then

$$(\bar{x}, f(\bar{x})) \in \mathrm{epi}\, f \subset \liminf_{i\to\infty} \mathrm{epi}\, f_i,$$

so that there exists $(x_i, r_i)\in\mathrm{epi}\, f_i$ satisfying $\lim_{i\to\infty}(x_i,r_i) = (\bar{x}, f(\bar{x}))$. It follows that

$$\inf f = f(\bar{x}) = \lim_{i \to \infty} r_i \geq \limsup_{i \to \infty} f_i(x_i) \geq \limsup_{i \to \infty} \inf f_i. \qquad (5.1.13)$$

On the other hand, let $x_i \in \operatorname{argmin} f_i \subset E$. Since E is compact there exists a subsequence (i_k) of the natural numbers such that for some $x \in E$, $x = \lim_{k \to \infty} x_{i_k}$ and

$$\lim_{k \to \infty} f_{i_k}(x_{i_k}) = \liminf_{i \to \infty} \inf f_i.$$

Thus,

$$(x, \liminf_{i \to \infty} \inf f_i) \in \limsup_{i \to \infty} \operatorname{epi} f_i \subset \operatorname{epi} f,$$

so that

$$\liminf_{i \to \infty} \inf f_i \geq f(x) \geq \inf f. \qquad (5.1.14)$$

Combining inequalities (5.1.13) and (5.1.14) we have

$$\lim_{i \to \infty} \inf f_i = \inf f.$$

Finally, let $\bar{x} \in \limsup \operatorname{argmin} f_i$ so that there exists a subsequence (i_k) of the natural numbers and $x_{i_k} \in \operatorname{argmin} f_{i_k}$ such that $\bar{x} = \lim_{k \to \infty} x_{i_k}$. Since

$$\limsup_{i \to \infty} \operatorname{epi} f_i \subset \operatorname{epi} f$$

we have $(\bar{x}, \limsup_{k \to \infty} f_{i_k}(x_{i_k})) \in \operatorname{epi} f$ so that

$$\limsup_{k \to \infty} f_{i_k}(x_{i_k}) \geq f(\bar{x}).$$

Now consider any $x \in \operatorname{dom} f$. Then

$$(x, f(x)) \in \operatorname{epi} f \subset \liminf_{i \to \infty} \operatorname{epi} f_i$$

so that there exists a sequence $(y_i, r_i) \in \operatorname{epi} f_i$ converging to $(x, f(x))$. It follows that

$$f(x) = \lim_{i \to \infty} r_i \geq \limsup_{i \to \infty} f_i(y_i)$$
$$\geq \limsup_{k \to \infty} f_{i_k}(x_{i_k})) \geq f(\bar{x}).$$

Since $x \in \operatorname{dom} f$ is arbitrary, $\bar{x} \in \operatorname{argmin} f$.   ●

By carefully examining the proof we can see that the condition that $\operatorname{dom} f$ and $\operatorname{dom} f_i$ are contained in a compact subset E of X is not needed in establishing inclusion (5.1.12). However, without this condition, (5.1.11) is false (Exercise 5.1.13).

5.1.4 Continuity of Multifunctions

The basic definition is given below.

Definition 5.1.15 (Continuity of Multifunction) *Let X and Y be two Hausdorff topological spaces and let $F\colon X \to 2^Y$ be a multifunction. We say that F is upper (lower) semicontinuous at $\bar{x} \in X$ provided that for any open set U in Y with $F(\bar{x}) \subset U$, $(F(\bar{x}) \cap U \neq \emptyset)$,*

$$\{x \in X \mid F(x) \subset U\} \quad (\{x \in X \mid F(x) \cap U \neq \emptyset\})$$

is an open set in X. We say that F is continuous at $\bar{x}$ if it is both upper and lower semicontinuous at $\bar{x}$. We say that F is upper (lower) continuous on $S \subset X$ if it is upper (lower) continuous at every $x \in S$. We omit S when it coincides with the domain of F.

We will also need a sequential approach to limits and continuity of multifunctions. This is mainly for applications in the subdifferential theory because the corresponding topological approach often yields objects that are too big.

Definition 5.1.16 (Sequential Lower and Upper Limits) *Let X and Y be two Hausdorff topological spaces and let $F\colon X \to 2^Y$ be a multifunction. We define the sequential lower and upper limit of F at $\bar{x} \in X$ by*

$$\operatorname*{s-lim\,inf}_{x \to \bar{x}} F(x) := \bigcap \{\liminf_{i \to \infty} F(x_i) \mid x_i \to \bar{x}\}$$

and

$$\operatorname*{s-lim\,sup}_{x \to \bar{x}} F(x) := \bigcup \{\limsup_{i \to \infty} F(x_i) \mid x_i \to \bar{x}\}.$$

When

$$\operatorname*{s-lim\,inf}_{x \to \bar{x}} F(x) = \operatorname*{s-lim\,sup}_{x \to \bar{x}} F(x)$$

we call the common set the sequential limit of F at $\bar{x}$ and denote it by $\text{s-lim}_{x \to \bar{x}} F(x)$.

Definition 5.1.17 (Semicontinuity and Continuity) *Let X and Y be two Hausdorff topological spaces and let $F\colon X \to 2^Y$ be a multifunction. We say that F is sequentially lower (upper) semicontinuous at $\bar{x} \in X$ provided that*

$$F(\bar{x}) \subset \operatorname*{s-lim\,inf}_{x \to \bar{x}} F(x) \ (\operatorname*{s-lim\,sup}_{x \to \bar{x}} F(x) \subset F(\bar{x})).$$

When F is both upper and lower semicontinuous at $\bar{x}$ we say it is continuous at $\bar{x}$. In the notation introduced above,

$$F(\bar{x}) = \operatorname*{s-lim}_{x \to \bar{x}} F(x).$$

Clearly, when Y is a metric space the sequential and the topological (semi) continuity coincide.

The following example illustrates how the semicontinuity and continuity of multifunctions relate to that of functions.

Example 5.1.18 (Profile Mappings) Let X be a Banach space and let $f\colon X \to \mathbb{R}\cup\{+\infty\}$ be a function. Then the epigraphic profile of f, E_f is upper (lower) semicontinuous at $\bar{x}$ if and only if f is lower (upper) semicontinuous at $\bar{x}$. Consequently, E_f is continuous at $\bar{x}$ if and only if f is continuous.

Example 5.1.19 (Sublevel Set Mappings) Let X be a Banach space and let $f\colon X \to \mathbb{R}\cup\{+\infty\}$ be a lsc function. Then the sublevel set mapping $S(a) = f^{-1}((-\infty,a])$ is upper semicontinuous.

When X and Y are metric spaces we have the following characterizations of the sequential lower and upper limit.

Theorem 5.1.20 (Continuity and Distance Functions) *Let X and Y be two metric spaces and let $F\colon X \to 2^Y$ be a multifunction. Then F is sequentially lower (upper) semicontinuous at $\bar{x} \in X$ if and only if for every $y \in Y$, the distance function $x \to d(F(x); y)$ is upper (lower) semicontinuous. Consequently, F is continuous at $\bar{x}$ if and only if for every $y \in Y$, the distance function $x \to d(F(x); y)$ is continuous.*

Proof. This follows from Lemma 5.1.11. Details are left as Exercise 5.1.15.

●

5.1.5 Uscos and Cuscos

The acronym *usco* (*cusco*) stands for a (convex) upper semicontinuous nonempty valued compact multifunction. Such multifunctions are interesting because they describe common features of the maximal monotone operators, of the convex subdifferential and of the Clarke generalized gradient.

Definition 5.1.21 *Let X be a Banach space and let Y be a Hausdorff topological vector space. We say $F\colon X \to 2^Y$ is an usco (cusco) provided that F is a nonempty (convex) compact valued upper semicontinuous multifunction. An usco (cusco) is minimal if it does not properly contain any other usco (cusco).*

A particularly useful case is when $Y = X^*$ with its weak-star topology. In this case we use the terminology weak*-usco (-cusco).

Closed multifunctions and uscos have an intimate relationship.

Proposition 5.1.22 *Let X and Y be two Hausdorff topological spaces and let $F\colon X \to 2^Y$ be a multifunction. Suppose that F is an usco. Then it is closed. If in addition, range F is compact, then F is an usco if and only if F is closed.*

Proof. It is easy to check that if $F: X \to 2^Y$ is an usco, then its graph is closed (Exercise 5.1.16). Now suppose F is closed and range F is compact. Then clearly F is compact valued. We show it is upper semicontinuous. Suppose on the contrary that F is not upper semicontinuous at $\bar{x} \in X$. Then there exists an open set $U \subset Y$ containing $F(\bar{x})$ and a net $x_\alpha \to \bar{x}$ and $y_\alpha \in F(x_\alpha) \backslash U$ for each α. Since range F is compact, we can take subnet (x_β, y_β) of (x_α, y_α) such that $x_\beta \to \bar{x}$ and $y_\beta \to \bar{y} \notin U$. On the other hand it follows from F is closed that $\bar{y} \in F(\bar{x}) \subset U$, a contradiction.   ●

An important feature of an usco (cusco) is that it always contains a minimal one.

Proposition 5.1.23 (Existence of Minimal usco) *Let X and Y be two Hausdorff topological spaces and let $F: X \to 2^Y$ be an usco (cusco). Then there exists a minimal usco (cusco) contained in F.*

Proof. By virtue of of Zorn's lemma we need only show that any decreasing chain (F_α) of usco (cusco) maps contained in F in terms of set inclusion has a minimal element. For $x \in X$ define $F_0(x) = \bigcap F_\alpha(x)$. Since $F_\alpha(x)$ are compact, $F_0(x)$ is nonempty, (convex) and compact. It remains to show that F_0 is upper semicontinuous. Suppose that $x \in X$, U is open in Y and $F_0(x) \subset U$. Then $F_\alpha(x) \subset U$ for some α. Indeed, if each $F_\alpha(x) \backslash U$ were nonempty then the intersection of these compact nested sets would be a nonempty subset of $F_0(x) \backslash U$, a contradiction. By upper semicontinuity of F_α, there exists an open set V containing x such that $F_0(V) \subset F_\alpha(V) \subset U$.   ●

When $Y = \mathbb{R}$ the proposition below provides a procedure of constructing a minimal usco contained in a given usco.

Proposition 5.1.24 *Let X be a Hausdorff topological space and $F: X \to 2^{\mathbb{R}}$ an usco. For each $x \in X$, put $f(x) := \min\{r \mid r \in F(x)\}$. Let $G: X \to 2^{\mathbb{R}}$ be the closure of f (i.e., the set-valued mapping whose graph is the closure of the graph of f). Now put $g(x) := \max\{r \mid r \in G(x)\}$ for each $x \in X$. Finally let $H: X \to 2^{\mathbb{R}}$ be the closure of g. Then H is a minimal usco contained in F.*

Proof. Since the graph of F is closed, G is contained in F, and G is an usco as G is closed and F is an usco. For the same reason H is an usco contained in G.

To show that H is minimal, consider open sets $U \subset X$ and $W \subset \mathbb{R}$, such that there is some $w \in H(U) \cap W$. It is sufficient to find a nonempty open subset of U, whose image under H is entirely contained in W.

Fix some $\varepsilon < d(\mathbb{R} \backslash W; w)$. Since $w \in H(U)$, there is some $x \in U$ such that $g(x) \in (w - \varepsilon; w + \varepsilon)$. This means that $G(x) \subset (-\infty; w + \varepsilon)$ and by upper semi-continuity of G there is an open $V \subset U$, $V \ni x$, such that $G(V) \subset (-\infty; w + \varepsilon)$.

As $g(x) \in (w - \varepsilon, w + \varepsilon)$, there is some $x' \in V$ with $f(x') \in (w - \varepsilon, w + \varepsilon)$. This means that $F(x') \subset (w - \varepsilon, +\infty)$ and by upper semi-continuity of F there is an open $V' \subset V$, $V' \ni x'$, such that $F(V') \subset (w - \varepsilon, +\infty)$.

Now $H(V') \subset F(V') \cap G(V) \subset (w - \varepsilon, w + \varepsilon) \subset W$. Thus H is a minimal usco. $\bullet$

Maximal monotone operators, in particular, subdifferentials of convex functions provide interesting examples of w*-cuscos. We leave the verification of the following example as a guided exercise (Exercise 5.1.17).

Example 5.1.25 Let X be a Banach space, let $F \colon X \to 2^{X^*}$ be a maximal monotone multifunction and let S be an open subset of $\mathrm{dom}\, F$. Then the restriction of F to S is a w*-cusco.

To further explore the relationship of maximal monotone multifunctions and cuscos we need to extend the notion of maximal monotone multifunctions to arbitrary set.

Definition 5.1.26 (Maximal Monotone on a Set) *Let X be a Banach space, let $F \colon X \to 2^{X^*}$ be a monotone multifunction and let S be a subset of X. We say that F is maximal monotone in S provided the monotone set*

$$\mathrm{graph}\, F \cap (S \times X^*) := \{(x, x^*) \in S \times X^* \mid x \in S \text{ and } x^* \in F(x)\}$$

is maximal under the set inclusion in the family of all monotone sets contained in $S \times X^*$.

It turns out that a monotone cusco on an open set is maximal.

Lemma 5.1.27 *Let X be a Banach space, let $F \colon X \to 2^{X^*}$ be a monotone multifunction and let S be an open subset of X. Suppose that $S \subset \mathrm{dom}\, F$ and F is a w*-cusco on S. Then F is maximal monotone in S.*

Proof. We need only show that if $(y, y^*) \in S \times X^*$ satisfies

$$\langle y^* - x^*, y - x \rangle \geq 0 \text{ for all } x \in S, x^* \in F(x), \qquad (5.1.15)$$

then $y^* \in F(y)$. If not, by the separation theorem there exists $z \in X \backslash \{0\}$ such that $F(y) \subset \{z^* \in X^* \mid \langle z^*, z \rangle < \langle y^*, z \rangle\} = W$. Since W is weak* open and F is w*-upper semicontinuous on S, there exists an $h > 0$ with $B_h(y) \subset S$ such that $F(B_h(y)) \subset W$. Now, for $t \in (0, h/\|z\|)$, we have $y + tz \in B_h(y)$, and therefore $F(y + tz) \subset W$. Applying (5.1.15) to any $u^* \in F(y + tz)$ we get

$$0 \leq \langle y^* - u^*, y - (y + tz) \rangle = -t \langle y^* - u^*, z \rangle,$$

which implies $\langle u^*, z \rangle \geq \langle y^*, z \rangle$, that is $u^* \notin W$, a contradiction. $\bullet$

As a corollary we have

Corollary 5.1.28 *Let X be a Banach space, let $F\colon X \to 2^{X^*}$ be a maximal monotone multifunction and let S be an open subset of X. Suppose that $S \subset \mathrm{dom}\, F$. Then F is maximal monotone in S.*

Proof. By Example 5.1.25 the maximal monotonicity of F implies that F is a w*-cusco on S, so the result follows from Lemma 5.1.27.   ●

Now we can prove the interesting relation that a maximal monotone multifunction on an open set is a minimal cusco.

Theorem 5.1.29 (Maximal Monotonicity and Minimal cusco) *Let X be a Banach space, let S be an open subset of X and let F be a maximal monotone multifunction in S. Then F is a minimal w*-cusco.*

Proof. We know by Example 5.1.25 that F is a w*-cusco. Suppose that $G\colon S \to 2^{X^*}$ is a w*-cusco and $\mathrm{graph}\, G \subset \mathrm{graph}\, F$. By Lemma 5.1.27, G is maximal monotone, and therefore $G = F$.   ●

Note that a maximal monotone multifunction need not be a minimal usco. The following example clarifies the difference whose easy proof is left as Exercise 5.1.18.

Example 5.1.30 Define monotone multifunctions F_0, F_1 and F_2 from $\mathbb{R} \to 2^{\mathbb{R}}$ by
$$F_0(x) = F_1(x) = F_2(x) = \mathrm{sgn}\, x \text{ if } x \neq 0,$$
while
$$F_0(0) = \{-1\}, F_1(0) = \{-1, 1\} \text{ and } F_2(0) = [-1, 1].$$
Then $\mathrm{graph}\, F_0 \subset \mathrm{graph}\, F_1 \subset \mathrm{graph}\, F_2$, and they are all distinct. The multifunction F_2 is maximal monotone and minimal cusco, F_1 is minimal usco and F_0 does not have a closed graph.

5.1.6 Monotone Operators and the Fitzpatrick Function

Throughout this subsection, $(X, \|\cdot\|)$ is a reflexive Banach space with dual X^* and $T\colon X \to 2^{X^*}$ is maximal monotone. The *Fitzpatrick function F_T*, associated with T, is the proper closed convex function defined on $X \times X^*$ by
$$F_T(x, x^*) := \sup_{y^* \in Ty} [\langle y^*, x \rangle + \langle x^*, y \rangle - \langle y^*, y \rangle]$$
$$= \langle x^*, x \rangle + \sup_{y^* \in Ty} \langle x^* - y^*, y - x \rangle.$$
Since T is maximal monotone
$$\sup_{y^* \in Ty} \langle x^* - y^*, y - x \rangle \geq 0$$

and the equality holds if and only if $x^* \in Tx$, it follows that

$$F_T(x, x^*) \geq \langle x^*, x \rangle \tag{5.1.16}$$

with equality holding if and only if $x^* \in Tx$. Thus, we capture much of a maximal monotone multifunction via an associated convex function.

Using only the Fitzpatrick function and the decoupling lemma we can prove the following fundamental result remarkably easily.

Theorem 5.1.31 (Rockafellar) *Let X be a reflexive Banach space and let $T: X \to 2^{X^*}$ be a maximal monotone operator. Then* range$(T + J) = X^*$. *Here J is the* duality map *defined by $J(x) := \partial \|x\|^2/2$.*

Proof. The Cauchy inequality and (5.1.16) implies that for all x, x^*,

$$F_T(x, x^*) + \frac{\|x\|^2 + \|x^*\|^2}{2} \geq 0. \tag{5.1.17}$$

Applying the decoupling result of Lemma 4.3.1 to (5.1.17) we conclude that there exists a point $(w^*, w) \in X^* \times X$ such that

$$0 \leq F_T(x, x^*) - \langle w^*, x \rangle - \langle x^*, w \rangle$$
$$+ \frac{\|y\|^2 + \|y^*\|^2}{2} + \langle w^*, y \rangle + \langle y^*, w \rangle \tag{5.1.18}$$

Choosing $y \in -Jw^*$ and $y^* \in -Jw$ in inequality (5.1.18) we have

$$F_T(x, x^*) - \langle w^*, x \rangle - \langle x^*, w \rangle \geq \frac{\|w\|^2 + \|w^*\|^2}{2}. \tag{5.1.19}$$

For any $x^* \in Tx$, adding $\langle w^*, w \rangle$ to both sides of the above inequality and noticing $F_T(x, x^*) = \langle x^*, x \rangle$ we obtain

$$\langle x^* - w^*, x - w \rangle \geq \frac{\|w\|^2 + \|w^*\|^2}{2} + \langle w^*, w \rangle \geq 0. \tag{5.1.20}$$

Since (5.1.20) holds for all $x^* \in Tx$ and T is maximal we must have $w^* \in Tw$. Now setting $x^* = w^*$ and $x = w$ in (5.1.20) yields

$$\frac{\|w\|^2 + \|w^*\|^2}{2} + \langle w^*, w \rangle = 0,$$

which implies $-w^* \in Jw$. Thus, $0 \in (T + J)w$. Since the argument applies equally well to all translations of T, we have range$(T + J) = X^*$ as required. ●

There is a tight relationship between nonexpansive mappings and monotone operators in Hilbert spaces, as stated in the next lemma.

Lemma 5.1.32 *Let H be a Hilbert space. Suppose that P and T are two multifunctions from subsets of H to 2^H whose graphs are related by the condition $(x, y) \in \operatorname{graph} P$ if and only if $(v, w) \in \operatorname{graph} T$ where $x = w + v$ and $y = w - v$. Then*

(i) *P is nonexpansive (and single-valued) if and only if T is monotone.*
(ii) *$\operatorname{dom} P = \operatorname{range}(T + I)$.*

Proof. Exercise 5.1.29. ●

This very easily leads to the Kirszbraun–Valentine theorem [161, 254] on the existence of nonexpansive extensions to all of Hilbert space of nonexpansive mappings on subsets of Hilbert space. The proof is left as a guided exercise.

Theorem 5.1.33 (Kirszbraun–Valentine) *Let H be a Hilbert space and let D be a non-empty subset of H. Suppose that $P\colon D \to H$ is a nonexpansive mapping. Then there exists a nonexpansive mapping $\widehat{P}\colon H \to H$ defined on all of H such that $\widehat{P}|_D = P$.*

Proof. Exercise 5.1.30. ●

Alternatively [226], one may directly associate a convex Fitzpatrick function F_P with a non-expansive mapping P, and thereby derive the Kirszbraun–Valentine theorem, see Exercise 5.1.31.

5.1.7 Commentary and Exercises

Multifunctions or set-valued functions have wide applications and have been the subject of intensive research in the past several decades. Our purpose in this short section is merely to provide minimal preliminaries and some interesting examples. Aubin and Frankowska's monograph [8] and Klein and Thompson's book [162] are excellent references for readers who are interested in this subject.

The subdifferential for convex functions is the first generalized differential concept that leads to a multifunction. It has many nice properties later generalized to the classes of usco and cusco multifunctions. The usco and cusco also relate to other concepts of generalized derivative such as the Clarke generalized gradient. Our discussion on usco and cusco here largely follows those in [56, 70, 221].

Maximal monotone operators are generalizations of the convex subdifferential—though they first flourished in partial differential equation theory. Rockafellar's result in Theorem 5.1.31 is in [230]. The original proofs were very extended and quite sophisticated—they used tools such as Brouwer's fixed point theorem and Banach space renorming theory. As with the proof of the local boundedness of Theorem 5.1.8, ultimately the result is reduced to

much more accessible geometric convex analysis. These proofs well illustrate
the techniques of variational analysis: using a properly constructed auxiliary
function, the variational principle with decoupling in the form of a sandwich
theorem and followed by an appropriate decoding of the information. Simon
Fitzpatrick played a crucial role in this process by constructing the auxiliary
functions. The proof of Theorem 5.1.8 follows [40]. The short proof of Theorem
5.1.31 is a reworking of that of [241] given in [67] using the Fitzpatrick function
discovered in [120]. The technique in the proof of Theorem 5.1.31 becomes
much more powerful when we view the Cauchy inequality as a special case
of the Fenchel–Young inequality for a general convex function. A beautiful
application is the proof of maximality of the sum of two maximal monotone
operators in reflexive spaces [32, 33, 240] (see also guided Exercises 5.1.44,
5.1.45 and 5.1.46).

Exercise 5.1.1 Let F be a multifunction from X to Y.

(i) Show that $\operatorname{dom} F = \operatorname{range} F^{-1}$ and $\operatorname{range} F = \operatorname{dom} F^{-1}$.
(ii) Show that $F(x) = \{y \in Y \mid (x,y) \in \operatorname{graph} F\}$ and $F^{-1}(y) = \{x \in X \mid (x,y) \in \operatorname{graph} F\}$.

Exercise 5.1.2 Let X and Y be Hausdorff topological spaces and let $F \colon X \to 2^Y$ be a multifunction.

(i) Show that if F is closed then it is closed valued.
(ii) Construct a closed valued multifunction whose graph is not closed.

Exercise 5.1.3 Let X be a metric space and let $f \colon X \to \mathbb{R} \cup \{+\infty\}$ be a lsc function. Show that $\operatorname{graph} E_f = \operatorname{epi} f$.

Exercise 5.1.4 Let X be a Banach space and let $f \colon X \to \mathbb{R} \cup \{+\infty\}$ be a function. Suppose that, for any $x^* \in X^*$, $x \to f(x) + \langle x^*, x \rangle$ is quasi-convex. Show that f is a convex function. Hint: Choose x^* such that $f(x) + \langle x^*, x \rangle = f(y) + \langle x^*, y \rangle$.

Exercise 5.1.5 Let X be a Banach space and let $F \colon X \to 2^{X^*}$ be a maximal monotone multifunction. Show that F is convex valued and closed.

Exercise 5.1.6 Prove the representations of the lower and upper limits of sequence of subsets in (5.1.3) and (5.1.4).

Exercise 5.1.7 Prove Lemma 5.1.10.

Exercise 5.1.8 Prove that in a metric space the sequential lower and upper limits of a sequence of subsets are always closed sets. Give an example showing that this is not the case in a general Hausdorff topological space.

Exercise 5.1.9 Prove the equivalence of (5.1.7) and (5.1.8) in Lemma 5.1.11.

Exercise 5.1.10 (Limits of Monotone and Sandwiched Sequences) Let (F_i) be a sequence in a metric space Y.

(i)   Suppose that (F_i) is monotone increasing, i.e, $F_i \subset F_{i+1}$ for $i = 1, 2, \ldots$. Then $\lim_{i \to \infty} F_i = \text{cl} \bigcup_{i=1}^{\infty} F_i$.

(ii) Suppose that (F_i) is monotone decreasing, i.e., $F_{i+1} \subset F_i$ for $i = 1, 2, \ldots$. Then $\lim_{i \to \infty} F_i = \bigcap_{i=1}^{\infty} \text{cl} \, F_i$.

(iii) Suppose that $F_i \subset G_i \subset H_i$ and $\lim_{i \to \infty} F_i = \lim_{i \to \infty} H_i = G$. Then $\lim_{i \to \infty} G_i = G$.

Exercise 5.1.11 (Lower Semicontinuity of Epi-limits) Let X be a metric space and let $f_i \colon X \to \mathbb{R}$ be a sequence of lsc functions. Then both e-$\liminf_{i \to \infty} f_i$ and e-$\limsup_{i \to \infty} f_i$ are lsc functions. Therefore, e-$\lim_{i \to \infty} f_i$ is a lsc function when exists.

Exercise 5.1.12 (Characterization of Epi-limits) Prove Lemma 5.1.13.

Exercise 5.1.13 Construct an example on $X = \mathbb{R}$ showing that without the condition that dom f and dom f_i belong to a compact subset of X, the conclusion (5.1.11) in Theorem 5.1.14 is false.

Exercise 5.1.14 Prove the claim in Example 5.1.18.

Exercise 5.1.15 Prove Theorem 5.1.20.

Exercise 5.1.16 Let $F \colon X \to 2^Y$ be an usco. Show that graph F is a closed subset of $X \times Y$.

Exercise 5.1.17 Verify Example 5.1.25. Hint: By Exercise 5.1.5 F is convex valued and closed. The upper semicontinuity of F follows from Theorem 5.1.8 and Proposition 5.1.22.

Exercise 5.1.18 Verify the claims in Example 5.1.30.

Exercise 5.1.19 Construct a multifunction F from $\mathbb{R}$ to $\mathbb{R}^2$ whose projections into $\mathbb{R}$ are both minimal usco mappings yet F itself is not. Hint: Let $F(x) = \{(\text{sgn}(x); \text{sgn}(x))\}$ for $z \neq 0$, while

$$F(0) = \{(-1; -1), (-1; 1), (1; -1), (1; 1)\}.$$

Exercise 5.1.20 Construct a minimal usco contained in a given usco $F \colon Z \to \mathbb{R}^N$.

Exercise 5.1.21 Deduce that every maximal monotone mapping on a reflexive space which is *coercive* (in the sense that $\inf_{x^* \in Tx} \langle x^*, x \rangle / \|x\| \to \infty$ with $\|x\| \to \infty$) is surjective, by considering the sequence $(T + \frac{1}{i} J)$. Hint: It helps to know that maximal monotone operators (and so their inverses) are sequentially *demi-closed*, that is $x_i \to_* x, y_i \to y, y_i \in Tx_i$ implies $y \in Tx$. This is neatly proved via the Fitzpatrick function.

In a non-reflexive space this fails badly. Indeed the existence of surjective, coercive subgradient mappings forces the space to be reflexive, [121].

Exercise 5.1.22 Show in finite dimensions that a single-valued surjective monotone mapping is *weakly coercive*, meaning that $\|Tx\| \to \infty$ when $\|x\| \to \infty$.

Exercise 5.1.23 Compute the Fitzpatrick function of T when T is a linear maximal monotone mapping.

Exercise 5.1.24 Compute the Fitzpatrick function of $T + S$ when T is maximal monotone and S is a skew bounded linear mapping.

***Exercise 5.1.25** Suppose T is maximal monotone and *skew* – that is, both T and $-T$ are monotone on X. Suppose, on translating if need be that $0 \in T(0)$ and $\mathrm{dom}(T)$ is a dense absorbing set.

Show that in any Banach space, a maximal monotone skew mapping whose domain has non-empty interior extends to a bounded skew affine mapping on the whole space. Hint: Show that $T(x) \subset K(x) := \{x^* \mid F_T(x, x^*) \le 0\}$, so that K is a convex multifunction. Now check that $K(0) = \{0\}$. Deduce that K is single valued, and therefore $T(x) = K(x)$ on $\mathrm{dom}(T)$.

Exercise 5.1.26 Supposing T is maximal monotone and skew, show that $\mathrm{dom}(T)$ is affine.

***Exercise 5.1.27** Determine when a C^1 monotone mapping, T, whose domain is open, can be written as $T = f' + S$ where f is a twice differentiable convex function and S is a skew and bounded linear mapping. Hint: (i) the gradient of T is a linear monotone mapping, and so can be written as $P(x) + S(x)$ where P is positive semi-definite, and (ii) the skew monotone part is linear by Exercise 5.1.25. It remains to determine when P is a Hessian.

***Exercise 5.1.28** Monotone mappings such that $T + J$ is surjective are called *hypermonotone*. Show that a hypermonotone mapping on a reflexive space is maximal monotone as soon as J and J^{-1} are both injective, but not necessarily more generally. In Hilbert space this result is due to Minty [194]. Deduce that T is hypermonotone as soon as $T + \alpha J$ is surjective for some $\alpha > 0$.

Exercise 5.1.29 Prove Lemma 5.1.32.

Exercise 5.1.30 Prove Theorem 5.1.33 as follows:

(i)   Associate P to a monotone function T as in Lemma 5.1.32.
(ii) Extend T to a maximal monotone multifunction $\widehat{T}$.
(iii) Define $\widehat{P}$ from $\widehat{T}$ using Lemma 5.1.32 and use Rockafellar's theorem to assert $\mathrm{dom}(\widehat{P}) = \mathrm{range}(\widehat{T} + I) = H$.
(iv) Check that $\widehat{P}$ is indeed an extension of P.

***Exercise 5.1.31** Use Lemma 5.1.32 to explicitly define a convex Fitzpatrick function associated with a nonexpansive mapping, and determine its properties.

Exercise 5.1.32 Let H be a Hilbert space and let $T\colon H \to 2^H$ be a monotone multifunction. Show that $Q := (I + T^{-1})^{-1}$ is nonexpansive. Moreover, if T is maximal monotone then $\operatorname{dom} Q = H$. Hint: $\operatorname{dom} Q = \operatorname{range}(I + T^{-1})$.

Exercise 5.1.33 (Resolvents) Let H be a Hilbert space with $T\colon H \to 2^H$ a maximal monotone multifunction. For $\lambda > 0$, show that the *resolvent* $R_\lambda := (I + \lambda T)^{-1}$ is everywhere defined, with range in the domain of T and non-expansive. Deduce that the *Yosida approximate* $T_\lambda(x) := TR_\lambda$ is an everywhere defined, $(1/\lambda)$-Lipschitz and maximal monotone mapping.

Show for x in the domain of T that $T_\lambda(x)$ converges to the minimal norm member of Tx. What happens when Tx is empty?

Non-expansivity is very definitely a Hilbert space property, but the Yosida approximate remains useful more generally (as in the next exercise) [96]. Hint: Supposing $x^* \in Tx$ and $x_i^* \in T_\lambda(x_i)$ we have $\langle x_i^* - x^*, x_i \rangle \le 0$. Thus $\limsup_{i \to \infty} \|x_i\| = \inf \|Tx\|$. Now use demi-closure.

Exercise 5.1.34 For a maximal monotone operator T in Hilbert space, show that $T_\lambda(x) = \left(T^{-1} + \lambda I\right)^{-1}(x)$ for all x in the space. Hint: for each x the righthand side is nonempty and a subset of the left.

*Exercise 5.1.35** Show that the domain—and hence range—of a maximal monotone operator on a reflexive space is *semi-convex* – that is, has a convex closure. It is unknown whether this holds in arbitrary Banach space [240], but see Exercise 5.1.41. Hint: Without loss assume 0 is in the closure of $\operatorname{conv} \operatorname{dom}(T)$. Fix $y \in \operatorname{dom}(T)$, $y^* \in T(y)$, and use inequality (5.1.19) for T/i to write

$$F_{T/i}(y, y^*/i) - \langle w_i^*, y \rangle - \frac{1}{i}\langle w_i, y^* \rangle \ge \|w_i\|^2,$$

where $\|w_i\| = \|w_i^*\|$ and $w_i \in \operatorname{dom}(T)$. Since $F_{T/i}(y, y^*/i) = \langle y, y* \rangle/i \to 0$ as $i \to \infty$ we deduce that $\sup_i \|w_i\| < \infty$. Thus (w_i^*) has a weak cluster point w^*. In particular,

$$d^2_{\operatorname{dom}(T)}(0) \le \liminf_i \|w_i\|^2 \le \inf_{y \in \operatorname{dom}(T)} \langle -w^*, y \rangle$$

$$= \inf_{y \in \operatorname{conv} \operatorname{dom}(T)} \langle -w^*, y \rangle \le \|w^*\| \, d_{\operatorname{conv} \operatorname{dom}(T)}(0) = 0.$$

We have actually shown that $\operatorname{cl} \operatorname{conv} \operatorname{dom}(T) \subset \operatorname{cl} \operatorname{dom}(T)$ and so $\operatorname{cl} \operatorname{dom}(T)$ is convex as required. What does this proof technique allow you to deduce in a non-reflexive Banach space?

*Exercise 5.1.36** (Maximality of the Sum) Let T and U be maximal monotone operators on a Hilbert space, H, and let $\lambda > 0$ be given.

(i)   Show that $\operatorname{range}(T_\lambda + U + \mu I) = H$, for $\mu > 1/\lambda$.
(ii) Deduce that $T_\lambda + U$ is maximal monotone.
(iii) Show that $T + U$ is maximal monotone when $\operatorname{dom}(U) \cap \operatorname{int}(\operatorname{dom} T) \ne \emptyset$.

Hint: (i) For any $y \in H$, the mapping

$$x \mapsto (U + \mu I)^{-1}[y - T_\lambda(x)]$$

is a Banach contraction. (iii) We may suppose $0 \in T(0) \cap U(0)$ and that 0 is interior to the domain of T. Let $\lambda_i \downarrow 0$. Note that $0 \in T_{\lambda_i}(0)$. Show that the solutions $t_i \in T_{\lambda_i}(x_i), u_i \in U(x_i)$ with $y = t_i + u_i + x_i$ yield a Cauchy sequence (x_i) as follows:

$$\langle x_i - x_j, x_i - x_j \rangle \leq -\langle t_i - t_j, \lambda_i t_i - \lambda_j t_j \rangle \leq 2(\lambda_i + \lambda_j) \sup \|t_k\|^2.$$

Use monotonicity and the fact that the domains intersect to show $\|x_i\| \leq \|y\|$. Now use the interiority hypothesis and the consequent local boundedness at 0 of the monotone operator T to show (t_i) remains bounded and also has a weakly convergent subsequence. Conclude that (x_i) converges in norm.

Finish by taking limits and using demi-closedness.

Note that everything has been reduced to Rockafellar's theorem and so to the Hahn–Banach theorem. An extension of this proof will work in arbitrary reflexive space, but step (i) must be replaced by a finite dimensional approximation argument.

Exercise 5.1.37 Show that for a closed convex set C in a Banach space and $\lambda > 0$ one has
$$(\partial \iota_C)_{2\lambda} = \partial \iota_C \,\square\, \lambda \|\cdot\|^2 = \lambda \, d_C^2(x).$$

∗**Exercise 5.1.38** (Monotone Variational Inequalities) Let T be a maximal monotone operator on a Banach space and let C be a closed convex subset of X.

(i) Show that the solution of the monotone variational inequality:

$$\mathrm{VI}\,(T, C) \quad \begin{cases} \text{there exist } x \in C \text{ and } t^* \in T(x) \\ \text{such that } \langle t^*, c - x \rangle \geq 0 \text{ for all } c \in C \end{cases}$$

is equivalent to the monotone inclusion

$$0 \in (T + \partial \iota_C)(x).$$

(ii) In particular, if T is coercive on C and the sum $T + \partial \iota_C$ is maximal monotone for which Exercise 5.1.36 gives conditions, then the variational inequality has a solution.

(iii) Specialize this to the cases when T is coercive and (a) $C = i B_X$, as $i \to \infty$, or (b) C is a closed convex cone – a so-called complementarity problem.

(iv) Consider two monotone operators T and U on X and Y respectively. Show that $M(x, y) := (Tx, Uy)$ is monotone on $X \times Y$ and is maximal if and only if both T and U are. Denote the diagonal convex set by $\Delta := \{(x, y) \in X \times Y \mid x = y\}$. Check that $0 \in \text{range}(T + U)$ if and only if $\mathrm{VI}\,(M, \Delta)$ has solution.

*Exercise 5.1.39 (Transversality I) Let T be maximal monotone operator on a Hilbert space, H, and let C be a non-empty closed convex subset of H.

(i) Show that when T is coercive on C the condition

$$0 \in \text{core}\,[\text{dom}(T) - C] \qquad (5.1.21)$$

implies VI (T, C) has a solution.

(ii) This remains true in a reflexive Banach space.

Hint: By Exercise 5.1.36, VI $(T_{1/i}, C)$ has a solution:

$$x_i \in C, \; t_i \in T(x_i - \frac{1}{i} t_i), \qquad \inf_{c \in C} \langle t_i, c - x \rangle \geq 0.$$

Condition (5.1.21) and the Baire category theorem imply that for some $N > 0$ one has $0 \in \text{cl}[T^{-1}(NB_H) - C \cap NB_H]$. This and coercivity of T suffice to show, much as in Exercise 5.1.36, that (x_i) and (t_i) remain bounded as i goes to infinity. Thence, (x_i) is norm convergent and one may to move to the limit.

*Exercise 5.1.40 (Transversality II) Let T and U be maximal monotone operators on a Hilbert space.

(i) Use Exercises 5.1.38 and 5.1.39 to show that

$$0 \in \text{core}[\text{dom}(T) - \text{dom}(U)]$$

implies $T + U$ is maximal monotone.

(ii) This remains true in a reflexive Banach space.

*Exercise 5.1.41 (Ranges) (i) Prove that a Banach space X is reflexive if the interior of range(∂f) is convex for each strongly coercive continuous convex function f on X. Hint: Suppose X is nonreflexive and $p \in X$ with $\|p\| = 5$ and $p^* \in Jp$ where J is the duality map. Define

$$f(x) := \max\left\{ \frac{1}{2}\|x\|^2, \|x - p\| - 12 + \langle p^*, x \rangle, \|x + p\| - 12 - \langle p^*, x \rangle \right\}$$

for $x \in X$. By the max-formula, we have, for $x \in B_X$,

$$\partial f(p) = B_{X^*} + p^*, \quad \partial f(-p) = B_{X^*} - p^*, \quad \partial f(x) = Jx \quad (5.1.22)$$

using inequalities like $\|p - p\| - 12 + \langle p^*, p \rangle = 13 > \frac{25}{2} = \frac{1}{2}\|p\|^2$.

Moreover, $f(0) = 0$ and $f(x) > \frac{1}{2}\|x\|$ for $\|x\| > 1$, thus $\|x^*\| > \frac{1}{2}$ if $x^* \in \partial f(x)$ and $\|x\| > 1$. Combining this with (5.1.22) shows

$$\text{range}(\partial f) \cap \frac{1}{2} B_{X^*} = \text{range}(J) \cap \frac{1}{2} B_{X^*}.$$

Let U_{X^*} denote the open unit ball in X^*. Now James' theorem gives us points $x^* \in \frac{1}{2} U_{X^*} \setminus \text{range}(J)$, thus $U_{X^*} \setminus \text{range}(\partial f) \neq \emptyset$. However, from (5.1.22)

$$U_{X^*} \subset \mathrm{conv}((p^* + U_{X^*}) \cup (-p^* + U_{X^*})) \subset \mathrm{conv}\,\mathrm{int}\,\mathrm{range}(\partial f)$$

so range(∂f) has non-convex interior.

(ii) Deduce the following:

Theorem 5.1.34 *A normed linear space X is reflexive if and only if every continuous convex function f on X has int range(∂f) convex.*

(iii) Observe that the easiest explicit example lies in the space c_0 of null sequences endowed with the supremum norm. One may use

$$f(x) := \|x - e_1\|_\infty + \|x + e_1\|_\infty \tag{5.1.23}$$

where e_1 is first unit vector. Then

$$\mathrm{int}\,\mathrm{range}(\partial f) = \{U_{\ell_1} + e_1\} \cup \{U_{\ell_1} - e_1\}$$

$$\mathrm{cl}\,\mathrm{int}\,\mathrm{range}(\partial f) = \{B_{\ell_1} + e_1\} \cup \{B_{\ell_1} - e_1\}$$

both of which are far from convex.

(iv) Compute the closure of the range of the subgradient.

*Exercise 5.1.42** Let X be a Banach space and let $S, T : X \to 2^{X^*}$ be monotone operators. Suppose that

$$0 \in \mathrm{core}[\mathrm{conv}\,\mathrm{dom}(T) - \mathrm{conv}\,\mathrm{dom}(S)].$$

Prove that there exist $r, c > 0$ such that, for any $x \in \mathrm{dom}(T) \cap \mathrm{dom}(S)$ and $t^* \in T(x)$ and $s^* \in S(x)$,

$$\max(\|t^*\|, \|s^*\|) \le c\,(r + \|x\|)(r + \|t^* + s^*\|).$$

Hint: Consider the convex lower semi-continuous function

$$\sigma_T(x) := \sup_{z^* \in T(z)} \frac{\langle z^*, x - z\rangle}{1 + \|z\|}.$$

This is a refinement of the function we used to prove local boundedness of monotone operators. First show that (i) conv dom(T) $\subset$ dom σ_T, and that (ii) $\bigcup_{i=1}^\infty [\{x \mid \sigma_S(x) \le i, \|x\| \le i\} - \{x \mid \sigma_T(x) \le i, \|x\| \le i\}] = X$. Reference: [258].

*Exercise 5.1.43** Let X be a Banach space and let $S, T : X \to 2^{X^*}$ be maximal monotone operators. Suppose that

$$0 \in \mathrm{core}[\mathrm{conv}\,\mathrm{dom}(T) - \mathrm{conv}\,\mathrm{dom}(S)].$$

Prove that, for any $x \in \mathrm{dom}(T) \cap \mathrm{dom}(S)$, $T(x) + S(x)$ is a w^*-closed subset of X^*.

Hint: In view of the Krein-Smulian theorem it is enough to prove that every bounded w^*-convergent net in $T(x) + S(x)$ has its limits in $T(x) + S(x)$. This can be done by using the estimate in Exercise 5.1.42. Reference: [258].

*Exercise 5.1.44 Let X be a reflexive Banach space. Prove that a monotone mapping $T : X \to 2^X$ is maximal if and only if the mapping $T(\cdot + x) + J$ is surjective for all x in X. References: [33, 240].

*Exercise 5.1.45 Prove the following theorem.

Theorem 5.1.35 *Let X be a reflexive space, let T be maximal and let f be closed and convex. Suppose that*

$$0 \in \text{core}\{\text{conv dom}(T) - \text{conv dom }\partial(f)\}.$$

Then

(a) $\partial f + T + J$ *is surjective.*
(b) $\partial f + T$ *is maximal monotone.*
(c) ∂f *is maximal monotone.*

Hint: Consider the Fitzpatrick function $F_T(x, x^*)$ and further introduce $f_J(x) := f(x) + 1/2\|x\|^2$. Let $G(x, x^*) := -f_J(x) - f_J^*(-x^*)$. Observe that

$$F_T(x, x^*) \geq \langle x, x^* \rangle \geq G(x, x^*)$$

pointwise thanks to the *Fenchel-Young inequality*. Now apply the decoupling result in Lemma 4.3.1 and Exercise 5.1.44.

*Exercise 5.1.46 Deduce the following result in [240] as a corollary of Theorem 5.1.35.

Theorem 5.1.36 *Let X be a reflexive Banach space, let $T, S : X \to 2^X$ be maximal monotone operators. Suppose that*

$$0 \in \text{core}[\text{conv dom}(T) - \text{conv dom}(S)].$$

Then $T + S$ is maximal monotone.

Hint: Apply Theorem 5.1.35 to $T(x, y) := (T_1(x), T_2(y))$ and the indicator function $f(x, y) = \iota_{\{x'=y'\}}(x, y)$.

Exercise 5.1.47 (Gossez' Example [130]) Define $T_N := 2^{-N} J_{\ell^1} - S$ for $N = 1, 2, \ldots$ where $S : \ell_1 \to \ell_\infty$ is a continuous linear map given by

$$(Sx)_n := -\sum_{k<n} x_k + \sum_{k>n} x_k, \quad \text{for all } x = (x_k) \in \ell_1, n = 1, 2, \ldots.$$

Show that T_N is a coercive maximal monotone operator with full domain whose range for N large has a non-convex closure.
Hint: We record that $S : \ell_1 \mapsto \ell_\infty$ is a skew bounded linear operator, for which S^* is not monotone but $-S^*$ is. Moreover, $e := (1, 1, \ldots, 1, \ldots) \notin \text{cl range}(S)$. [12, 15]. To see that $\Gamma_N := \text{cl range}(T_N)$ is not convex, first note that Γ_N is homogeneous. Hence, it is impossible that $\text{cl range}(T_N) = \ell^\infty$ for infinitely

many $N > 0$ as this forces $e \in \mathrm{cl}\,\mathrm{range}(S)$. Thus, if convex, Γ_N must eventually be a norm closed proper subspace in ℓ^∞. Fix such an N with Γ_N proper. There then is some $0 \neq \mu \in (\ell^\infty)^*$ with

$$2^{-N}\langle x, S^*\mu \rangle = 2^{-N}\langle Sx, \mu \rangle = \langle J(x), \mu \rangle$$

for all $x \in \ell^1$. By considering the image of $\{te_1 - te_2 \mid t > 0\}$, we may derive that for all $\overline{m} \in \ell^\infty$, the set Γ_N contains points of the form $(y_1, y_2, \overline{m})$, and so for all $z \in \ell^\infty$ we have $\langle z, \mu \rangle = \langle P(z), \mu \rangle$ where $P(z) := (z_1, z_2, \overline{0})$. Thus, $\mu = P^*(\mu) \in \ell^1$. Select a bounded net $x_a \to^* \mu, x_a \in \ell_1, \|x_a\|_1 \leq \|\mu\|_1$, so $x_a \to \mu$ in ℓ_1, by the Kadec property. Thus,

$$0 \geq 2^{-N}\langle \mu, S^*\mu \rangle = 2^{-N} \lim_{a \to \infty} \langle x_a, S^*\mu \rangle = \liminf_{a \to \infty} \langle J(x_a), \mu \rangle$$
$$= \langle J(\mu), \mu \rangle = \|\mu\|^2 > 0,$$

which is a contradiction.

5.2 Subdifferentials as Multifunctions

5.2.1 Clarke's Generalized Gradient

We define the Clarke subdifferential and related normal cone concepts following Clarke's original three-step approach. First, we define the Clarke subdifferential for Lipschitz functions by using the Clarke directional derivative. Then, we generate the corresponding normal cone with the Clarke subdifferential for distance functions. Finally, we define the Clarke subdifferential for a lsc function f using the normal cone to the epigraph of f.

Definition 5.2.1 (Clarke Directional Derivative) *Let X be a Banach space and let $f \colon X \to \mathbb{R}$ be a locally Lipschitz function. We define the Clarke directional derivative of f at $\bar{x}$ in the direction h by*

$$f^\circ(\bar{x}; h) := \limsup_{t \to 0+,\, y \to x} \frac{f(y + th) - f(y)}{t}.$$

The following proposition is easy to check.

Proposition 5.2.2 *Let X be a Banach space and let $f \colon X \to \mathbb{R}$ be Lipschitz with a Lipschitz constant L near $\bar{x}$. Then the function $h \to f^\circ(\bar{x}; h)$ is finite, positively homogeneous, subadditive and satisfies*

$$|f^\circ(\bar{x}; h)| \leq L\|h\|.$$

Proof. Exercise 5.2.2.   ●

Definition 5.2.3 (Clarke Subdifferential) *Let X be a Banach space and let $f\colon X \to \mathbb{R}$ be a locally Lipschitz function. We define the Clarke subdifferential of f at $\bar{x}$ by*

$$\partial_C f(\bar{x}) := \{x^* \in X^* \mid \langle x^*, h \rangle \le f^\circ(\bar{x}; h) \ \text{ for all } h \in X\}.$$

We can show that f° is the support function of $\partial_C f$.

Proposition 5.2.4 *Let X be a Banach space and let $f\colon X \to \mathbb{R}$ be Lipschitz with a Lipschitz constant L near $\bar{x}$. Then*

(i) $x \to \partial_C f(x)$ *is a nonempty, convex, weak*-compact subset of X^* and $\|x^*\| \le L$ for every $x^* \in \partial_C f(\bar{x})$.*
(ii) *For every $h \in X$,*

$$f^\circ(\bar{x}; h) = \max\{\langle x^*, h \rangle \mid x^* \in \partial_C f(\bar{x})\}.$$

Proof. Conclusion (i) follows directly from the definition and Alaoglu's theorem (for the weak* compactness). To prove (ii) observe that for any $h \in X$, $f^\circ(\bar{x}; h)$ is no less than the given maximum by the definition. Suppose that for some h, $f^\circ(\bar{x}; h)$ exceeds the maximum. Then by the Hahn–Banach Extension Theorem of Theorem 4.3.7 (with the linear subspace being the span of h) there exists a linear functional $x^* \in X^*$ majorized by $f^\circ(\bar{x}; \cdot)$ and agreeing with it at h. It follows that $x^* \in \partial_C f(\bar{x})$, and therefore $f^\circ(\bar{x}; h) > \langle x^*, h \rangle = f^\circ(\bar{x}; h)$, a contradiction. $\bullet$

Thus, properties of $\partial_C f$ can often be derived through corresponding properties of f°. We illustrate this method by three examples.

Proposition 5.2.5 (Optimality Condition) *Let X be a Banach space and let $f\colon X \to \mathbb{R}$ be a locally Lipschitz function. Suppose that f attains a local minimum at $\bar{x}$. Then*

$$0 \in \partial_C f(\bar{x}).$$

Proof. We need only check by definition that when $\bar{x}$ is a local minimum of f, $f^\circ(\bar{x}; h) \ge 0$ for any $h \in X$ (Exercises 5.2.1). $\bullet$

Theorem 5.2.6 (Sum Rule) *Let X be a Banach space and let $f_n\colon X \to \mathbb{R}, n = 1, \dots, N$ be locally Lipschitz functions. Then*

$$\partial_C \Big(\sum_{n=1}^{N} f_n \Big)(\bar{x}) \subset \sum_{n=1}^{N} \partial_C f_n(\bar{x}).$$

Proof. It suffices to prove the case when $N = 2$ and the general case follows by induction. Since the support function on the left- and right-hand sides (evaluated at h) are, respectively, $(f_1 + f_2)^\circ(\bar{x}; h)$ and $f_1^\circ(\bar{x}; h) + f_2^\circ(\bar{x}; h)$, this follows readily from Proposition 5.2.2. $\bullet$

Theorem 5.2.7 (Cusco Property) *Let X be a Banach space and let $f: X \to \mathbb{R}$ be a locally Lipschitz function. Then $\partial_C f$ is a weak* cusco.*

Proof. In view of Proposition 5.2.4 we need only show that $\partial_C f$ is an upper semicontinuous multifunction. We show that $(x, h) \to f^\circ(x; h)$ is an upper semicontinuous function which implies the conclusion. Let L be a Lipschitz constant of f near $\bar{x}$ and let (x_i) and (h_i) be arbitrary sequences converging to $\bar{x}$ and $\bar{h}$, respectively. By the definition of the upper limit, for each i there exist $y_i \in X$ and $t_i > 0$ such that $\|y_i - x_i\| + t_i < 1/i$, and

$$f^\circ(x_i; h_i) - \frac{1}{i} \le \frac{f(y_i + t_i h_i) - f(y_i)}{t_i}$$
$$= \frac{f(y_i + t_i \bar{h}) - f(y_i)}{t_i} + \frac{f(y_i + t_i h_i) - f(y_i + t_i \bar{h})}{t_i}.$$

Note that the last term is bounded in magnitude by $L\|h_i - \bar{h}\|$ due to the Lipschitz property of f. Taking limits as $i \to \infty$ we have

$$\limsup_{i \to \infty} f^\circ(x_i; h_i) \le f^\circ(\bar{x}; \bar{h}),$$

which establishes the upper semicontinuity.   ●

The Clarke normal cone to a closed set is defined through the distance function to the set – a Lipschitz function with Lipschitz constant 1. The Clarke tangent cone is defined as the polar of the Clarke normal cone.

Definition 5.2.8 (Clarke Normal and Tangent Cones) *Let X be a Banach space and let S be a closed subset of X. We define the Clarke normal cone of S at $\bar{x}$ by*

$$N_C(S; \bar{x}) := \mathrm{cl}^* \bigcup_{\lambda \ge 0} \partial_C d(S; \bar{x}).$$

Here cl^* stands for the weak-star closure. We define the Clarke tangent cone of S at $\bar{x}$ by

$$T_C(S; \bar{x}) := N_C(S; \bar{x})^\circ = \{v \in X \mid \langle v^*, v \rangle \le 0, v^* \in N_C(S; \bar{x})\}.$$

Combining the optimality condition of Proposition 5.2.5, the sum rule of Theorem 5.2.6 and the definition of the Clarke normal cone, we have the following necessary optimality conditions for a constrained minimization problem.

Theorem 5.2.9 *Let X be a Banach space, let S be a closed subset of X and let $f: X \to \mathbb{R}$ be a locally Lipschitz function. Suppose that f attains a local constrained minimum over S at $\bar{x} \in S$. Then*

$$0 \in \partial_C f(\bar{x}) + N_C(S; \bar{x}).$$

Proof. Using the exact penalization in Example 3.0.3 there exists a constant μ such that $\bar{x}$ is a local minimum of $f + \mu d_S$. Applying Proposition 5.2.5 and Theorem 5.2.6 we have $0 \in \partial_C f(\bar{x}) + \mu \partial_C d(S; \bar{x})$. Since $\mu \partial_C d(S; \bar{x}) \subset N_C(S; \bar{x})$, throwing away the information about the size of μ, the conclusion follows.   ●

The Clarke subdifferential and singular subdifferential for a general lsc function are defined through the normal cone to its epigraph.

Definition 5.2.10 (Clarke Subdifferential for lsc Functions) *Let X be a Banach space and let $f\colon X \to \mathbb{R}$ be a lsc function. We define the Clarke subdifferential and singular subdifferential of f at $\bar{x}$ by*

$$\partial_C f(\bar{x}) := \{x^* \in X^* \mid (x^*, -1) \in N_C(\text{epi}\, f; (\bar{x}, f(\bar{x})))\}$$

and

$$\partial_C^\infty f(\bar{x}) := \{x^* \in X^* \mid (x^*, 0) \in N_C(\text{epi}\, f; (\bar{x}, f(\bar{x})))\},$$

respectively.

Unlike the Fréchet subdifferential, the Clarke subdifferential or singular subdifferential is defined everywhere. However, it is "coarse" compared to the Fréchet subdifferential.

Example 5.2.11 Consider the absolute value function $f(x) := |x|$ on $\mathbb{R}$. Then

$$\partial_C f(0) = \partial_C(-f)(0) = [-1, 1].$$

We can see in this example that the Clarke subdifferential does not distinguish between the absolute value function and its negative at the crucial point where the minimum (respectively, maximum) occurs. In contrast $\partial_F f(0) = [-1, 1]$ while $\partial_F(-f)(0) = \emptyset$. With a little more work we can produce the phenomenon of Example 5.2.11 at every point of an interval.

Example 5.2.12 (Rockafellar) Let E be a measurable set in $[0, 1]$ with the property that the intersection of any nonempty open interval (a, b) in $[0, 1]$ with both E and its complement has positive measure; such sets are sometimes termed *ubiquitous* [244]. Let χ_E be the characteristic function of E (i.e., $\chi_E(x) = 1$ if $x \in E$ and 0 otherwise) and define

$$f(x) = \int_0^x \chi_E(t)\, dt = \lambda(E \cap [0, x]),$$

with λ Lebesgue measure on the line. Then, for all $x \in (0, 1)$

$$\partial_C f(x) = [0, 1].$$

In fact we know by Lebesgue's version of the Fundamental theorem of calculus that for almost all $y \in (0, 1)$, f is differentiable and $f'(y) = \chi(y)$. For any such

y, we have $\chi_E(y) \in \partial_C f(y)$. Since $\partial_C f$ is a cusco, we have $[0, 1] \subset \partial_C f(x)$ for any $x \in (0, 1)$, by the property of E. On the other hand it is easy to check that $f^\circ(x; 1) \leq 1$, and $f^\circ(x; -1) \lesssim 0$ follows from the equality $f(y \pm h) - f(y) = \int_y^{y \pm h} \chi_E(t)dt$. Thus, $\partial_C f(x) \equiv [0, 1]$.

As a caution, the Clarke subdifferential is not a generalization of the usual Fréchet differentiability. This is clarified by the following example.

Example 5.2.13 Let

$$f(x) := \begin{cases} x^2 \sin(1/x) & \text{if } x \neq 0, \\ 0 & \text{if } x = 0. \end{cases}$$

Then, $f'(0) = 0$ while $\partial_C f(0) = [-1, 1]$.

One way to resolve the difficulty that the Clarke subdifferential may sometimes – often – be too coarse is to consider functions for which the Clarke directional derivative does coincide with the usual directional derivative. This leads to the following definition.

Definition 5.2.14 (Regularity) *Let X be a Banach space and let $f \colon X \to \mathbb{R}$ be a locally Lipschitz function. We say that f is (Clarke) regular at $\bar{x}$ provided that $f'(\bar{x}; h)$ exists and agrees with $f^\circ(\bar{x}; h)$.*

Clearly if a function f is C^1 in a neighborhood of $\bar{x}$ then it is regular at $\bar{x}$. In fact, this is true for points of *strict differentiability* (Exercise 5.2.4) – it is this property that the Clarke derivative is really generalizing. Another important class of functions with the regularity property is the class of convex or concave functions. Thus, regularity captures both strictly differentiable functions and convex functions.

Theorem 5.2.15 (Regularity of Convex Functions) *Let X be a Banach space and let $f \colon X \to \mathbb{R} \cup \{+\infty\}$ be a lsc convex function. Suppose that $\bar{x} \in \mathrm{core}(\mathrm{dom}\, f)$. Then f is regular at $\bar{x}$.*

Proof. By Theorem 4.1.8 $\mathrm{core}(\mathrm{dom}\, f) = \mathrm{int}(\mathrm{dom}\, f)$. By Theorem 4.1.3 and Proposition 4.2.4 f is locally Lipschitz at $\bar{x}$ and $f'(\bar{x}; h)$ exists for all $h \in X$. Denoting the local Lipschitz constant of f near $\bar{x}$ by K and choosing a $\delta > 0$, we know

$$\begin{aligned} f^\circ(\bar{x}; h) &= \lim_{\varepsilon \to 0+} \sup_{\|x - \bar{x}\| \leq \varepsilon \delta} \sup_{0 < t < \varepsilon} \frac{f(x + th) - f(x)}{t} \\ &\leq \lim_{t \to 0+} \sup_{\|x - \bar{x}\| \leq t \delta} \frac{f(x + th) - f(x)}{t} \\ &\leq \lim_{t \to 0+} \frac{f(\bar{x} + th) - f(\bar{x})}{t} + 2K\delta \\ &= f'(\bar{x}; h) + 2K\delta. \end{aligned}$$

Letting $\delta \to 0$, we deduce $f^\circ(\bar{x}; h) \leq f'(\bar{x}; h)$. The opposite inequality follows directly from the definition.   ●

5.2.2 Representation of the Clarke Subdifferential

In a Fréchet smooth Banach space, the Clarke subdifferential can be represented as the convex closure of the weak-star sequential limit of the Fréchet subdifferential. Let us start with Lipschitz functions.

Theorem 5.2.16 (Representation of the Clarke Subdifferential: Lipschitz Case) *Let X be a Fréchet smooth Banach space and let $f \colon X \to \mathbb{R}$ be a locally Lipschitz function. Then*

$$\partial_C f(\bar{x}) = \mathrm{cl}^* \mathrm{conv} \left\{ \underset{i \to \infty}{\mathrm{w}^*\text{-}\lim} \, x_i^* \mid x_i^* \in \partial_F f(x_i), x_i \to \bar{x} \right\}.$$

Here $\mathrm{w}^*\text{-}\lim_{i \to \infty} x_i^*$ signifies limit in the weak-star topology.

Proof. We need show that the support function for both sides is the same, i.e.,

$$f^\circ(\bar{x}; h) = \sup \left\{ \langle x^*, h \rangle \mid x^* = \underset{i \to \infty}{\mathrm{w}^*\text{-}\lim} \, x_i^*, x_i^* \in \partial_F f(x_i), x_i \to \bar{x} \right\}.$$

Since $\partial_F f(x_i) \subset \partial_C f(x_i)$, and, by Theorem 5.2.7, $\partial_C f$ is upper semicontinuous we have

$$f^\circ(\bar{x}; h) \geq \sup \left\{ \langle x^*, h \rangle \mid x^* = \underset{i \to \infty}{\mathrm{w}^*\text{-}\lim} \, x_i^*, x_i^* \in \partial_F f(x_i), x_i \to \bar{x} \right\}.$$

To prove the opposite inequality, choose $y_i \to \bar{x}$ and $t_i \to 0+$ such that

$$f^\circ(\bar{x}; h) = \lim_{i \to \infty} \frac{f(y_i + t_i h) - f(y_i)}{t_i}.$$

By the approximate mean value theorem of Theorem 3.4.7, for any $\varepsilon > 0$ and each i, there exist $x_i \in B_{\varepsilon t_i}([y_i, y_i + t_i h])$ and $x_i^* \in \partial_F f(x_i)$ such that

$$\langle x_i^*, t_i h \rangle \geq f(y_i + t_i h) - f(y_i) - t_i \varepsilon$$

or

$$\langle x_i^*, h \rangle \geq \frac{f(y_i + t_i h) - f(y_i)}{t_i} - \varepsilon.$$

Since f is Lipschitz, (x_i^*) is bounded by the Lipschitz constant of f (Theorem 3.4.8). Thus, without loss of generality we may assume that (x_i^*) weak-star converges to some x^*. Taking limits in the last inequality we have

$$\sup \left\{ \langle x^*, h \rangle \mid x^* = \underset{i \to \infty}{\mathrm{w}^*\text{-}\lim} \, x_i^*, x_i^* \in \partial_F f(x_i), x_i \to \bar{x} \right\} \geq f^\circ(\bar{x}; h) - \varepsilon.$$

Since $\varepsilon > 0$ is arbitrary we have

$$\sup \left\{ \langle x^*, h \rangle \mid x^* = \underset{i\to\infty}{\text{w}^*\text{-lim}}\, x_i^*, x_i^* \in \partial_F f(x_i), x_i \to \bar{x} \right\} \geq f^\circ(\bar{x}; h).$$

●

The key to representing the normal cone is the following lemma.

Lemma 5.2.17 *Let X be a Fréchet smooth Banach spaces and let S be a closed subset of X. Suppose $x \in X$ and $x^* \in \partial_F d(S; x)$. Then for any $\varepsilon > 0$, there are $s \in S$ and $s^* \in N_F(S; s)$ such that*

$$\|x - s\| \leq d(S; x) + \varepsilon \text{ and } \|s^* - x^*\| < \varepsilon.$$

Proof. If $x \in S$ then $\partial_F d(S; x) \subset N_F(S; x)$ and the conclusion holds trivially for $x = s$. Now consider the case when $x \notin S$. Let g be a C^1 function such that $g'(x) = x^*$ and $d_S - g$ attains a minimum at x. Choose $\eta \in (0, \min\{1, \varepsilon/3, d(S; x)\})$ such that

$$\|g'(z) - g'(x)\| < \varepsilon/3, \quad \text{for all } z \in B_\eta(x). \tag{5.2.1}$$

Choose $v \in S$ satisfying

$$\|v - x\| < d(S; x) + \eta^2$$

and define $f(y, u) := \|u - y\| - g(y)$. Then

$$\begin{aligned} f(v, x) &= \|v - x\| - g(x) \leq d(S; x) - g(x) + \eta^2 \\ &\leq \inf_{y\in X}(d(S; y) - g(y)) + \eta^2 \\ &\leq \inf_{y\in X, u\in S} f(y, u) + \eta^2. \end{aligned}$$

By the Ekeland variational principle of Theorem 2.1.2 there exists $s \in S$ and $z \in X$ satisfying $\|s - v\| < \eta$, $\|z - x\| < \eta$ (and hence $\|s - x\| < d(S; x) + \varepsilon$) such that

$$f(z, s) \leq f(v, x)$$

and

$$f(z, s) \leq f(y, u) + \eta(\|y - z\| + \|u - s\|), \quad \text{for all } u \in S, y \in X.$$

Fixing $y = z$ we have that

$$u \to \iota_S(u) + \|z - u\| - g(z) + \eta\|u - s\|$$

attains minimum at $u = s$, and therefore

$$\| \cdot \|'(z - s) \in N_F(S; s) + \eta B_{X^*}. \tag{5.2.2}$$

Similarly fixing $u = s$ we have that

$$y \to \|y - s\| - g(y) + \eta\|y - z\|$$

attains minimum at $y = z$, and hence

$$0 \in \| \cdot \|'(z - s) - g'(z) + \eta B_{X^*}. \tag{5.2.3}$$

Combining (5.2.1), (5.2.2) and (5.2.3) we have that there exists $s^* \in N_F(S; s)$ such that

$$\|s^* - x^*\| < \varepsilon.$$

●

Combining Theorem 5.2.16 and Lemma 5.2.17 we obtain a representation of the Clarke normal cone in terms of the adjacent Fréchet normal cones.

Theorem 5.2.18 (Representation of the Clarke Normal Cone) *Let X be a Fréchet smooth Banach space and let S be a closed subset of X. Then*

$$N_C(S; \bar{x}) = \text{cl}^* \text{conv}\{\underset{i\to\infty}{\text{w}^*\text{-lim}}\, x_i^* \mid x_i^* \in N_F(S; x_i), S \ni x_i \to \bar{x}\}.$$

Proof. Exercise 5.2.6   ●

Now we are ready to discuss the representation of the Clarke subdifferential of general lsc functions.

Theorem 5.2.19 (Representation of the Clarke Subdifferential and Singular Subdifferential) *Let X be a Fréchet smooth Banach space and let $f \colon X \to \mathbb{R}$ be a lsc function. Then*

$$\partial_C f(\bar{x}) = \text{cl}^* \text{conv}\Big\{\underset{i\to\infty}{\text{w}^*\text{-lim}}\, x_i^* \,\Big|\, x_i^* \in \partial_F f(x_i),$$

$$(x_i, f(x_i)) \to (\bar{x}, f(\bar{x}))\Big\} + \partial_C^\infty f(\bar{x}),$$

and

$$\partial_C^\infty f(\bar{x}) = \text{cl}^* \text{conv}\Big\{\underset{i\to\infty}{\text{w}^*\text{-lim}}\, \lambda_i x_i^* \,\Big|\, x_i^* \in \partial_F f(x_i),$$

$$(x_i, f(x_i)) \to (\bar{x}, f(\bar{x})), \lambda_i \to 0_+\Big\}.$$

Proof. Combine Theorem 5.2.18 and the representations of Fréchet normal vectors in Theorem 3.1.8 and singular Fréchet normal vectors in Theorem 3.3.6.   ●

5.2.3 Limiting Subdifferentials and Calculus

The representation of Clarke subdifferentials and normal cones shows that in a Fréchet smooth Banach space they can be viewed as the convex sequential upper semicontinuous closure of the Fréchet subdifferential and the Fréchet normal cone. The convexification brings about the nice characterization in terms of the Clarke directional derivatives, which leads to much flexibility in discussing the properties of the Clarke subdifferential and normal cone. On the other hand this convexification process also adds additional elements to the natural sequential limits of the Fréchet subdifferential and makes the Clarke subdifferential larger than desired in some situations, such as those illustrated in Example 5.2.11. Naturally, one may wonder whether omitting the convexification in this process will still yield subdifferentials and a normal cone that have a reasonable enough calculus to be useful. This is the goal of the current subsection.

Definition 5.2.20 (Limiting and Singular Subdifferential) *Let X be a Banach space and let $f : X \to \mathbb{R} \cup \{+\infty\}$ be a lsc function. Define*

$$\partial_L f(x) := \left\{ \operatorname*{w^*-lim}_{i \to \infty} x_i^* : x_i^* \in \partial_F f(x_i),\, (x_i, f(x_i)) \to (x, f(x)) \right\},$$

and

$$\partial^\infty f(x) := \left\{ \operatorname*{w^*-lim}_{i \to \infty} t_i x_i^* : x_i^* \in \partial_F f(x_i),\, (x_i, f(x_i)) \to (x, f(x)), t_i \to 0_+ \right\}$$

and call $\partial_L f(x)$ and $\partial^\infty f(x)$ the limiting subdifferential and singular subdifferential of f at x, respectively.

Definition 5.2.21 (Limiting Normal Cone) *Let X be a Banach space and let S be a closed subset of X. Define*

$$N_L(S; x) := \left\{ \operatorname*{w^*-lim}_{i \to \infty} x_i^* : x_i^* \in N_F(S; x_i),\, S \ni x_i \to x \right\}$$

and call $N_L(S; x)$ the limiting normal cone of S at x.

The representations in Theorems 5.2.16, 5.2.18 and 5.2.19 now can be rewritten as:

Theorem 5.2.22 (Limiting and Clarke Subdifferentials) *Let X be a Fréchet smooth Banach space, let $f : X \to \mathbb{R} \cup \{+\infty\}$ be a lsc function and let S be a closed subset of X. Then*

$$\partial_C f(x) = \operatorname{cl}^* \operatorname{conv}[\partial_L f(x) + \partial^\infty f(x)],$$

$$\partial_C^\infty f(x) = \operatorname{cl}^* \operatorname{conv} \partial^\infty f(x),$$

and

$$N_C(S; x) = \operatorname{cl}^* \operatorname{conv} N_L(S; x).$$

Proof. Exercise 5.2.8.   ●

Thus, the Clarke subdifferential and normal cone are convex weak-star closure of the limiting subdifferential and the limiting normal cone.

Clearly, $\partial_F f(x) \subset \partial_L f(x)$ and $N_F(S; x) \subset N_L(S; x)$ (Exercise 5.2.7). As a consequence the limiting subdifferential preserves the necessary minimization condition (Exercise 5.2.9). The limiting subdifferential is more accurate in that it can distinguish a maximum and a minimum as illustrated by the following example.

Example 5.2.23 Let $f(x) = |x| \colon \mathbb{R} \to \mathbb{R}$. Then

$$\partial_L f(x) = \begin{cases} 1 & \text{if } x > 0, \\ -1 & \text{if } x < 0, \\ [-1, 1] & \text{if } x = 0, \end{cases}$$

and

$$\partial_L(-f)(x) = \begin{cases} -1 & \text{if } x > 0, \\ 1 & \text{if } x < 0, \\ \{-1, 1\} & \text{if } x = 0. \end{cases}$$

A natural method of deriving calculus results for the limiting subdifferential and normal cone is by taking limits in the corresponding approximate calculus for the Fréchet subdifferential and the Fréchet normal cone. We illustrate this method by deriving a sum rule for the limiting subdifferential in finite dimensional spaces.

Theorem 5.2.24 (Limiting Sum Rule) *Let X be a finite dimensional Banach space. Let $f_1, \ldots, f_N \colon X \to \mathbb{R} \cup \{+\infty\}$ be lsc functions such that $\sum_{n=1}^{N} f_n$ attains a local minimum at $\bar{x}$. Then, either*

(A1) $$0 \in \sum_{n=1}^{N} \partial_L f_n(\bar{x});$$

or there exist $u_n \in \partial^\infty (f_n)(\bar{x}), n = 1, \ldots, N$ not all zero such that

(A2) $$0 = \sum_{n=1}^{N} u_n.$$

Proof. By the approximate local sum rule of Theorem 3.3.1, for each i there exist $(x_n^i, f_n(x_n^i)) \in (\bar{x}, f_n(\bar{x})) + \frac{1}{i} B_{X \times R}$ and $\xi_n^i \in \partial_F f_n(x_n^i)$ such that

$$\left\| \sum_{n=1}^{N} \xi_n^i \right\| < \frac{1}{i}. \tag{5.2.4}$$

Define $t_i := \sum_{n=1}^{N} \|\xi_n^i\|$. We consider the following two cases.

Case 1. The sequence (t_i) is bounded. Then without loss of generality we may assume that (ξ_n^i) converges to ξ_n for $n = 1, \ldots, N$. It is obvious that $\xi_n \in \partial_L f_n(\bar{x})$. Upon taking limits in (5.2.4) we obtain $0 = \sum_{n=1}^{N} \xi_n \in \sum_{n=1}^{N} \partial f_n(\bar{x})$.

Case 2. The sequence (t_i) is unbounded. Then without loss of generality we may assume that $t_i \to \infty$ and (ξ_n^i/t_i) converges to u_n. Then $u_n \in \partial^\infty f_n(\bar{x})$ by the definition of the singular limiting subdifferential. Dividing (5.2.4) by t_i and taking limits we obtain $0 = \sum_{n=1}^{N} u_n$. Since $\sum_{n=1}^{N} \|\xi_n^i\|/t_i = 1$ we conclude that $\sum_{n=1}^{N} \|u_n\| = 1$, and therefore not all u_n are 0.   ●

Using a similar approach we can establish a limiting multiplier rule and a limiting extremal principle. Consider the minimization problem $\mathcal{P}$ in Subsection 3.3.4. Using the notation τ_n introduced there we have:

Theorem 5.2.25 (Limiting Multiplier Rule) *Let X be a finite dimensional Banach space, let S be a closed subset of X and let f_n be lsc for $n = 0, 1, \ldots, N$ and continuous for $n = N+1, \ldots, M$. Suppose that $\bar{x}$ is a local solution of problem $\mathcal{P}$. Then either there exist $\mu_n \geq 0$, $n = 0, \ldots, M$ satisfying $\sum_{n=0}^{M} \mu_n = 1$ such that*

(A1) $0 \in \displaystyle\sum_{m \in \{n : \mu_n > 0\}} \mu_m \partial_L(\tau_m f_m)(\bar{x}) + \sum_{m \in \{n : \mu_n = 0\}} \partial^\infty(\tau_m f_m)(\bar{x}) + N_L(S; \bar{x}),$

or

(A2) *there exist $u_n \in \partial^\infty(\tau_n f_n)(\bar{x})$, $n = 0, 1, \ldots, M$ and $u_{M+1} \in N_L(S; \bar{x})$ not all zero such that*

$$0 = \sum_{n=0}^{M+1} u_n.$$

Proof.   Exercise 5.2.11.   ●

When f_n's are smooth functions we can recover the Karush–Kuhn–Tucker conditions. Theorem 5.2.25 is also a quite general result that can be used to derive many subdifferential calculus rules. The following is an example.

Theorem 5.2.26 (Limiting Chain Rule) *Let X be a finite dimensional Banach space, let $f \colon \mathbb{R}^M \to \mathbb{R} \cup \{+\infty\}$ and $f_n \colon X \to \mathbb{R} \cup \{+\infty\}, n = 1, \ldots, N$ be lsc functions and let $f_n \colon X \to \mathbb{R}, n = N+1, \ldots, M$ be continuous functions. Suppose that $f(f_1, \ldots, f_M)$ attains a minimum at $\bar{x}$. Then either:*

(A1) *there exist $u_n \in \partial^\infty(\tau_n f_n)(\bar{x})$, $n = 1, \ldots, M$ not all zero such that*

$$0 = \sum_{n=1}^{M} u_n.$$

or there exist $\mu = (\mu_1, \ldots, \mu_M) \in \partial f(f_1, \ldots, f_M)(\bar{x})$ *such that*

$$\text{(A2)} \qquad 0 \in \sum_{m \in \{n : \mu_n \neq 0\}} \mu_m \partial_L(\tau_m f_m)(\bar{x}) + \sum_{m \in \{n : \mu_n = 0\}} \partial^\infty(\tau_m f_m)(\bar{x}).$$

Proof. We leave the proof as a guided exercise (Exercise 5.2.12). ●

Theorem 5.2.27 (Limiting Extremal Principle) *Let X be a finite dimensional Banach space and let $S_1, \ldots, S_N \subset X$ be an extremal system of fixed sets as in Definition 3.7.5 with an extremal point $\bar{x}$. Then there exists $x_n^* \in N_L(S_n; \bar{x})$ such that*

$$\sum_{n=1}^N \|x_n^*\| = 1 \quad \text{and} \quad \sum_{n=1}^N x_n^* = 0.$$

Proof. Exercise 5.2.14. ●

It is not hard to see that in a finite dimensional space the limiting subdifferentials and the limiting normal cone are closed.

Proposition 5.2.28 *Let X be a finite dimensional Banach space, let $f \colon X \to \mathbb{R} \cup \{+\infty\}$ be a lsc function and let S be a closed subset of X. Then $\partial_L f$, $\partial^\infty f$ and N_L are closed multifunctions.*

Proof. We prove $\partial_L f$ is a closed multifunction and leave the other two as Exercise 5.2.16.

Since X is finite dimensional the weak-star and norm topology of X^* are the same. Let $x_i \to x$ and $\partial_L f(x_i) \ni x_i^* \to x^*$. Then for each i it follows from the definition of the limiting subdifferential that there exist $y_i \in B_{1/i}(x_i)$ and $y_i^* \in \partial_F f(y_i)$ such that $y_i^* \in B_{1/i}(x_i^*)$. Clearly, $y_i \to x$ and $y_i^* \to x^*$, and therefore $x^* \in \partial_L f(x)$. ●

5.2.4 Additional Examples and Counterexamples

Analyzing the proof of the limiting sum rule of Theorem 5.2.24 we can see that in an infinite dimensional space we will not be able to guarantee that not all u_n are 0. When all the u_n's are 0 the alternative (A2) is trivial. In fact, most of the limiting results fail in infinite dimensional spaces. There are also other complications in infinite dimensional spaces. We present several examples here to illustrate pathological situations that may occur in an infinite dimensional space.

Our first example shows that the limiting sum rule does not hold in infinite dimensional spaces.

Example 5.2.29 (Failure of Limiting Sum Rule) Consider a Hilbert space H with two closed subspaces M_1 and M_2 such that $M_1^\perp + M_2^\perp$ is dense in H but not closed and $M_1^\perp \cap M_2^\perp = \{0\}$.

Define $f_1 := \delta_{M_1}$ and $f_2 := \delta_{M_2} + \langle v, \cdot \rangle$ where $-v \in H \backslash (M_1^\perp + M_2^\perp)$. Since $M_1^\perp + M_2^\perp$ dense implies that $M_1 \cap M_2 = \{0\}$, $f_1 + f_2$ attains a minimum at 0. However, it is easy to check that $\partial f_1(0) = \partial^\infty f_1(0) = M_1^\perp$, $\partial f_2(0) = M_2^\perp + v$ and $\partial^\infty f_2(0) = M_2^\perp$. Thus,

$$0 \notin \partial f_1(0) + \partial f_2(0)$$

and

$$\partial^\infty f_1(0) \cap (-\partial^\infty f_2(0)) = \{0\}$$

or equivalently

$$0 \in \partial^\infty f_1(0) + \partial^\infty f_2(0)$$

holds only in the trivial case.

As a concrete example of the basic construction let $H := \ell_2$ and denote the standard unit vectors by $\{e_i\}$. Suppose (α_i) is a sequence of positive real numbers with $1 > \alpha_i \geq \sqrt{1 - \frac{1}{i^2}}$. Define $M_1 := \mathrm{cl\,span}\{e_{2i}\}_{i=1}^\infty$ and $M_2 := \mathrm{cl\,span}\{\sqrt{1 - \alpha_i^2} e_{2i-1} - \alpha_i e_{2i}\}_{i=1}^\infty$. Then we can directly verify that

$$M_1^\perp := \mathrm{cl\,span}\{f_1, f_2, \dots\} \quad \text{and} \quad M_2^\perp := \mathrm{cl\,span}\{g_1, g_2, \dots\}$$

where $f_i := e_{2i-1}$, $g_i := \alpha_i e_{2i-1} + \sqrt{1 - \alpha_i^2} e_{2i}$.

Then for any $x = \sum_{i=1}^\infty x_i e_i \in H$, we have

$$\sum_{i=1}^{2k} x_i e_i = \sum_{i=1}^{k} \left(x_{2i-1} - \frac{x_{2i}\alpha_i}{\sqrt{1 - \alpha_i^2}} \right) f_i + \sum_{i=1}^{k} \frac{x_{2i}}{\sqrt{1 - \alpha_i^2}} g_i \in M_1^\perp + M_2^\perp.$$

Therefore, $M_1^\perp + M_2^\perp$ is dense in H. We can show by a similar argument that $M_1 + M_2$ is dense in H, which implies that $M_1 \cap M_2 = 0$. It remains to show that $M_1^\perp + M_2^\perp \neq H$. Consider

$$v := \sum_{n=1}^\infty \sqrt{1 - \alpha_i^2} e_{2i}.$$

If $v = y + z$ with $y \in M_1^\perp$ and $z \in M_2^\perp$ then $y = \sum_{i=1}^\infty y_i f_i$ and $z = \sum_{i=1}^\infty z_i g_i$ because $\{f_i\}$ and $\{g_i\}$ are orthonormal bases for $M_1^\perp$ and $M_2^\perp$, respectively. Then we must have $z_i = 1$ and $y_i = z_i \alpha_i = \alpha_i \to 1$ which is impossible.

Since Theorem 5.2.25 implies Theorem 5.2.26 and the latter implies the limiting sum rule (Exercises 5.2.12 and 5.2.13), Example 5.2.29 also shows that these two results fail in infinite dimensional spaces.

Our next example shows that in an infinite dimensional space the limiting normal cone fails to be a closed multifunction in general. Since the limiting normal cone is the limiting subdifferential of an indicator function this also provides an example for the limiting subdifferential.

Example 5.2.30 (Nonclosed Normal Cone) Let $H := \ell_2$ and denote the standard basis by $\{e_i\}$. We define a closed subset S of H by

$$S := \{s(e_1 - je_j) + t(je_1 - e_k) \mid k > j > 1, \ s, t \geq 0\} \cup \{te_1 \mid t \geq 0\}.$$

Then $N_L(S;0)$ is not closed since (i) $N_L(S;0) \ni e_1 + j^{-1}e_j \to e_1$ and (ii) $e_1 \notin N_L(S;0)$.

It is easy to check that S is a closed set and that (i) holds. We leave them as Exercises 5.2.17 and 5.2.18. We will concentrate on verifying (ii). Suppose not; then there are $x_i \to 0$ and $x_i^* \in N_F(S;0)$ such that x_i^* weakly converges to e_1. Suppose some $x_i = t_i e_1$ for $t_i \geq 0$. Put $u(r) := x_i + re_1$ for $r > 0$. We have $u(r) \in S$, and therefore

$$0 \geq \limsup_{r \to 0+} \left\langle x_i^*, \frac{u(r) - x_i}{\|u(r) - x_i\|} \right\rangle = \limsup_{r \to 0+} \left\langle x_i^*, \frac{re_1}{\|re_1\|} \right\rangle = \langle x_i^*, e_1 \rangle.$$

On the other hand, (x_i^*) weakly converging to e_1 implies $\langle x_i^*, e_1 \rangle \to 1$, so that only finitely many x_i can be of the form $x_i = t_i e_1$ for $t_i \geq 0$. So all but finitely many x_i are necessarily of the form $s(e_1 - je_j) + t(je_1 - e_k)$ where $k > j > 1$, $s, t \geq 0$.

Now let $x_i = s(e_1 - je_j) + t(je_1 - e_k)$ where $k = k(i) > j = j(i) > 1, s = s(i) \geq 0$ and $t = t(i) \geq 0$. Considering $u(r) = x_i + r(je_1 - e_k) \in S$ we get

$$0 \geq \limsup_{r \to 0+} \left\langle x_i^*, \frac{u(r) - x_i}{\|u(r) - x_i\|} \right\rangle$$

$$= \limsup_{r \to 0+} \left\langle x_i^*, \frac{r(je_1 - e_k)}{\|r(je_1 - e_k)\|} \right\rangle$$

$$= \left\langle x_i^*, \frac{je_1 - e_k}{\|je_1 - e_k\|} \right\rangle$$

so that

$$\langle x_i^*, e_1 \rangle \leq \langle x_i^*, j^{-1}e_k \rangle, \tag{5.2.5}$$

while considering $u(r) = x_i + r(e_1 - je_j) \in S$ we get

$$0 \geq \limsup_{r \to 0+} \left\langle x_i^*, \frac{u(r) - x_i}{\|u(r) - x_i\|} \right\rangle$$

$$= \limsup_{r \to 0+} \left\langle x_i^*, \frac{r(e_1 - je_j)}{\|r(e_1 - je_j)\|} \right\rangle$$

$$= \left\langle x_i^*, \frac{e_1 - je_j}{\|e_1 - je_j\|} \right\rangle$$

so that

$$\langle x_i^*, e_1 \rangle \leq \langle x_i^*, je_j \rangle. \tag{5.2.6}$$

Letting $i \to \infty$ in (5.2.5) we obtain

$$1 \leq \liminf \langle x_i^*, j(i)^{-1} e_{k(i)} \rangle.$$

If the $j(i)$'s are unbounded then this shows the sequence (x_i^*) is unbounded, contradicting its weak convergence. Therefore, in the sequence of (x_i) we have only finitely many $j(i)$. But then (5.2.5) contradicts (x_i^*) weakly converges to e_1.

We have seen in Theorem 5.1.29 that a maximal monotone operator, in particular the subdifferential of a proper convex lsc function, is a minimal cusco in the interior of its domain. The next example show that the restriction to the interior of the domain is necessary.

Example 5.2.31 Again let $X = \ell_2$ and let $\{e_i\}$ be the standard basis of ℓ_2. Define

$$y_{p,i} := \frac{1}{p}(e_p + e_{p^i}), \quad y_{p,i}^* := e_p + (p-1)e_{p^i}$$

for prime numbers p and $i = 2, 3, \ldots$. Then we have

$$\langle y_{p,i}^*, y_{q,j} \rangle = \begin{cases} 0 & \text{if } p \neq q, \\ 1/p & \text{if } p = q, i \neq j, \\ 1 & \text{if } p = q, i = j. \end{cases}$$

Further, define $f \colon X \to \mathbb{R}$ by

$$f(x) := \max(\langle e_1, x \rangle + 1, \sup\{\langle y_{p,i}, x \rangle \mid p \text{ prime } i \geq 2\})$$

so that f is a proper lsc convex function on X. Then $f(0) = f(y_{p,i}) = 1$, $f(-e_1) = 0$ and $f(x) \geq \langle y_{p,i}^*, x \rangle$ for all $x \in X$ and p prime, $i \geq 2$, which implies $y_{p,i}^* \in \partial f(y_{p,i})$. In fact,

$$f(x) - f(y_{p,i}) = f(x) - 1 \geq \langle y_{p,i}^*, x \rangle - 1 = \langle y_{p,i}^*, x - y_{p,i} \rangle, \quad \text{for all } x \in X.$$

We also have $0 \notin \partial f(0)$, since $0 \in \partial f(0)$ is equivalent to $f(x) - f(0) \geq 0$ for all $x \in X$, which is not true for $x = -e_1$. Thus, $(0,0)$ is not in the graph of ∂f. Now we can show that the graph of ∂f is not closed in the norm $\times$ weak* topology by checking that $(0,0)$ is in the norm $\times$ weak* closure of the set

$$\{(y_{p,i}, y_{p,i}^*) \mid p \text{ prime}, i \geq 2\} \subset \operatorname{graph} \partial f.$$

Finally we examine a generic version of Example 5.2.12 and its application. Fix a bounded subset A in a Banach space X. Let C be a w*-compact convex subset of X^*. Consider

$$\mathcal{N}_C := \{f \mid f \colon A \to \mathbb{R} \text{ and } f(x) - f(y) \leq \sigma_C(x - y) \text{ for all } x, y \in A\}.$$

Note that C is norm bounded and $f \in \mathcal{N}_C$ implies that f is Lipschitz with a Lipschitz constant $L = \sup\{\|x^*\| \mid x^* \in C\}$ (Exercise 5.2.32).

Lemma 5.2.32 *Let X be a Banach space, let A be a bounded subset of X and let C be a w^*-compact convex subset of X^*. Then the metric space $(\mathcal{N}_C, \rho)$ is complete, where*

$$\rho(f,g) := \sup_{x \in A} |f(x) - g(x)|.$$

Proof. Assume (f_i) is Cauchy. Then (f_i) converges pointwise to some f on A. Because $f_i(x) - f_i(y) \le \sigma_C(x-y)$, we have $f(x) - f(y) \le \sigma_C(x-y)$ for all $x, y \in A$, so $f \in \mathcal{N}_C$. We now show (f_i) converges uniformly to f on A. For every $\varepsilon > 0$ there exists $N > 0$ such that if $i, j \ge N$ we have $|f_i(x) - f_j(x)| \le \varepsilon$ for all $x \in A$. Letting $j \to \infty$, we obtain $\rho(f_i, f) \le \varepsilon$. That is, (f_i) converges to f uniformly on A. $\qquad\bullet$

Thus, in $(\mathcal{N}_C, \rho)$ the Baire category theorem applies and the generic version of Example 5.2.12 can be stated.

Theorem 5.2.33 *Let X be a Banach space, let A be a bounded subset of X and let C be a w^*-compact convex subset of X^*. Then in $(\mathcal{N}_C, \rho)$, the set*

$$\{f \in \mathcal{N}_C|\ \partial_C f \equiv C \text{ on } A\},$$

is a residual set.

Proof. Fix $x \in A$ and $v \in X$. Consider

$$G_k := \left\{ f \in \mathcal{N}_C \,\Big|\, \frac{f(x+tv) - f(x)}{t} - \sigma_C(v) > -\frac{1}{k} \text{ for some } 0 < t < \frac{1}{k} \right\}.$$

(a) G_k is open in $\mathcal{N}_C$: Let $f_0 \in G_k$. Then for some $0 < t < 1/k$ we have

$$\frac{f_0(x+tv) - f_0(x)}{t} - \sigma_C(v) > -1/k. \tag{5.2.7}$$

Let $\rho(f, f_0) < \varepsilon$ and $f \in \mathcal{N}_C$. Consider

$$\frac{f(x+tv) - f(x)}{t} - \sigma_C(v)$$
$$= \frac{(f(x+tv) - f(x)) - (f_0(x+tv) - f_0(x))}{t} + \frac{f_0(x+tv) - f_0(x)}{t} - \sigma_C(v)$$
$$\ge -\frac{|f(x+tv) - f_0(x+tv)| + |f(x) - f_0(x)|}{t} + \frac{f_0(x+tv) - f_0(x)}{t} - \sigma_C(v)$$
$$\ge -\frac{2\varepsilon}{t} + \frac{f_0(x+tv) - f_0(x)}{t} - \sigma_C(v).$$

The last expression is greater than $-1/k$ by equation (5.2.7). We may set ε sufficiently small such that

$$\frac{-2\varepsilon}{t} + \frac{f_0(x+tv) - f_0(x)}{t} - \sigma_C(v) > -\frac{1}{k}.$$

Thus, the same t may be used, and so $B(f_0, \varepsilon) \subset G_k$.

(b) G_k is dense in $\mathcal{N}_C$: With $f \in \mathcal{N}_C$, for every $\varepsilon > 0$ we verify that the open ball $B(f, 3\varepsilon)$ contains a point of G_k. Define $h \colon X \to R$ by $h(\tilde{x}) := f(x) - \varepsilon + \sigma_C(\tilde{x} - x)$, which is in $\mathcal{N}_C$ (Exercise 5.2.33), and set

$$h_1 := \min\{f, h\}, \quad h_2 := \max\{f - 2\varepsilon, h_1\}.$$

Then $h_1 \in \mathcal{N}_C$, as is h_2 (Exercise 5.2.34). Since $h_1 \leq f$ and $f - 2\varepsilon \leq f$, we have $f - 2\varepsilon \leq h_2 \leq f$. As $f, \sigma_C(\cdot - x)$ are continuous at x, for $0 < \delta < \varepsilon/2$ sufficiently small we have for $y \in B_\delta(x)$,

$$f(x) - \frac{\varepsilon}{2} \leq f(y) \leq f(x) + \frac{\varepsilon}{2} \tag{5.2.8}$$

and

$$-\frac{\varepsilon}{2} \leq \sigma_C(y - x) \leq \frac{\varepsilon}{2}. \tag{5.2.9}$$

Now for $\tilde{x} \in B_\delta(x)$ we have

$$h(\tilde{x}) \leq f(x) - \varepsilon + \varepsilon/2 \leq f(x) - \frac{\varepsilon}{2} \leq f(\tilde{x}),$$

and so $h_1 = \min\{f, h\} = h$ on $B_\delta(x)$. On the other hand, on $B_\delta(x)$ by equations (5.2.9) and (5.2.8) we have

$$h_1(\tilde{x}) = h(\tilde{x}) \geq f(x) - \varepsilon - \varepsilon/2 \geq f(x) - \frac{3\varepsilon}{2},$$

and

$$f(\tilde{x}) - 2\varepsilon \leq f(x) - \frac{3\varepsilon}{2},$$

and so $h_2 = h_1 = h$ on $B_\delta(x)$. Choosing $0 < t < 1/k$ sufficiently small such that $x + tv \in B_\delta(x)$, we have

$$\frac{h_2(x + tv) - h_2(x)}{t} = \frac{h(x + tv) - h(x)}{t}$$
$$= \frac{f(x) - \varepsilon + \sigma_C(tv) - (f(x) - \varepsilon)}{t} = \sigma_C(v),$$

which shows $h_2 \in G_k$ while h_2 is arbitrarily close to f.

(c) Since G_k is open and dense in $\mathcal{N}_C$,

$$G_{x,v} := \bigcap_{k=1}^{\infty} G_k,$$

is a dense G_δ in $\mathcal{N}_C$. If $f \in G_{x,v}$, then for every k we can find $0 < t_k < 1/k$ such that

$$\frac{f(x + t_k v) - f(x)}{t_k} - \sigma_C(v) > -\frac{1}{k},$$

and taking the limit we derive

$$f^\circ(x; v) \geq \limsup_{t\downarrow 0} \frac{f(x + tv) - f(x)}{t} \geq \sigma_C(v).$$

(d) Now let $\{v_k\}$ be a norm dense countable set in E. For each v_k, the set G_{x,v_k} is a dense G_δ set in $\mathcal{N}_C$. Hence,

$$G_x := \bigcap_{k=1}^{\infty} G_{x,v_k}$$

is also a dense G_δ set in $\mathcal{N}_C$.

Given $f \in G_x$ we note that for each k we have

$$f^\circ(x; v_k) \geq \sigma_C(v_k).$$

Because $f^\circ(x; \cdot)$ and $\sigma_C(\cdot)$ are Lipschitz, we deduce

$$f^\circ(x; v) \geq \sigma_C(v),$$

for every $v \in X$.

(e) Finally let $\{x_k\}$ be a norm dense countable set in A. Since each G_{x_k} is a dense G_δ set in $\mathcal{N}_C$, the set

$$G := \bigcap_{k=1}^{\infty} G_{x_k}$$

is also a dense G_δ set in $\mathcal{N}_C$. For $f \in G$, and each positive integer k we have $f^\circ(x_k, v) \geq \sigma_C(v)$ for all $v \in E$. Since $f^\circ(x; v)$ is upper semicontinuous in x, we obtain

$$f^\circ(x; v) \geq \sigma_C(v),$$

for every $x \in A$ and $v \in X$.

Since $f \in \mathcal{N}_C$, for every $v \in E$, we have

$$f^\circ(x; v) = \limsup_{\substack{y \to x \\ t\downarrow 0}} \frac{f(y + tv) - f(y)}{t} \leq \limsup_{\substack{y \to x \\ t\downarrow 0}} \frac{\sigma_C(y + tv - y)}{t}$$

$$= \limsup_{t\downarrow 0} \frac{\sigma_C(tv)}{t} = \sigma_C(v). \tag{5.2.10}$$

Then for $f \in G$, we have $f^\circ(x; v) = \sigma_C(v)$ for every $x \in A$ and $v \in X$. Dually, $\partial_C f \equiv C$ on A. ●

Corollary 5.2.34 *Let X be a Banach space, let A be a bounded subset of X and let B^* be the unit ball of X^*. Then in the metric space of nonexpansive functions on A with the uniform metric, the set*

$$\{f \mid \partial_C f \equiv B^* \text{ on } A\}$$

is a residual set.

Proof. Noting that

$$\mathcal{N}_{B^*} := \{f \mid f \colon A \mapsto \mathbb{R} \text{ is nonexpansive with respect to } \|\cdot\|\},$$

the Corollary follows directly from Theorem 5.2.33.   ●

When $X = \mathbb{R}$, this provides a generic version of Example 5.2.12. The result in Theorem 5.2.33 also holds for unbounded A. Details are left as guided Exercise 5.2.35. These results showed that most – in the Baire category sense – Lipschitz functions have maximal Clarke subdifferentials and therefore contain no information beyond the Lipschitz constant. It is interesting to observe the following recast of a function with maximal Clarke subdifferential in mathematical economics.

Example 5.2.35 Cornet (see [157]) formalized a nonsmooth *marginal price rule* in mathematical economics by establishing that given a closed *production set Y*, the *price $p \in N_C(Y; y)$* for all y in the boundary of Y. Take $f \colon X \to \mathbb{R}$ with $\partial_C f(x) \equiv C$. The Clarke normal cone and tangent cone to the epigraph of f at $(x, f(x))$ are then constant multifunctions:

$$T_C(\text{epi } f; (x, f(x))) = \{(v, \beta) \mid \sigma_C(v) \le \beta\}$$

and

$$N_C(\text{epi } f; (x, f(x))) = \bigcup_{\lambda \ge 0} \lambda[C, -1].$$

For every $(x, r) \in \text{epi } f$ and $(v, \beta) \in T_C(\text{epi } f; (x, f(x)))$ we have

$$f(x + v) \le f(x) + \sigma_C(v) \le r + \beta,$$

thus $\text{epi } f + T_C(\text{epi } f ; \cdot) \subset \text{epi } f$. In $\mathbb{R}^N$, if we take $0 \in C \subset \mathbb{R}^{N-1}$ with N extreme points $v_1, \ldots, v_N$ such that

$$\langle (v_n, -1), (v_m, -1) \rangle = 0 \quad \text{for } n \ne m,$$

then $N_C(\text{epi } f; (x, f(x)))$ is the closed convex cone generated by $(v_1, -1), \ldots,$ $(v_N, -1)$ which is linearly isometric to $\mathbb{R}^N_+$. Thus $T_C(\text{epi } f; (x, f(x)))$ is linearly isometric to $-\mathbb{R}^N_+$. Then $\text{epi } f$ is isometric to a closed set $Y \subset \mathbb{R}^N$ such that $N_C(Y ; y) = \mathbb{R}^N_+$ for y in the boundary of Y and $Y - \mathbb{R}^N_+ \subset Y$ ("free disposal"). Thus the marginal rule generically imposes no restriction on the price vector.

5.2.5 Commentary and Exercises

The Clarke generalized gradient was introduced in Clarke's thesis [80], a pioneering work that marked the beginning of nonsmooth analysis, although other earlier efforts such as that of Pshenichnii [223] on quasi-derivatives have also been very influential. The exposition here largely follows [84]. Representations of the Clarke generalized gradient in terms of simpler smooth subdifferentials are the result of efforts of many researchers [58, 48, 45, 61, 62, 146]. Rockafellar laid down much of the theory for lsc functions in a series of seminal papers [232, 233, 234, 235].

Example 5.2.12 is due to Rockafellar [235]. It shows that although that Clarke's subdifferential is weak* cusco (Theorem 5.2.7), it is far from a minimal one. The generic versions of Rockafellar's example (and its generalizations in [157]) in Theorem 5.2.33 and Corollary 5.2.34 are taken from Borwein and Wang [65].

Mordukhovich introduced the limiting subdifferential and developed its calculus [195, 198]. This has the advantage of being smaller and more accurate in many applications. Other subdifferential constructions in the same spirit have also been proposed in [143, 193, 250, 263, 264]. The difference between these alternative limiting subdifferentials and the Clarke subdifferential is vividly illustrated in the recent paper by Borwein, Borwein and Wang [28]: even in $\mathbb{R}^N$ for $N > 1$ the limiting subdifferential and the Clarke subdifferential for Lipschitz function may differ almost everywhere. For functions on $\mathbb{R}$ a precise relationship between the Clarke subdifferential and the limiting subdifferential was established by Borwein and Fitzpatrick in [41]. In particular, on the line they differ at most countably. While the limiting subdifferential and normal cone are more accurate, these objects are less "regular" compared to the Clarke subdifferential and normal cone as reflected by examples and counterexamples in subsection 5.2.4. For the construction of these examples and related literature we refer to [69] (Example 5.2.29), [44] (Examples 5.2.30 and 5.2.31) and [65] (Theorem 5.2.33, Corollary 5.2.34 and Example 5.2.35).

An alternative idea of looking for "small" subdifferentials is to search for classes of functions whose Clarke subdifferentials are minimal cuscos. We refer the interested readers to [43, 57, 31, 70] and the references therein.

When discussing the geometry of a set, the variational technique naturally leads to an emphasis on its normal cones. However, in the development of nonsmooth optimality conditions, tangent cones and their duality with normal cones in the tradition of convex analysis still play an important role. The guided Exercise 5.2.21 provides a glimpse into the various tangent cones and their relationship with the normal cones we have examined. Historically necessary optimality conditions for constrained minimization problems were developed in terms of the tangent vectors first. Different ways of defining the tangent cones will lead to differential strength of the necessary optimality conditions. This is illustrated in Exercise 5.2.27. Although in the nonsmooth setting it is clear now that one can often formulate more precise necessary optimality conditions in terms of the normal cones for constrained minimization problems, the tangent cone perspective is still useful in providing intuitive understanding of the geometry of these problems.

Exercise 5.2.1 Let X be a Banach space and let $f\colon X \to \mathbb{R}$ be a locally Lipschitz function. Suppose that f attains a local minimum at $\bar{x}$. Prove that $f^\circ(\bar{x}; h) \geq 0$ for any $h \in X$.

Exercise 5.2.2 Prove Proposition 5.2.2. Reference: [84, p. 26].

Exercise 5.2.3 Verify Example (5.2.13).

Exercise 5.2.4 Let X be a Banach space and let $f\colon X \to \mathbb{R}$ be a function. We say that f is strictly differentiable at x provided that $f'(x)$ exists and for any $\varepsilon > 0$ there exists a $\delta > 0$ such that

$$\left| \frac{f(z + th) - f(z)}{t} - \langle f'(x), h \rangle \right| < \varepsilon$$

for all $t \in (0, \delta)$, $\|z - x\| < \delta$ and $h \in B_X$. Prove that $\partial_C f(x)$ is a singleton if and only if f is strictly differentiable at x.

Exercise 5.2.5 Show that $\partial_F f(x) \subset \partial_C f(x)$.

Exercise 5.2.6 Prove Theorem 5.2.18.

Exercise 5.2.7 Verify that $\partial_F f(x) \subset \partial_L f(x)$ and $N_F(S; x) \subset N_L(S; x)$.

Exercise 5.2.8 Prove Theorem 5.2.22.

Exercise 5.2.9 Let X be a Banach space and let $f\colon X \to \mathbb{R} \cup \{+\infty\}$ be a lsc function. Suppose that f attains a local minimum at $\bar{x}$. Show that $0 \in \partial_L f(\bar{x})$.

Exercise 5.2.10 Verify Example 5.2.23.

Exercise 5.2.11 Take limits in the weak approximate multiplier rule of Theorem 3.3.8 to prove Theorem 5.2.25.

Exercise 5.2.12 Deduce the limiting chain rule of Theorem 5.2.26 from Theorem 5.2.25. Hint: Note the fact that $f(f_1, \ldots, f_M)$ attains a minimum at $\bar{x}$ implies that $(\bar{x}, f_1(\bar{x}), \ldots, f_M(\bar{x}))$ is a solution of the minimization problem:

$$\text{minimize} \quad f(y_1, \ldots, y_M) \tag{5.2.11}$$
$$\text{subject to} \quad f_n(x) \leq y_n, n = 1, \ldots, N,$$
$$f_n(x) = y_n, n = N + 1, \ldots, M.$$

Exercise 5.2.13 Deduce the limiting sum rule of Theorem 5.2.24 from Theorem 5.2.26.

Exercise 5.2.14 Prove Theorem 5.2.27.

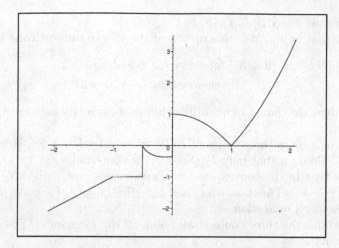

Fig. 5.1. Determine the various subdifferentials.

Exercise 5.2.15 (Extremal Principle and Convex Separation) Let X be a finite dimensional Banach space and let S_1, S_2 be closed convex subsets of X. Suppose that $S_1 \cap S_2 = \{\bar{x}\}$.

(i)   Show that $\bar{x}$ is an extremal point of the extremal system of fixed sets (S_1, S_2).
(ii) Deduce a convex separation theorem from the limiting extremal principle of Theorem 5.2.27.

Exercise 5.2.16 Prove that $\partial^\infty f$ and N_L in Proposition 5.2.28 are closed multifunctions.

Exercise 5.2.17 Verify that the set S defined in Example 5.2.30 is closed.

Exercise 5.2.18 Verify that $N_L(S; 0) \ni e_1 + j^{-1}e_j \to e_1$ for S defined in Example 5.2.30. Hint: Check that, for $1 < j < k$, $e_{j,k} = e_1 + j^{-1}e_j + je_k \in N_F(S; k^{-1}(je_1 - e_k))$ and $(e_{j,k})$ converges weakly to $e_1 + j^{-1}e_j$ as $k \to \infty$.

Exercise 5.2.19 Supplement the details for Example 5.2.31.

Exercise 5.2.20 The lower semicontinuous function drawn in Figure 5.1 is continuously differentiable except at $z := -1, -1/2, 0, 1$. Determine its Fréchet, Limiting and Clarke subdifferentials graphically at each such z.

*Exercise 5.2.21** The classical or *Bouligand tangent cone*, also called the *contingent cone*, is

$$T_B(S; x) := \Big\{ d \in X \mid d = \lim_{i \to \infty} \frac{x_i - x}{t_i}, t_i \downarrow 0, x_i \to_S x \Big\},$$

and the *pseudotangent cone* is its closed convex hull, $T_P(S\,;x) := \overline{\mathrm{co}}\,T_B(S;x)$.

(i) Show that $T_C(S; x) \subseteq T_B(S; x) \subseteq T_P(S; x)$.

(ii) Show that an intrinsic description of the Clarke tangent cone is

$$T_C(S; x) := \{d \in X \mid \text{for every } t_i \downarrow 0 \text{ and } x_i \to_S x,$$
$$\text{there exists } s_i \to_S x \text{ with } (s_i - x_i)/t_i \to d\}.$$

Note how the change of quantifiers imposes convexity and destroys monotonicity.

(iii) Compute these three cones at $(0, 0)$ when $S := \{0\} \times \mathbb{R} \cup \mathbb{R} \times \{0\}$ (the axes). Confirm that only $T_B(S; (0, 0))$ is nontrivial.

(iv) Show that in the convex case all three cones agree with $-N_F(S; x)^\circ = \overline{cl}\, \mathbb{R}^+(S - x)$. (The three cones also agree in the case of a smooth manifold as discussed in Section 7.)

(v) Determine the three cones at all points of the *Pacman set* given in polar coordinates by

$$P(\sigma) := \{(r, \theta) \mid 0 \leq r \leq 1, |\theta| \geq \sigma\}.$$

The set $P(1/5)$ is shown in Figure 5.2.

(vi) Observe that the Clarke and pseudo-tangent cones are necessarily convex while the contingent cone need not be. Note also that the Clarke cone is not monotone – it may decrease in size as the set grows. Finally, the Bouligand cone is often not convex while the pseudotangent cone may often be "too" big.

Exercise 5.2.22 (Representation of the Clarke Tangent Cone) Let X be a Banach space and let S be a closed subset of X. Show that for any $\bar{x} \in S$,

$$T_C(S; \bar{x}) = \{h \in X \mid d_S^\circ(\bar{x}; h) \leq 0\}.$$

Exercise 5.2.23 (Representation of the Contingent Cone) A similar characterization exists for the contingent cone. Recall that the Dini directional derivative of a locally Lipschitz function $f : X \to \mathbb{R}$ at x in the direction $h \in X$ is defined by

$$f^-(x; h) := \liminf_{t \to 0+} \frac{f(x + th) - f(x)}{t}.$$

Let X be a Banach space and let S be a closed subset of X. Show that, for any $\bar{x} \in S$,

$$T_B(S; \bar{x}) = \{h \in X \mid d_S^-(\bar{x}; h) \leq 0\}$$

*Exercise 5.2.24 (Tangential Regularity) Let X be a Banach space and let S be a closed subset of X. We say S is *regular* at $\bar{x} \in S$ provided that $T_C(S; \bar{x}) = T_B(S; \bar{x})$. In terms of the Dini directional derivative, regularity of a locally Lipschitz function f at x is equivalent to $f^\circ(x; h) = f^-(x; h)$ for all $h \in X$. Show that if d_S is regular at $\bar{x} \in S$ then S is regular at $\bar{x}$ and the converse holds when X is finite dimensional. Reference: [75].

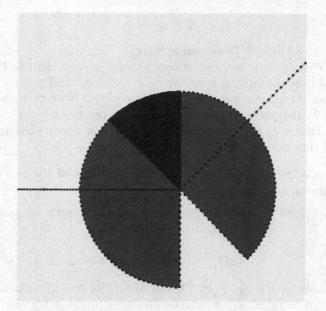

Fig. 5.2. A Pacman set, star and normal cone.

*Exercise 5.2.25 Show that the converse of the conclusion in Exercise 5.2.24 fails when X is infinite dimensional. Hint: Let (e_i) be an infinite sequence in the unit sphere of X satisfying $\|e_i - e_j\| \geq \delta, i \neq j$ for some $\delta > 0$. Define

$$S := \{0\} \cup \left\{ 4^{-i}(e_0 + \frac{1}{4}e_i) \mid i = 1, 2, \dots \right\}.$$

Then verify that $T_B(S;0) = T_C(S;0) = \{0\}$, yet $d_S^-(0;e_0) \leq 1/4 < 1 = d_S^{\circ}(0;e_0)$. Reference: [39].

Exercise 5.2.26 Consider the problem of minimizing a Fréchet differentiable function f over a closed set S. Show that a necessary condition for a local optimum to occur at $\bar{x}$ is

$$f'(\bar{x}) \in -T_P(S;\bar{x})^{\circ}$$

and that this is sufficient when f is convex and S is *pseudoconvex* at $\bar{x}$ in the sense that $S - \bar{x} \subseteq T_P(S;\bar{x})$.

Exercise 5.2.27 Consider the problem of minimizing $f(x,y) = x^2 + y$ over $S := \{0\} \times \mathbb{R} \cup \mathbb{R} \times \{0\} \subset \mathbb{R}^2$.

The necessary optimality condition

$$f'(x) \in -T_P(S;x)^{\circ} = -T_B(S;x)^{\circ}$$

clearly implies the necessary condition $f'(x) \in -N_C(S;x)$. Alternatively, we can derive the above necessary optimality condition as well as a tighter one

$$f'(x) \in -N_L(S; x)$$

from $f'(x) \in -N_F(S; x)$ (Proposition 3.1.7).

Check that $(0, 0)$ is not a solution to the above problem and show that, of the candidate conditions, only $f'(x) \in -T_P(S; x)^o$ and $f'(x) \in -N_F(S; x)$ can rule out $(0, 0)$ as a candidate for optimality. (Note that the necessary conditions in the previous exercise actually apply to Gâteaux differentiable functions. Hence, examples similar to this one can also be constructed with function f that are not so "nice".)

*Exercise 5.2.28** Prove the following representation of the contingent cone of the preimage of a surjective function.

Theorem 5.2.36 (Liusternik) *Let* $g: \mathbb{R}^N \to \mathbb{R}^M$ *be a* C^1 *mapping. Suppose that* $g'(\bar{x}): \mathbb{R}^N \to \mathbb{R}^M$ *is surjective. Then*

$$T_B(g^{-1}(g(\bar{x})); \bar{x}) = \operatorname{Ker} g'(\bar{x}).$$

Hint: Fix a vector $v \in \operatorname{Ker} g'(\bar{x})$. Choose any $N \times (N - M)$ matrix D making the matrix $A = (g'(\bar{x}), D)$ invertible. Define a function $h(x) := (g(x), Dx)$, and for small real $\delta > 0$ define

$$p(t) := h^{-1}(h(\bar{x}) + tAv).$$

Prove that p is well-defined when δ is small and that p is C^1 with :

(i) $p(0) = \bar{x}$.
(ii) $p'(0) = v$.
(iii) $g(p(t)) = g(\bar{x})$ for all small t.

Thence, deduce that $v \in \operatorname{Ker} g'(\bar{x})$ if and only if for some $\delta > 0$ there exists a C^1 function $p: (-\delta, \delta) \to \mathbb{R}^N$ satisfying the three conditions above.

Exercise 5.2.29 (Lagrange Multiplier Rule) Consider

$$\text{minimize } f(x) \text{ subject to } \quad g(x) = 0, \tag{5.2.12}$$

where $f: \mathbb{R}^N \to \mathbb{R}$ and $g: \mathbb{R}^N \to \mathbb{R}^M$ are C^1 mappings. Suppose that $\bar{x}$ is a solution to the constrained minimization problem (5.2.12) and suppose that $g'(\bar{x}): \mathbb{R}^N \to \mathbb{R}^M$ is surjective. Show that there exists $\lambda \in \mathbb{R}^M$ such that

$$f'(\bar{x}) + \langle \lambda, g'(\bar{x}) \rangle = 0.$$

Hint: Note that problem (5.2.12) is equivalent to minimizing f over $S = g^{-1}(0) = g^{-1}(g(\bar{x}))$. Then use Exercise 5.2.21 and Theorem 5.2.36.

Exercise 5.2.30 (Ubiquity) Show that the existence of a function as in Example 5.2.12 implies the existence of a ubiquitous set. Use the Baire category theorem to help construct a ubiquitous set.

*Exercise 5.2.31 The star of a set S is the set of points $x \in S$ such that $[x, s] \subset S$ for all $s \in S$.

(i)   Show that in any Banach space,

$$\text{star}(S) = \bigcap_{x \in S} x + T_B(S; x) = \bigcap_{x \in S} x + T_C(S; x).$$

(ii) If X has a Fréchet renorm, show

$$\text{star}(S) = \bigcap_{x \in S} x + T_P(S; x).$$

Deduce that a closed set in such a space is pseudoconvex at all its points if and only if it is convex. Reference: [56].

(iii) Verify these claims for the set in Figure 5.2.

Exercise 5.2.32 Let C be a w*-compact convex subset of X^*. Show that C is norm bounded and $f \in \mathcal{N}_C$ implies that f is Lipschitz with a Lipschitz constant $L = \sup\{\|x^*\| \mid x^* \in C\}$.

Exercise 5.2.33 Show that for every $x \in A$, the function $\sigma_C(\cdot - x)$ belongs to $\mathcal{N}_C$.

Exercise 5.2.34 Let Γ be a set and let $f_\gamma \in \mathcal{N}_C$ for $\gamma \in \Gamma$. Show that $\sup_{\gamma \in \Gamma} f_\gamma$ and $\inf_{\gamma \in \Gamma} f_\gamma$ belong to $\mathcal{N}_C$ when they are finite everywhere.

*Exercise 5.2.35 Generalize the results in Theorem 5.2.33 to an unbounded set A. Hint: Consider:

$$\mathcal{X}_C := \{f : X \mapsto \mathbb{R} \mid f(x) - f(y) \leq \sigma_C(x - y) \text{ for } x, y \in X\}.$$

Define the metric of uniform convergence on bounded sets, $\tilde{\rho}$, on $\mathcal{X}_C$ by

$$\tilde{\rho}(f, g) := \sum_{i=1}^{\infty} \frac{1}{2^i} \frac{\rho_i(f, g)}{1 + \rho_i(f, g)} \quad \text{where } \rho_i(f, g) := \sup_{x \in iB} |f(x) - g(x)|.$$

For the metric $\tilde{\rho}$, $f \to g$ if and only if $f \to g$ on iB in the metric ρ_i for every i. In an entirely standard fashion, we may verify that $(\mathcal{X}_C, \tilde{\rho})$ is complete. For fixed x, v with f_0 and t as in equation (5.2.7), there exists some integer $I > 0$ such that $\|x\| \leq I, \|x + tv\| \leq I$. By definition of $\tilde{\rho}$,

$$\frac{1}{2^I} \frac{\rho_I(f, f_0)}{1 + \rho_I(f, f_0)} \leq \tilde{\rho}(f, f_0).$$

For $\tilde{\rho}(f, f_0) < \varepsilon$, we have $\rho_I(f, f_0) \leq (2^I \varepsilon)/(1 - 2^I \varepsilon)$. Thus for small ε, the same argument in (a) applies to guarantee G_k being open. The arguments in (b)–(e) still apply. Hence we can get

Corollary 5.2.37 *Let X be a Banach space, let A be a bounded subset of X and let C be a w*-compact convex subset of X^*. Then in $(\mathcal{X}_C, \tilde{\rho})$, the set*

$$\{f \in \mathcal{X}_C \mid \partial_C f \equiv C \text{ on } X\}$$

is residual.

5.3 Distance Functions

We discuss distance functions in Hilbert spaces to illustrate the various variational and nonsmooth analysis techniques.

5.3.1 Distance Functions as Differences of Convex Functions

Many nice properties of a distance function are due to the fact that it is the difference of two convex functions (a DC function), one of which is smooth. To prove this fact we need the following technical lemma.

Lemma 5.3.1 *Let X be a Hilbert space, let $G \subset X$ be an open convex set and let g be a Fréchet differentiable function whose derivative g' is Lipschitz on G with a Lipschitz constant L. Then the function $f(x) := L\|x\|^2/2 - g(x)$ is convex on G.*

Proof. Let $a \in G$ and let $v \in X$ be a unit vector. Define $h(t) := f(a + tv)$. Then, for any $t_2 > t_1$ and $a + t_1 v, a + t_2 v \in G$ we have

$$
\begin{aligned}
h'(t_2) - h'(t_1) &= L\langle a + t_2 v, v\rangle - L\langle a + t_1 v, v\rangle \\
&\quad - \langle g'(a + t_2 v), v\rangle + \langle g'(a + t_1 v), v\rangle \\
&\geq L(t_2 - t_1) - \|g'(a + t_2 v) - g'(a + t_1 v)\| \geq 0.
\end{aligned}
$$

Therefore h is convex on the interval $\{t \mid a + tv \in G\}$ and consequently f is convex on G $\qquad\bullet$

Now we can prove that a distance function on a Hilbert space is always a DC function.

Theorem 5.3.2 (Distance Functions Are DC) *Let X be a Hilbert space and let $S \subset X$ be a closed set. Then, d_S is locally the difference of a C^1 convex function and a convex function on $X \backslash S$.*

Proof. Let $\bar{x} \in X \backslash S$ and let $G = \{x \in X \mid \|x - \bar{x}\| < d_S(\bar{x})/2\}$. Then, for any $y \in S$, $x \to \|x - y\|' = (x - y)/\|x - y\|$ is Lipschitz with a Lipschitz constant $L = 4/d_S(\bar{x})$ (Exercise 5.3.1). By Lemma 5.3.1 each function $x \to L\|x\|^2/2 - \|x - y\|, y \in S$ is convex on G, and therefore the function

$$
c(x) := L\|x\|^2/2 d_S(x) = \sup\{L\|x\|^2/2 - \|x - y\| \mid y \in S\}
$$

is continuous and convex on G. Now $d_S(x) = L\|x\|^2/2 - c(x)$, as was to be shown. $\qquad\bullet$

5.3.2 The Clarke Subdifferential of a Distance Function

We have seen that the Clarke subdifferential of a Lipschitz function is always a cusco but not necessarily a minimal one. Now we have

Theorem 5.3.3 (Minimality of the Clarke Subdifferential of a Distance Function) *Let X be a Hilbert space and let $S \subset X$ be a closed set. Then, $\partial_C(-d_S)$ is a minimal w^*-cusco on $X \backslash S$. Consequently, $\partial_C(-d_S)$ is a minimal w^*-cusco on X.*

Proof. Let c be the continuous convex function in the proof of Theorem 5.3.2. Then

$$\partial_C(-d_S)(x) = \partial c(x) - Lx.$$

Since ∂c is a minimal w^*-cusco on $X \backslash S$ (Theorem 5.1.7, Corollary 5.1.28 and Theorem 5.1.29) so is $\partial_C(-d_S)$. ●

5.3.3 Closest Points

Let X be a Banach space and let $S \subset X$ be a closed set. Consider any $y \notin S$. We say $x \in S$ is a closest point to y in S provided that $\|y - x\| = d_S(y)$. It turns out that in a Hilbert space X at any point $y \in X \backslash S$ where $\partial_F d_S(y) \neq \emptyset$, y has a closest point in S.

Theorem 5.3.4 (Subdifferential of the Distance Function and Closest Point) *Let X be a Hilbert space and let $S \subset X$ be a closed set. Suppose that $x \notin S$ and $x^* \in \partial_F d_S(x)$. Then there exists $\bar{x} \in S$ such that*

(i) *every minimizing sequence (x_i) in S of $\inf\{\|s - x\| \mid s \in S\}$ converges to $\bar{x}$, so that $\bar{x}$ is the unique closest point of x in S;*

(ii) *the distance function d_S is Fréchet differentiable at $\bar{x}$ and $x^* = d_S'(x) = (x - \bar{x})/\|x - \bar{x}\|$; and*

(iii) *$x^* \in N_F(S; \bar{x})$.*

Proof. Let g be a C^1 function such that $g'(x) = x^*$ and $d_S - g$ attains a minimum 0 at x. It is not hard to check the identity (Exercise 5.3.2)

$$d_S^2(y) - d_S^2(x) = 2d_S(x)(d_S(y) - d_S(x)) + (d_S(y) - d_S(x))^2.$$

It follows that

$$d_S^2(y) - d_S^2(x) \geq 2d_S(x)(d_S(y) - d_S(x)) \geq 2d_S(x)(g(y) - (g(x)),$$

and therefore $2d_S(x)x^* \in \partial_F d_S^2(x)$. Now by Proposition 3.1.3 we have

$$d_S^2(y) - d_S^2(x) \geq \langle 2d_S(x)x^*, y - x \rangle + o(\|y - x\|). \tag{5.3.1}$$

Let (x_i) be a minimizing sequence of $\inf\{\|s - x\| \mid s \in S\}$ in S. Then there exists a sequence of positive numbers $\varepsilon_i \to 0$ such that

$$d_S^2(x) \geq \|x_i - x\|^2 - \varepsilon_i^2. \tag{5.3.2}$$

Combining (5.3.1) and (5.3.2) we have

$$
\begin{aligned}
\langle 2d_S(x)x^*, y - x \rangle &\leq d_S^2(y) - d_S^2(x) + o(\|y - x\|) \\
&\leq \|x_i - y\|^2 - \|x_i - x\|^2 + \varepsilon_i^2 + o(\|y - x\|) \\
&= \langle 2x_i - y - x, x - y \rangle + \varepsilon_i^2 + o(\|y - x\|) \\
&= \langle 2(x_i - x), x - y \rangle + \|y - x\|^2 + \varepsilon_i^2 + o(\|y - x\|)
\end{aligned}
$$

or

$$\langle 2(d_S(x)x^* - x + x_i), y - x \rangle \leq \|y - x\|^2 + \varepsilon_i^2 + o(\|y - x\|). \tag{5.3.3}$$

For any $v \in B_X$, setting $y = x + \varepsilon_i v$ in (5.3.3) and dividing by ε_i we have

$$\langle 2(d_S(x)x^* - x + x_i), v \rangle \leq \varepsilon_i(\|v\| + 1) + o(1).$$

Since $v \in B_X$ is arbitrary, it follows that

$$\|d_S(x)x^* - x + x_i\| \to 0$$

or $x_i \to \bar{x} \in S$ where

$$\bar{x} = x - d_S(x)x^*. \tag{5.3.4}$$

Clearly $\bar{x}$ is a unique minimizer of $\inf\{\|s - x\| \mid s \in S\}$, which verifies (i).

Since $d_S^2(x) = \|x - \bar{x}\|^2$ and for any $y \in X$, $d_S^2(y) \leq \|y - \bar{x}\|^2$, we have

$$
\begin{aligned}
d_S^2(x) - d_S^2(y) &\geq \|x - \bar{x}\|^2 - \|y - \bar{x}\|^2 \\
&= \langle x + y - 2\bar{x}, x - y \rangle \\
&= \langle 2(x - \bar{x}), x - y \rangle - \|x - y\|^2 \\
&= \langle 2d_S(x)x^*, x - y \rangle - \|x - y\|^2. \tag{5.3.5}
\end{aligned}
$$

Thus, $2d_S(x)x^* \in \partial_F d_S^2(x)$ (Exercise 5.3.3). It follows from Exercise 3.1.4 that d_S^2 is Fréchet differentiable at x and $(d_S^2)'(x) = 2d_S(x)x^*$. Using identity

$$\frac{d_S^2(x + tv) - d_S^2(x)}{t} = \frac{d_S(x + tv) - d_S(x)}{t}(d_S(x + tv) + d_S(x))$$

we have that both sides converge to $\langle 2d_S(x)x^*, v \rangle$ uniformly as $t \to 0+$ for v in bounded sets. Since $d_S(x + tv) + d_S(x)$ converges to $2d_S(x) > 0$ uniformly for v in bounded sets as $t \to 0+$, it follows that the difference quotient

$$\frac{d_S(x + tv) - d_S(x)}{t}$$

converges to $\langle x^*, v \rangle$ uniformly as $t \to 0+$ for v in bounded sets. Thus, $d_S'(x) = x^*$. By (5.3.4) we have $x^* = (x - \bar{x})/d_S(x) = (x - \bar{x})/\|x - \bar{x}\|$, this verifies (ii).

Finally, observing that function $s \to \iota_S(s) + \|s - x\|^2$ attains minimum at $\bar{x}$ we have $2(x - \bar{x}) \in N_F(S; \bar{x})$. Hence $x^* = (x - \bar{x})/\|x - \bar{x}\| \in N_F(S; \bar{x})$, since this set is a cone. The theorem is proved.   ●

Since $\partial_F d_S$ is nonempty on a dense set of X we have the following corollary.

Corollary 5.3.5 *Let X be a Hilbert space and let $S \subset X$ be a closed subset. Then d_S is attained on a dense subset of X.*

Proof. Since $\partial_F d_S$ is nonempty on a dense subset of $X \backslash S$, d_S is attained on every such point by Theorem 5.3.4. Clearly, d_S is attained on every point of S.   ●

Closest points are interesting in approximation theory. Ideally given a set S of "nice" elements in a Hilbert space X. We hope for each point $x \in X$ there is a unique closest point in S. A *norm* closed set that has the above property is called a *Chebyshev set*. We can easily deduce from Theorem 5.3.4 that every closed convex set in a Hilbert space is a Chebyshev set.

Corollary 5.3.6 *Let X be a Hilbert space and let C be a closed convex subset of X. Then C is a Chebyshev set.*

Proof. Note that the distance function d_C is convex and everywhere finite. It follows from Theorem 4.2.8 that $\partial_F d_C(x) = \partial d_C(x) \neq \emptyset$ for all $x \in X$. Thus, the conclusion follows directly from Theorem 5.3.4.   ●

5.3.4 Commentary and Exercises

Distance functions associated with sets comprise a most important class of nonsmooth functions. They closely reflect the properties of the corresponding sets and norms and have very nice properties. Besides being DC-functions – in the Hilbert case – they are also *nonexpansive*, i.e., Lipschitz with a Lipschitz constant 1, and can be represented as an inf-convolution (Exercises 5.3.4). They play important roles in exact penalization (see Example 3.0.3) and in the relationship between functions and sets (see Exercises 5.2.24, 5.2.25 and Section 5.5). Theorem 5.3.4 and Corollary 5.3.5 actually work beyond the Hilbert space settings. Whether the converse of Corollary 5.3.6 holds in infinite dimensions is a long standing open problem. Much research has been done in this area and we refer the interested reader to [126, 259]. Exercises 5.3.8–5.3.12 introduce the notion of a *Chebyshev sun* and use this to prove that a closed Chebyshev set in Euclidean space is necessarily convex. Infinite dimensional versions are given in Chapter 12 of [97], where it is conjectured that there is a non-convex Chebyshev set in any infinite dimensional Hilbert space.

Exercise 5.3.1 Let $\bar{x} \in X \backslash S$ and let $G = \{x \in X \mid \|x - \bar{x}\| < d(S; \bar{x})/2\}$. Given $y \in S$, prove that $x \to \|x - y\|' = (x - y)/\|x - y\|$ is Lipschitz with a Lipschitz constant $4/d(S; \bar{x})$. Hint: Note that for any $x \in G$ and $y \in S$, $\|x - y\| > d(S; \bar{x})/2$.

Exercise 5.3.2 Verify the identity

$$d_S^2(y) - d_S^2(x) = 2d_S(x)(d_S(y) - d_S(x)) + (d_S(y) - d_S(x))^2.$$

Exercise 5.3.3 Verify that inequality (5.3.2) implies $2d_S(x)x^* \in \partial^F d_S^2(x)$.

Exercise 5.3.4 Verify that $d_S = d_{\overline{S}} = \iota_{\overline{S}} \square \|\cdot\|$ and by Exercise 1.3.7 is a nonexpansive function (a Lipschitz function with rank 1).

Exercise 5.3.5 (Proximal Normal Cone) Let S be a closed subset of a Hilbert space X. If $x \notin S$ and $s \in S$ satisfies $d_S(x) = \|x - s\|$, i.e., s is a closest point to x in S, then any nonnegative multiple of $x - s$ is a *proximal normal vector* to S at s.

The set of all proximal normal vectors to S at s is denoted $N_P(S; s)$.

(i)   Show that $N_P(S; s)$ is a (typically non-convex) cone containing zero, called the *proximal normal cone* of S at s.
(ii) Show that $N_P(S; s) = \{0\}$ when $s \in \text{int } S$.
(iii) Construct an example with $N_P(S; s) = \{0\}$ for some $s \in \text{bd}(S)$.
(iv) Show that $N_P(S; s) \neq \{0\}$ densely in $\text{bd}(S)$.
(v)   Prove that $x^* \in N_P(S; \bar{x})$ if and only if there exists $r > 0$ such that

$$\langle x^*, s - \bar{x} \rangle \leq r\|s - \bar{x}\|^2, \quad \text{for all } s \in S.$$

Conclude that $N_P(S; \bar{x}) \subset N_F(S; \bar{x})$, usually properly.

Exercise 5.3.6 Show that Theorem 5.3.4 (iii) can be strengthened to

$$x^* \in N_P(S; \bar{x}).$$

*Exercise 5.3.7** (Proximal Normal Formula) Let S be a closed subset of a Hilbert space X and let $\bar{x} \in S$. Use Exercise 5.3.6 to prove the following *proximal normal formula*

$$N_C(S; \bar{x}) = \overline{\text{conv}}\Big\{ \lim_{i \to \infty} x_i^* \mid x_i^* \in N_P(S; x_i), x_i \to_S \bar{x} \Big\}$$

by recalling $N_C(S; \bar{x}) = \bigcup_{r \geq 0} r \partial_C d_S(\bar{x})$ and showing that

$$\partial_C d_S(\bar{x}) = \overline{\text{conv}}\Big\{ \lim_{i \to \infty} x_i^* \mid x_i^* \in N_P(S; x_i), x_i \to_S \bar{x} \Big\}.$$

Exercise 5.3.8 Let K be a Chebyshev subset of an inner product space. Show that

$$P_K[x + t(P_K(x) - x)] = P_K(x),$$

for $1 \geq t \geq 0$, and $x \in X$.

*Exercise 5.3.9 A subset K of an inner product space X is a *sun* if

$$P_K[x + t(x - P_K(x))] = P_K(x),$$

for all $t \geq 0$ and $x \in X$.

Show that the following coincide for a set.

(i) K is convex.
(ii) K is a sun.
(iii) P_K is nonexpansive.

Hint: Show (iii) implies (i) implies (ii) implies $P_K(x) = P_{[k,P_K(x)]}(x)$ for each k in K, which implies (iii).

*Exercise 5.3.10 Let X be a finite dimensional inner product space. Show that P_K is continuous when K is a Chebyshev set in X.

*Exercise 5.3.11 (Convexity of Sets, I) Let X be a finite dimensional inner product space. Show that a set in X is a Chebyshev set if and only if it is closed and convex. Hint: By Exercise 5.3.9 it suffices to show that every set K is a sun.

(i)   Suppose K is not a sun. Then there is a point $x \notin K$ such that

$$t_0 = \sup\{t \geq 0 \,|\, P_K(x + t(x - P_K(x)) = P_K(x)\} < \infty$$

   is attained.
(ii) Denote $x(t) := x + t(x - P_K(x))$ and note that $x_0 := x(t_0)$ is the point on the ray farthest from $P_K(x)$ such that $P_K(x(t)) = P_K(x)$.
(iii) Pick $\varepsilon > 0$ such that $B_\varepsilon(x_0) \cap K = \emptyset$. Define a self map Q on the closed convex set $B_\varepsilon(x_0)$ by

$$Q(y) := x_0 + \varepsilon \frac{x_0 - P_K(y)}{\|x_0 - P_K(y)\|}.$$

   Use Exercise 5.3.10 to show Q is well defined and continuous.
(iv) By Brouwer's fixed point theorem, Q has a fixed point x_1 in B_ε'.
(v)   Show that x_0 is a convex combination of x_1 and $P_K(x_1)$ and then use Exercise 5.3.8 to conclude that $P_K(x_0) = P_K(x_1)$.
(vi) Use (iv) to show that x_1 is farther from $P_K(x)$ than x_0 on the ray $x + t(x - P_K(x))$, $t \geq 0$. This contradicts the construction of x_0.

The same proof works for bounded relatively compact sets in Hilbert space with Schauder's theorem used instead of Brouwer's – one needs to show $Q(B_\varepsilon)$ is relatively compact. Indeed, for any non convex set K there must be an uncountable spanning set of points at which P_K fails to be continuous.

An alternative finite dimensional approach is as follows.

*Exercise 5.3.12 (Convexity of Sets, II) Let X be a finite dimensional inner product space.

(i) Show that a closed set K in X is a set if and only if d_K^2 is everywhere Fréchet differentiable.

(ii) Show that when d_K^2 is everywhere Fréchet differentiable one has $d_K = d_K^{**} = d_{\operatorname{conv} K}$, and so K is convex.

5.4 Coderivatives of Multifunctions

5.4.1 Definitions and Examples

Coderivatives are convenient derivative-like objects for multifunctions. We define them via normal cones to the graph of the multifunctions and we emphasize studying coderivatives by related subdifferentials of the indicator function for the graph of the multifunction.

Definition 5.4.1 (Coderivatives) *Let X and Y be Banach spaces, let $F\colon X \to 2^Y$ be a closed multifunction and let $(\bar{x}, \bar{y}) \in \operatorname{graph} F$. The* Fréchet coderivative (limiting coderivative) *of F at $(\bar{x}, \bar{y})$ is defined by:*

$$D_F^* F(\bar{x}; \bar{y})(y^*) := \{x^* \in X^* \mid (x^*, -y^*) \in N_F(\operatorname{graph} F; (\bar{x}, \bar{y}))\}.$$

$$\left(D_L^* F(\bar{x}; \bar{y})(y^*) := \{x^* \in X^* \mid (x^*, -y^*) \in N_L(\operatorname{graph} F; (\bar{x}, \bar{y}))\}\right).$$

Since F is a multifunction, in general, when discussing coderivatives we must specify the point $\bar{y} \in F(\bar{x})$ to avoid ambiguity. If F is a single valued function then we must have $\bar{y} = F(\bar{x})$ and in this case we will use the notation $D_F^* F(\bar{x})$ $(D_L^* F(\bar{x}))$ instead of $D_F^* F(\bar{x}; F(\bar{x}))$ $(D_L^* F(\bar{x}; F(\bar{x})))$. When F is a C^1 function both the Fréchet and the limiting coderivative of F coincide with the dual of the Fréchet derivative of F (Exercise 5.4.1). Thus, coderivatives are natural generalizations of the derivative concept for functions.

We have seen that the coderivative is defined through the subdifferential of the indicator function of the graph of the multifunction. On the other hand, given a lsc function, its subdifferentials are completely characterized by the coderivatives of its profile mapping.

Proposition 5.4.2 *Let X be a Banach space and let $f\colon X \to \mathbb{R} \cup \{+\infty\}$ be a lsc function. Then*

$$\partial_F f(x) = D_F^* E_f(x; f(x))(1),$$

$$\partial_L f(x) = D_L^* E_f(x; f(x))(1),$$

and

$$\partial^\infty f(x) = D_L^* E_f(x; f(x))(0).$$

Proof.   Exercise 5.4.2.   ●

To effectively use coderivatives of multifunctions we also need to develop calculus rules. These calculus rules may be established by reducing them to calculus for subdifferentials of indicator functions of the graphs of the corresponding multifunctions. As with the approximate calculus for functions, the calculus for the Fréchet coderivative can be derived in weak (accurate up to a weak-star neighborhood) and strong (accurate up to a norm neighborhood) forms.

5.4.2 Weak Fréchet Coderivative Calculus

We start with the weak coderivative calculus. It applies to general closed multifunctions.

Theorem 5.4.3 (Weak Coderivative Sum Rule) *Let X and Y be Fréchet-smooth Banach spaces. Let $F_n, n = 1, \ldots, N$ and $F = \sum_{n=1}^{N} F_n$ be closed multifunctions from X into Y, and $\bar{y} \in \sum_{n=1}^{N} F_n(\bar{x})$. Fix arbitrary $\bar{y}_n \in F_n(\bar{x}), n = 1, \ldots, N$ with $\bar{y} = \sum_{n=1}^{N} \bar{y}_n$. Suppose $x^* \in D_F^* F(\bar{x}; \bar{y})(y^*)$. Then for any $\varepsilon > 0$ and any weak-star neighborhoods, U and V, of the origins in X^* and Y^* respectively, there exist $(x_n, y_n) \in (\text{graph } F_n) \cap B_\varepsilon((\bar{x}, \bar{y}_n)), y_n^* \in y^* + V, n = 1, \ldots, N$ and $x_n^* \in D_F^* F_n(x_n; y_n)(y_n^*)$ with $\max_{n=1,\ldots,N}(\|x_n^*\|, \|y_n^*\|) \times \text{diam}((x_1, y_1), \ldots, (x_N, y_N)) < \varepsilon$ such that*

$$x^* \in \sum_{n=1}^{N} x_n^* + U. \tag{5.4.1}$$

Proof. Let $x^* \in D_F^* F(\bar{x}, \bar{y})(y^*)$. Then there exists a C^1 function g on $X \times Y$ with $(x^*, -y^*) = g'(\bar{x}, \bar{y})$ such that

$$(x, y) \to \iota_{\text{graph}(\sum_{n=1}^{N} F_n)}(x, y) - g(x, y)$$

attains a local minimum 0 at $(\bar{x}, \bar{y})$. Since

$$\sum_{n=1}^{N} \iota_{\text{graph} F_n}(x, y_n) \geq \iota_{\text{graph} F}\left(x, \sum_{n=1}^{N} y_n\right)$$

and

$$\sum_{n=1}^{N} \iota_{\text{graph} F_n}(\bar{x}, \bar{y}_n) = \iota_{\text{graph} F}(\bar{x}, \bar{y}) = 0,$$

the function

$$(x, y_1, y_2, \ldots, y_N) \to \sum_{n=1}^{N} \iota_{\text{graph} F_n}(x, y_n) - g\left(x, \sum_{n=1}^{N} y_n\right)$$

attains a local minimum at $(\bar{x}, \bar{y}_1, \bar{y}_2, \ldots, \bar{y}_N)$. Thus

$$(x^*, -y^*, \ldots, -y^*) \in \partial_F\left(\sum_{n=1}^{N} \iota_{\text{graph} F_n}(\bar{x}, \bar{y}_n)\right).$$

Invoking the weak local approximate sum rule of Theorem 3.3.3, there exist $x_n \in B_\varepsilon(\bar{x})$ and $y_n \in B_\varepsilon(\bar{y}_n)$, as well as $y_n^* \in Y^*$ and $x_n^* \in D_F^* F(x_n; y_n)(y_n^*)$ with

$$\max_{n=1,\ldots,N}(\|x_n^*\|, \|y_n^*\|) \times \mathrm{diam}((x_1, y_1), \ldots, (x_N, y_N)) < \varepsilon$$

such that

$$(x^*, -y^*, \ldots, -y^*) \in (x_1^*, -y_1^*, 0, \ldots, 0) + (x_2^*, 0, -y_2^*, 0, \ldots, 0)$$
$$+ \cdots + (x_2^*, 0, \ldots, 0, -y_N^*) + U \times V \times \cdots \times V.$$

Therefore one has $y_n^* \in y^* + V, n = 1, \ldots, N$ and (5.4.1). This completes the proof of the theorem.   ●

Let X, Y, and Z be Banach spaces and let $G \colon X \to 2^Y$ and $F \colon Y \to 2^Z$ be arbitrary closed multifunctions. We define the *composition* of F and G by

$$(F \circ G)(x) := F(G(x)) = \bigcup_{y \in G(x)} F(y). \qquad (5.4.2)$$

A chain rule follows from a similar argument.

Theorem 5.4.4 (Weak Coderivative Chain Rule) *Let X, Y and Z be Fréchet-smooth Banach spaces, let $G \colon X \to 2^Y$ and $F \colon Y \to 2^Z$ be closed multifunctions, and let $\bar{y} \in G(\bar{x})$ and $\bar{z} \in F(\bar{y})$. Suppose $x^* \in D_F^*(F \circ G)(\bar{x}; \bar{z})(z^*)$. Then, for any $\varepsilon > 0$ and any weak-star neighborhoods, U, V and W of the origins in X^*, Y^* and Z^* respectively, there exist $x_2 \in B_\varepsilon(\bar{x})$, $y_n \in B_\varepsilon(\bar{y}_n), n = 1, 2$ and $z_1 \in B_\varepsilon(\bar{z})$, as well as $x_2^* \in X^*, y_n^* \in Y^*, n = 1, 2$ and $z_1^* \in Z^*$ satisfying $y_1^* - y_2^* \in V$, $z_1^* \in z^* + W$, $y_1^* \in D_F^* F(y_1; z_1)(z_1^*)$, and $x_2^* \in D_F^* G(x_2; y_2)(y_2^*)$ with*

$$\max(\|x_2^*\|, \|y_1^*\|, \|y_2^*\|, \|z_1^*\|) \times \|(x_1, y_1) - (x_2, y_2)\| < \varepsilon$$

such that

$$x^* \in x_2^* + U. \qquad (5.4.3)$$

Proof. Let $x^* \in D_F^*(F \circ G)(\bar{x}; \bar{z})(z^*)$. Then there exists a C^1 function g on $X \times Z$ with $(x^*, -z^*) = g'(\bar{x}, \bar{z})$ such that

$$(x, z) \to \iota_{\mathrm{graph}(F \circ G)}(x, z) - g(x, z)$$

attains a local minimum 0 at $(\bar{x}, \bar{z})$. Observe that

$$\iota_{\mathrm{graph} F}(y, z) + \iota_{\mathrm{graph} G}(x, y) \ge \iota_{\mathrm{graph}(F \circ G)}(x, z)$$

and

$$\iota_{\mathrm{graph} F}(\bar{y}, \bar{z}) + \iota_{\mathrm{graph} G}(\bar{x}, \bar{y}) = \iota_{\mathrm{graph}(F \circ G)}(\bar{x}, \bar{z}) = 0.$$

We conclude that $(\bar{x}, \bar{y}, \bar{z})$ is a local minimum of the function

$$(x, y, z) \to \iota_{\mathrm{graph}F}(y, z) + \iota_{\mathrm{graph}G}(x, y) - g(x, z).$$

Therefore

$$(x^*, 0, -z^*) \in \partial_F(\iota_{\mathrm{graph}F}(\bar{y}, \bar{z}) + \iota_{\mathrm{graph}G}(\bar{x}, \bar{y})).$$

Applying the Weak Local Approximate Sum Rule of Theorem 3.3.3, we can select $x_2 \in B_\varepsilon(\bar{x}), y_n \in B_\varepsilon(\bar{y})$, $n = 1, 2$, and $z_1 \in B_\varepsilon(\bar{z})$ as well as $x_2^*, y_1^*, y_2^*, z_1^*$ with

$$\max(\|x_2^*\|, \|y_1^*\|, \|y_2^*\|, \|z_1^*\|) \times \|(x_1, y_1) - (x_2, y_2)\| < \varepsilon$$

such that $y_1^* \in D_F^*F(y_1; z_1)(z_1^*)$, $x_2^* \in D_F^*G(x_2; y_2)(y_2^*)$ and

$$(x^*, 0, -z^*) \in (0, y_1^*, -z_1^*) + (x_2^*, -y_2^*, 0) + U \times V \times W.$$

Then we have $y_1^* - y_2^* \in V$, $z_1^* \in z^* + W$ and (5.4.3). ●

5.4.3 Strong Fréchet Coderivative Calculus

The strong calculus for the Fréchet coderivative can be established similarly by using the strong local approximate sum rule. Now the sequential uniform lower semicontinuity condition in Definition 3.3.17 comes into play. This condition is important here for two reasons. First it is stable when adding a "nice" function as is made precise in item (iii) of Exercise 3.3.5. For convenience we restate that result as a lemma below.

Lemma 5.4.5 *Let X be a Banach space and let $f_1, \ldots, f_N \colon X \to \mathbb{R} \cup \{+\infty\}$ be lsc functions. If $(f_1, \ldots, f_N)$ is sequentially uniform lower semicontinuous at x and $f_{N+1} \colon X \to \mathbb{R}$ is uniformly continuous around x then $(f_1, \ldots, f_N, f_{N+1})$ is sequentially uniform lower semicontinuous at x.*

Proof. Exercise 3.3.5 item (iii). ●

Secondly, the sequential uniform lower semicontinuity is equivalent to the following *general metric regularity condition* which is convenient to apply to the sum of indicator functions of sets. In what follows we will use $\mathcal{K}$ to denote the collection of nonnegative continuous functions ω on $[0, +\infty)$ with $\omega(0) = 0$.

Definition 5.4.6 (General Metric Regularity Qualification Condition) *Let X be a Banach space, let $f_1, \ldots, f_N \colon X \to \mathbb{R} \cup \{+\infty\}$ be lsc functions and let $\bar{x} \in \bigcap_{n=1}^N \mathrm{dom}(f_n)$. We say that $(f_1, \ldots, f_N)$ satisfies the general metric qualification condition at $\bar{x}$ provided that there is an $\omega \in \mathcal{K}$ such that*

$$d\Big(\mathrm{epi}\Big(\sum_{n=1}^N f_n\Big); (x, a)\Big) \leq \omega\Big(\sum_{n=1}^N d(\mathrm{epi}(f_n); (x, a_n))\Big)$$

for all x in a neighborhood of $\bar{x}$ and all a, a_n satisfying $a = \sum_{n=1}^N a_n$.

Lemma 5.4.7 *Let X be a Banach space, let $f_1, \ldots, f_N \colon X \to \mathbb{R} \cup \{+\infty\}$ be lsc functions and let $\bar{x} \in \bigcap_{n=1}^N \mathrm{dom}(f_n)$. Then $(f_1, \ldots, f_N)$ is sequentially uniformly lower semicontinuous at $\bar{x}$ if and only if it satisfies the general metric qualification condition at $\bar{x}$.*

Proof. Exercise 5.4.3.   ●

The form of this qualification condition for indicator functions of closed subsets is given in the following lemma.

Lemma 5.4.8 *Let X be a Banach space, let $S_1, \ldots, S_N$ be closed subsets of X and let $\bar{x} \in \bigcap_{n=1}^N S_n$. Then the indicator functions ι_{S_n} satisfy the general metric regularity condition at $\bar{x}$ if and only if there is an $\omega \in \mathcal{K}$ such that*

$$d\Big(\bigcap_{n=1}^N S_n; x \Big) \le \omega \Big(\sum_{n=1}^N d(S_n; x) \Big)$$

for all x in a neighborhood of $\bar{x}$.

Proof. Exercise 5.4.5.   ●

Geometrically, this says that the distance to the intersection is of the same order as the sum of the distances to the individual sets. Strong calculus for the coderivatives of multifunctions then can be established under this general metric regularity condition.

Theorem 5.4.9 (Strong Coderivative Sum Rule) *Let X and Y be Fréchet-smooth spaces, and let $F_n, n = 1, \ldots, N$, $F = \sum_{n=1}^N F_n$ be closed multifunctions from X into Y. Suppose $\bar{y} \in \sum_{n=1}^N F_n(\bar{x})$.*

Fix arbitrary $\bar{y}_n \in F_n(\bar{x}), n = 1, \ldots, N$ with $\bar{y} = \sum_{n=1}^N \bar{y}_n$. Let $x^* \in D_F^* F(\bar{x}; \bar{y})(y^*)$. Suppose that $\mathrm{graph} F_n, n = 1, \ldots, N$ satisfy the following general metric regularity condition: for any $(x, y_1, \ldots, y_N)$ sufficiently close to $(\bar{x}, \bar{y}_1, \ldots, \bar{y}_N)$,

$$d\big(T; (x, y_1, \ldots, y_N) \big) \le \omega \Big(\sum_{n=1}^N d(\mathrm{graph} F_n; (x, y_n)) \Big)$$

where $T := \{ (x, y_1, \ldots, y_N) \mid (x, y_n) \in \mathrm{graph} F_n, n = 1, \ldots, N \}$ and $\omega \in \mathcal{K}$. Then for any $\varepsilon > 0$, there exist $(x_n, y_n) \in (\mathrm{graph} F_n) \cap B_\varepsilon((\bar{x}, \bar{y}_n)), \|y_n^* - y^*\| < \varepsilon, n = 1, \ldots, N$ and $x_n^* \in D_F^* F_n(x_n; y_n)(y_n^*)$ with

$$\max_{n=1,\ldots,N} (\|x_n^*\|, \|y_n^*\|) \times \mathrm{diam}((x_1, y_1), \ldots, (x_N, y_N)) < \varepsilon$$

such that

$$\Big\| x^* - \sum_{n=1}^N x_n^* \Big\| < \varepsilon. \tag{5.4.4}$$

Proof. Let $x^* \in D_F^* F(\bar{x}; \bar{y})(y^*)$. Then there exists a C^1 function g on $X \times Y$ with $(x^*, -y^*) = g'(\bar{x}, \bar{y})$ such that $(x, y) \to \iota_{\mathrm{graph}(F)}(x, y) - g(x, y)$ attains a local minimum 0 at $(\bar{x}, \bar{y})$. Since

$$\sum_{n=1}^{N} \iota_{\mathrm{graph}F_n}(x, y_n) \geq \iota_{\mathrm{graph}(F)}\left(x, \sum_{n=1}^{N} y_n\right)$$

and

$$\sum_{n=1}^{N} \iota_{\mathrm{graph}F_n}(\bar{x}, \bar{y}_n) = \iota_{\mathrm{graph}F}(\bar{x}, \bar{y}) = 0,$$

the function

$$(x, y_1, y_2, \ldots, y_N) \to \sum_{n=1}^{N} \iota_{\mathrm{graph}F_n}(x, y_n) - g\left(x, \sum_{n=1}^{N} y_n\right)$$

attains a local minimum at $(\bar{x}, \bar{y}_1, \bar{y}_2, \ldots, \bar{y}_N)$. Since the graphs of $F_n, n = 1, \ldots, N$ satisfy the general metric regularity condition, $(\iota_{\mathrm{graph}\, F_1}, \ldots, \iota_{\mathrm{graph}\, F_N})$ is sequentially uniformly lower semicontinuous. Then Lemma 5.4.5 implies that $(\iota_{\mathrm{graph}\, F_1}, \ldots, \iota_{\mathrm{graph}\, F_N}, -g)$ is sequentially uniformly lower semicontinuous. Let $\varphi(x, y_1, \ldots, y_N) := g(x, \sum_{n=1}^{N} y_n)$. Then

$$\varphi'(\bar{x}, \bar{y}_1, \ldots, \bar{y}_N) = (x^*, -y^*, \ldots, -y^*).$$

Since g is C^1 there exists an $\varepsilon' < \varepsilon/2$ such that

$$\|(x, y_1, \ldots, y_N) - (x, \bar{y}_1, \ldots, \bar{y}_N)\| < \varepsilon'$$

implies that

$$\|\varphi'(\bar{x}, \bar{y}_1, \ldots, \bar{y}_N) - (x^*, -y^*, \ldots, -y^*)\| < \varepsilon/2.$$

Invoking the strong local approximate sum rule of Theorem 3.3.1 with ε' in place of ε and using $(-x^*, y^*, \ldots, y^*)$ to replace the gradient of φ at a point in the ε' neighborhood of $(\bar{x}, \bar{y}_1, \ldots, \bar{y}_N)$ with an error of at most $\varepsilon/2$ we conclude that there exist $x_n \in B_\varepsilon(\bar{x})$ and $y_n \in B_\varepsilon(\bar{y}_n)$, as well as $y_n^* \in Y^*$ and $x_n^* \in D_F^* F(x_n; y_n)(y_n^*)$, with

$$\max_{n=1,\ldots,N}(\|x_n^*\|, \|y_n^*\|) \times \mathrm{diam}((x_1, y_1), \ldots, (x_N, y_N)) < \varepsilon$$

such that

$$0 \in (-x^*, y^*, \ldots, y^*) + (x_1^*, -y_1^*, 0, \ldots, 0) + (x_2^*, 0, -y_2^*, 0, \ldots, 0) + \cdots$$
$$\cdots + (x_2^*, 0, \ldots, 0, -y_N^*) + \varepsilon B_{X^* \times Y^* \times \cdots \times Y^*}.$$

The conclusion of the theorem follows.   ●

A chain rule may be similarly derived.

Theorem 5.4.10 (Strong Coderivative Chain Rule) *Let X, Y and Z be Fréchet-smooth Banach spaces, let $G\colon X \to 2^Y$ and $F\colon Y \to 2^Z$ be closed multifunctions, and let $\bar{y} \in G(\bar{x})$ and $\bar{z} \in F(\bar{y})$. Suppose that $x^* \in D_F^*(F \circ G)(\bar{x}; \bar{z})(z^*)$ and suppose that $\mathrm{graph}F$ and $\mathrm{graph}G$ satisfy the following general metric regularity condition: for all (x, y, z) sufficiently close to $(\bar{x}, \bar{y}, \bar{z})$,*

$$d((x, y, z), T) \leq \omega(d(\mathrm{graph}G; (x, y)) + d(\mathrm{graph}F; (y, z)))$$

where $T := \{(x, y, z) \mid y \in G(x), z \in F(y)\}$ and $\omega \in \mathcal{K}$. Then for any $\varepsilon > 0$ there exist $x_2 \in B_\varepsilon(\bar{x})$, $y_n \in B_\varepsilon(\bar{y}_n), n = 1, 2$ and $z_1 \in B_\varepsilon(\bar{z})$, as well as $x_2^* \in X^*, y_n^* \in Y^*, n = 1, 2$ and $z_1^* \in Z^*$ satisfying $\|y_1^* - y_2^*\| < \varepsilon$, $\|z_1^* - z^*\| < \varepsilon$, $y_1^* \in D_F^* F(y_1; z_1)(z_1^*)$, and $x_2^* \in D_F^* G(x_2; y_2)(y_2^*)$ with

$$\max(\|x_2^*\|, \|y_1^*\|, \|y_2^*\|, \|z_1^*\|) \times \|(x_1, y_1) - (x_2, y_2)\| < \varepsilon$$

such that

$$\|x^* - x_2^*\| < \varepsilon. \tag{5.4.5}$$

Proof.   Exercise 5.4.6.   ●

5.4.4 Limiting Coderivative Calculus

Calculus for limiting coderivatives follows naturally. For this we need the following condition. A multifunction $F\colon X \to 2^Y$ is *lower semicompact* around $\bar{x}$ if there is a neighborhood U of $\bar{x}$ such that for any $x \in U$ and any sequence $x_i \to x$ with $F(x_i) \neq \emptyset$, there is a sequence $y_i \in F(x_i)$ containing a norm convergent subsequence.

Theorem 5.4.11 (Limiting Coderivative Sum Rule) *Let X and Y be finite dimensional Banach spaces, let F_1 and F_2 be closed multifunctions from X to Y and let $\bar{y} \in F_1(\bar{x}) + F_2(\bar{x})$. Suppose that the multifunction*

$$S(x, y) := \{(y_1, y_2) \mid y_1 \in F_1(x), y_2 \in F_2(x), y_1 + y_2 = y\}$$

is lower semicompact around $(\bar{x}, \bar{y})$ and suppose that the following constraint qualification condition holds

$$D_L^* F_1(\bar{x}; y_1)(0) \cap (-D_L^* F_2(\bar{x}; y_2)(0)) = \{0\}, \quad \text{for all } (y_1, y_2) \in S(\bar{x}, \bar{y}).$$

Then

$$D_L^*(F_1 + F_2)(\bar{x}, \bar{y})(y^*) \subset \bigcup_{(y_1, y_2) \in S(\bar{x}, \bar{y})} [D_L^* F_1(\bar{x}, y_1)(y^*) + D_L^* F_2(\bar{x}, y_2)(y^*)].$$

Proof. Suppose that $x^* \in D_L^*(F_1 + F_2)(\bar{x}, \bar{y})(y^*)$. Then

$$(x^*, -y^*) \in N_L(\text{graph } F_1 + F_2; (\bar{x}, \bar{y})).$$

Thus, there exist sequences $(x_i, y_i) \to (\bar{x}, \bar{y})$ and $x_i^* \in D_F^*(F_1 + F_2)(x_i, y_i)(y_i^*)$ such that $x_i^* \to x^*$ and $y_i^* \to y^*$. Since S is lower semicompact around $(\bar{x}, \bar{y})$, without loss of generality we may assume that there exists a sequence $(y_{i1}, y_{i2}) \in S(x_i, y_i)$ converging to $(y_1, y_2) \in S(\bar{x}, \bar{y})$.

By Theorem 5.4.3 and using the fact that the norm and weak-star topology coincide in finite dimensional Banach spaces, for each i, there exist $(y_{i1}^*, y_{i2}^*) \in B_{1/i}(y_i^*)$ and

$$x_{in}^* \in D_F^* F_n(x_i, y_{in})(y_{in}^*), n = 1, 2 \tag{5.4.6}$$

satisfying

$$\|x_i^* - x_{i1}^* - x_{i2}^*\| < 1/i. \tag{5.4.7}$$

If the sequence $(\|x_{i1}^*\|)$ is unbounded then taking a subsequence if necessary we may assume that $\|x_{i1}^*\| \to \infty$ and $(x_{i1}^*/\|x_{i1}^*\|)$ converges to a unit vector v. Dividing (5.4.7) by $\|x_{i1}^*\|$ and taking limits as $i \to \infty$ we obtain that $(x_{i2}^*/\|x_{i1}^*\|)$ converges to $-v$. It follows from (5.4.6) that

$$0 \neq v \in D_L^* F_1(\bar{x}; y_1)(0) \cap (-D_L^* F_2(x; y_2)(0)) - \{0\},$$

a contradiction to the constraint qualification condition. Thus, $(\|x_{i1}^*\|)$, and therefore $(\|x_{i2}^*\|)$ must be bounded. Taking subsequences if necessary we may assume that both x_{i1}^* and x_{i2}^* converge. Taking limits in (5.4.6) and (5.4.7) we have

$$x^* \in D_L^* F_1(\bar{x}, y_1)(y^*) + D_L^* F_2(\bar{x}, y_2)(y^*)$$

and the conclusion of the theorem follows.   ●

Similarly we can prove a limiting coderivative chain rule.

Theorem 5.4.12 (Limiting Coderivative Chain Rule) *Let X, Y and Z be finite dimensional Banach spaces and let $F: X \times Y \to 2^Z$ and $G: X \to 2^Y$ be closed multifunctions. Suppose that the multifunction*

$$M(x, z) := G(x) \cap F^{-1}(z) = \{y \in G(x) : z \in F(x, y)\}$$

is lower semicompact around $(\bar{x}, \bar{z})$ and suppose that for any $\bar{y} \in M(\bar{x}, \bar{z})$ the constraint qualification condition

$$\left.\begin{cases} (x^*, y^*) \in D_L^* F((\bar{x}, \bar{y}); \bar{z})(0) \\ \text{and} -x^* \in D_L^* G(\bar{x}; \bar{y})(y^*) \end{cases}\right\} \implies x^* = 0 \text{ and } y^* = 0$$

holds. Then, for all $z^* \in Z^*$,

$$D_L^*(F \circ G)(\bar{x}; \bar{z})(z^*) \subset$$

$$\bigcup_{\bar{y} \in M(\bar{x}, \bar{z})} \left[x_1^* + x_2^* : x_1^* \in D_L^* G(\bar{x}; \bar{y})(y^*), (x_2^*, y^*) \in D_L^* F((\bar{x}, \bar{y}); \bar{z})(z^*) \right].$$

Proof. Exercise 5.4.7.

5.4.5 Commentary and Exercises

Coderivatives were introduced by Mordukhovich in [196]. In [143] Ioffe used the term coderivative and systematically studied its properties. We follow [197, 202] for the definition of coderivatives through the normal cone of the graph of the multifunction under consideration. For a single valued C^1 function the Fréchet and limiting coderivatives reduce to the dual of the derivative (Exercise 5.4.1). However, this is not true for the limiting coderivative when the single valued function is merely Fréchet differentiable. In fact, in [28] it is shown that an uncountable number of Fréchet differentiable vector-valued Lipschitz functions differing by more than constants can share the same limiting coderivatives.

One can find alternative definitions of generalized derivative concepts for multifunctions using tangent cones in Aubin and Frankowska [8]. Calculus results for coderivatives may be found in [202, 207, 209] and the references therein. We emphasize that the relationship between coderivative calculus and that of the subdifferential is made through the careful use of indicator functions of the graphs of multifunctions.

Exercise 5.4.1 Let X and Y be Banach spaces and let $f \colon X \to Y$ be a C^1 function. Show that

$$D_F^* f(x) = D_L^* f(x) = [f'(x)]^*.$$

Exercise 5.4.2 Prove Proposition 5.4.2.

Exercise 5.4.3 Prove Lemma 5.4.7.

Exercise 5.4.4 Let X and Y be Fréchet smooth Banach spaces, let F be a closed multifunction from X to Y and let $f \colon X \to Y$ be a C^1 function. Show that

$$D_F^*(f + F)(\bar{x}, \bar{y})(y^*) = (f'(\bar{x}))^* y^* + D_F^* F(\bar{x}, \bar{y} - f(\bar{x}))(y^*)$$

and

$$D_L^*(f + F)(\bar{x}, \bar{y})(y^*) = (f'(\bar{x}))^* y^* + D_L^* F(\bar{x}, \bar{y} - f(\bar{x}))(y^*).$$

Reference: [207].

Exercise 5.4.5 Prove Lemma 5.4.8.

Exercise 5.4.6 Prove Theorem 5.4.4.

Exercise 5.4.7 Prove Theorem 5.4.12.

5.5 Implicit Multifunction Theorems

The implicit function theorem plays an important role in classic analysis for smooth mappings. In this subsection, we discuss its nonsmooth and multifunction counterpart which connects and unifies many concepts arising from diverse backgrounds, including open mapping theorems, metric regularity, pseudo-Lipschitz continuity for set-valued mappings, stability and solvability.

Let X, Y and Z be Banach spaces and let $F: X \times Y \to Z$ be a mapping. The classical implicit function theorem asserts that if $F(\bar{x}, \bar{y}) = 0$, F is smooth near $(\bar{x}, \bar{y})$ and $F_x(\bar{x}, \bar{y}): Y \to Z$ is bijective, then in a neighborhood of $(\bar{x}, \bar{y})$ the equation $F(x, y) = 0$ determines $x = x(y)$ as a function of y. Moreover, $x'(y) = -F_x(x(y), y)^{-1} F_y(x(y), y)$. Here, we consider the more general situation when $F: X \times Y \to 2^Z$ is a close-valued multifunction. We want to find conditions to ensure that the inclusion $0 \in F(x, y)$ determines x as a multifunction of y. In other words, defining $G(y) := \{x \in X \mid 0 \in F(x, y)\}$ we want to find conditions ensuring G to be (locally) nonempty. The coderivatives will replace the role of derivatives in these conditions.

5.5.1 Solvability Revisited

Set $f(x, y) := d(F(x, y); 0)$. It is clear that $0 \in F(x, y)$ if and only if $f(x, y) = 0$. Thus, we can study the implicit multifunction problem $0 \in F(x, y)$ by studying the solvability of the functional equation $f(x, y) = 0$. We will take this approach and first we discuss a refinement of the solvability results in subsection 3.6.2, including an estimate of the coderivative for the implicit multifunction.

Theorem 5.5.1 (Solvability) *Let X and Y be Fréchet smooth Banach spaces and let U be an open set in $X \times Y$. Suppose that $f: U \to \mathbb{R} \cup \{+\infty\}$ satisfies the following conditions:*

(i) *there exists $(\bar{x}, \bar{y}) \in U$ such that*

$$f(\bar{x}, \bar{y}) \leq 0;$$

(ii) *$y \to f(\bar{x}, y)$ is upper semicontinuous at $\bar{y}$;*
(iii) *for any fixed y near $\bar{y}$, $x \to f(x, y)$ is lower semicontinuous;*
(iv) *there exists a $\sigma > 0$ such that for any $(x, y) \in U$ with $f(x, y) > 0$, $\xi \in \partial_{F,x} f(x, y)$ implies that $\|\xi\| \geq \sigma$.*

Then there exist open sets $W \subset X$ and $V \subset Y$ containing $\bar{x}$ and $\bar{y}$ respectively such that

(a) *for any $y \in V$, $W \cap G(y) \neq \emptyset$;*

(b) *for any $y \in V$ and $x \in W$,*

$$d(x, G(y)) \leq \frac{f_+(x,y)}{\sigma},$$

where $f_+(x,y) := \max\{0, f(x,y)\}$;
(c) *for any $(x, y) \in W \times V$, $x \in G(y)$,*

$$D_F^* G(p;y)(x^*) = \{y^* \mid (-x^*, y^*) \in \text{cone } \partial_F f_+(x,y)\}.$$

Proof. Conclusions (a) and (b) were proven in the proof of Theorem 3.6.3. We need only establish estimate (c). Consider a pair (x, y) with $x \in W \cap G(y)$ and let $y^* \in D_F^* G(y;x)(x^*)$. Then

$$(y^*, -x^*) \in N_F(\text{graph } G; (x,y)) = \bigcup_{K>0} K\partial_F d(\text{graph } G; (x,y)).$$

By definition there exists a C^1 function g with $g'(y,x) = (y^*, -x^*)$ and a positive constant K such that for any $(u, v) \in Y \times X$, we have

$$g(u, v) \leq g(y, x) + Kd(\text{graph } G; (u,v))$$
$$\leq g(y, x) + Kd(G(v); v) \leq g(y, x) + (K/\sigma)f_+(u,v).$$

Thus $(K/\sigma)f_+(u, v) - g(u, v)$ attains a minimum at (y, x), i.e., $(-x^*, y^*) \in (K/\sigma)\partial_F f_+(x, y)$. This establishes

$$D_F^* G(y;x)(x^*) \subset \{y^* \mid (-x^*, y^*) \in \text{cone } \partial_F f_+(x,y)\}.$$

The reverse inclusion follows directly from the inequality $\iota_{\text{graph } G} \geq Kf_+$ for any $K > 0$. $\quad\bullet$

5.5.2 The Subdifferential of the Infimum Function

Using the relationship of nonemptiness of $G(y)$ and the solvability of $f(x,y) = d(F(x,y); 0) \leq 0$ we can deduce an implicit multifunction theorem from Theorem 5.5.1. The key is the relationship between the infinitesimal regularity coderivative condition for a multifunction F and that of the subdifferential condition for f. Since $f(x,y) = d(F(x,y); 0)$ can be written as an infimum function we need the following representation of the subdifferential of infimum. This result is also interesting on its own.

Theorem 5.5.2 (Representation of the Subdifferential of an Infimum) *Let X be a Fréchet smooth Banach space, let $f_a \colon X \to \mathbb{R} \cup \{+\infty\}, a \in A$ be a family of lsc function and define $f(x) := \inf_{a \in A} f_a(x)$. Suppose that $x^* \in \partial_F \underline{f}(x)$ where $\underline{f}$ is the lsc closure of f. Then there exists a function $\varphi \in K$ and a positive number M such that for any small $\varepsilon > 0$ and any (a, y) satisfying*

$y \in B_\varepsilon(x)$ and $f_a(y) < \underline{f}(x) + \varepsilon$, there exist $z \in B_{M\sqrt{\varepsilon}}(x)$ and $z^* \in \partial_F f_a(z)$ such that

$$f_a(z) < \underline{f}(x) + M\sqrt{\varepsilon}$$

and

$$\|z^* - x^*\| < \varphi(\varepsilon).$$

Proof. Let g be a C^1 function with $g'(x) = x^*$ such that $\underline{f} - g$ attains a minimum at x over $B_{2\delta}(x)$ and g is Lipschitz with a Lipschitz constant L on $B_{2\delta}(x)$. For $\varepsilon > 0$ sufficiently small, by the definition of (a, y) we have

$$(f_a(u) - g(u)) - (f_a(y) - g(y)) \geq -L\varepsilon, \quad \text{for all } u \in B_{\sqrt{\varepsilon}}(y). \quad (5.5.1)$$

Applying the multidirectional mean value inequality of Theorem 3.6.1, there exist $z \in B_{\sqrt{\varepsilon}}([y, B_{\sqrt{\varepsilon}}(y)]) \subset B_{2\sqrt{\varepsilon}}(x)$ and $z^* \in \partial_F f_a(z)$ such that

$$
\begin{aligned}
f_a(z) &< f_a(y) + (g(z) - g(y)) + \varepsilon \\
&< \underline{f}(x) + L\|z - y\| + 2\varepsilon \\
&< \underline{f}(x) + 2(L+1)\sqrt{\varepsilon}
\end{aligned}
\quad (5.5.2)
$$

and

$$\langle z^* - g'(z), u - y \rangle \geq -L\varepsilon, \quad \text{for all } u \in B_{\sqrt{\varepsilon}}(y). \quad (5.5.3)$$

It follows that

$$\|z^* - g'(z)\| \leq L\sqrt{\varepsilon}. \quad (5.5.4)$$

Set $M := 2(L+1)$ and

$$\varphi(\varepsilon) := L\sqrt{\varepsilon} + \sup\{\|g'(z) - g'(x)\| \mid z \in B_{2\sqrt{\varepsilon}}(x)\}.$$

Then φ has the required properties. Moreover, it follows from (5.5.4) that

$$\|z^* - x^*\| \leq \|g'(z) - g'(x)\| + L\sqrt{\varepsilon} \leq \varphi(\varepsilon), \quad (5.5.5)$$

which completes the proof.   ●

Notice that subdifferentiability implies lower semicontinuity; we have the following corollary.

Corollary 5.5.3 *Let X be a Fréchet smooth Banach space, let $f_a\colon X \to \mathbb{R} \cup \{+\infty\}, a \in A$ be a family of lsc functions and define $f(x) := \inf_{a \in A} f_a(x)$. Suppose that $x^* \in \partial_F f(x)$. Then there exists a function $\varphi\colon [0, \infty) \to [0, \infty)$ with $\varphi(t) \to 0$ as $t \to 0+$ and a positive number M such that for any small $\varepsilon > 0$ and any a satisfying $f_a(x) < f(x) + \varepsilon$, there exist $z \in B_{M\sqrt{\varepsilon}}(x)$ and $z^* \in \partial_F f_a(z)$ such that*

$$f_a(z) < f(x) + M\sqrt{\varepsilon}$$

and

$$\|z^* - x^*\| < \varphi(\varepsilon).$$

Proof. Exercise 5.5.1.   ●

5.5.3 Implicit Multifunction Theorems

We now use the tools developed in the previous sections to derive an implicit multifunction theorem. The idea is to view f as an infimum function.

$$f(x) = d(F(x); 0) = \inf\{\|y\| + \iota_{\operatorname{graph} F}(x, y)\}.$$

To state the relationship between the subdifferentials of $d(F(x); 0)$ and the coderivatives of $F(x)$ we need the concept of an *approximate projection*. For $\eta > 0$, we denote the η-approximate projection of v to S by $\operatorname{pr}_\eta(S; v) := \{s \in S \mid \|s - v\| \leq d(S; v) + \eta\}$.

Lemma 5.5.4 *Let X and Y be Fréchet smooth Banach spaces, let $U \subset X$ be an open set and let $F\colon U \to 2^Y$ be a close-valued upper semicontinuous multifunction. Let $f(x) := d(F(x); 0)$. Suppose*

(i) *for any $x \in U$ with $0 \notin F(x)$*

$$\sigma \leq \liminf_{\eta \to 0} \Big\{ \|x^*\| \ \Big| \ x^* \in D_F^* F(x'; y')(y^*), \|y^*\| = 1$$

$$\text{with } x' \in B_\eta(x), y' \in \operatorname{pr}_\eta(F(x'); 0) \Big\}.$$

Then

(ii) *for any $\bar{x} \in U$ with $f(\bar{x}) > 0$, $\xi \in \partial_F f(\bar{x})$ implies that $\|\xi\| \geq \sigma$.*

Proof. Consider

$$\varphi(x', y') = \|y'\| + \iota_{\operatorname{graph} F}(x', y').$$

Since F is a closed valued upper semicontinuous multifunction, the graph of F is a closed set. Thus, φ is lsc. Moreover, $f(x') = d(F(x'); 0) = \inf_{y' \in Y} \varphi(x', y')$. Let $\xi \in \partial_F f(x)$ where $f(x) > 0$. Then f is lsc at x. (In fact, when F is upper semicontinuous it is not hard to verify directly that f is lower semicontinuous.) Take η small enough so that $\|x' - x\| < \eta$ implies that $f(x') \geq f(x)/2 > 0$. Applying Theorem 5.5.2 with $\varepsilon = \eta$, we find (u_η, v_η) and $(\xi_\eta, \zeta_\eta) \in \partial_F \varphi(u_\eta, v_\eta)$ such that $\|x - u_\eta\| < \eta$,

$$0 < f(u_\eta) < \varphi(u_\eta, v_\eta) < f(u_\eta) + \eta \tag{5.5.6}$$

and

$$\|\xi_\eta - \xi\| < \eta, \quad \|\zeta_\eta\| < \eta. \tag{5.5.7}$$

It follows from (5.5.6) that $v_\eta \in \operatorname{pr}_\eta(F(u_\eta); 0)$. By the sum rule of Theorem 3.3.1 there exist (x_η, y_η), (x'_η, y'_η) close to (u_η, v_η) and a subgradient y^*_η of the norm function $\|y\|$ at the point y'_η such that $y_\eta \in \operatorname{pr}_\eta(F(x_\eta); 0)$, $\|y'_\eta\| > 0$ and

$$(\xi_\eta, \zeta_\eta) \in (0, y^*_\eta) + N_F(\operatorname{graph} F; (x_\eta, y_\eta)) + \eta(B_{X^*} \times B_{Y^*}),$$

i.e., there exists $(\xi'_\eta, \zeta'_\eta) \in \eta(B_{X^*} \times B_{Y^*})$ such that

$$\xi_\eta - \xi'_\eta \in D^*_F F(x_\eta, y_\eta)(y^*_\eta - \zeta_\eta + \zeta'_\eta). \tag{5.5.8}$$

Rewriting (5.5.8) as

$$(\xi_\eta - \xi'_\eta)/\|y^*_\eta - \zeta_\eta + \zeta'_\eta\| \in D^*_F F(x_\eta, y_\eta)((y^*_\eta - \zeta_\eta + \zeta'_\eta)/\|y^*_\eta - \zeta_\eta + \zeta'_\eta\|)$$

and noting that $\|y^*_\eta - \zeta_\eta + \zeta'_\eta\| \geq 1 - 2\eta$, it follows from assumption (i) that

$$\liminf_{\eta \to 0} \|\xi_\eta\| = \liminf_{\eta \to 0} \|\xi_\eta - \xi'_\eta\|/\|y^*_\eta - \zeta_\eta + \zeta'_\eta\| \geq \sigma.$$

Relation (5.5.7) then implies that

$$\|\xi\| \geq \sigma.$$

$\bullet$

Combining Theorem 5.5.1 and Lemma 5.5.4 we have the following implicit multifunction theorem.

Theorem 5.5.5 (Implicit Multifunction Theorem) *Let X, Y and Z be Fréchet smooth Banach spaces and let U be an open set in $X \times Y$. Suppose that $F : U \to 2^Z$ is a closed valued multifunction satisfying.*

(i)   *there exists $(\bar{x}, \bar{y}) \in U$ such that*

$$0 \in F(\bar{x}, \bar{y}),$$

(ii)   *$y \to F(\bar{x}, y)$ is lower semicontinuous at $\bar{y}$,*
(iii) *for any fixed y near $\bar{y}$, $x \to F(x, y)$ is upper semicontinuous, and*
(iv)   *there exists $\sigma > 0$ such that for any $(x, y) \in U$ with $0 \notin F(x, y)$*

$$\sigma \leq \liminf_{\eta \to 0} \Big\{ \|x^*\| \mid x^* \in D^*_F F(x', y; z')(y^*), \|y^*\| = 1$$

$$\text{with } x' \in B_\eta(x), z' \in \mathrm{pr}_\eta(F(x', y); 0) \Big\}.$$

Then there exist open sets $W \subset X$ and $V \subset P$ containing $\bar{x}$ and $\bar{y}$ respectively such that

(a) *for any $y \in V$, $W \cap G(y) \neq \emptyset$,*
(b) *for any $y \in V$ and $x \in W$, $d(G(y); x) \leq d(F(x, y); 0)/\sigma$, and*
(c) *for any $(x, y) \in W \times V$, $x \in G(y)$,*

$$D^*_F G(y; x)(x^*) = \{x^* \mid (-x^*, p^*) \in \mathrm{cone}\ \partial_F d(F(x, y); 0)\}.$$

Proof.   Exercise 5.5.2. $\bullet$

5.5.4 Open Covering, Regularity and Pseudo-Lipschitzian Properties

In this section we discuss the open covering, metric regularity and pseudo-Lipschitzian properties. These concepts are closely related, yet historically they arose in different contexts. We start with the definitions.

Definition 5.5.6 (Metric Regularity) *Let X and Y be Banach spaces and let $F\colon X \to 2^Y$ be a closed multifunction. We say that F is metrically regular at $(\bar{x}, \bar{y}) \in$ graph F with modulus r provided that there exist neighborhoods W and V of $\bar{x}$ and $\bar{y}$ respectively such that, for any $x \in W$ and $y \in V$,*

$$d(F^{-1}(y); x) \leq r d(F(x); y).$$

Definition 5.5.7 (Pseudo-Lipschitz) *Let X and Y be Banach spaces and let $F\colon X \to 2^Y$ be a closed multifunction. We say that F is pseudo-Lipschitz at $(\bar{x}, \bar{y}) \in$ graph F with rank r provided that there exist neighborhoods W and V of $\bar{x}$ and $\bar{y}$ respectively such that, for any $x_1, x_2 \in W$,*

$$F(x_2) \cap V \subset F(x_1) + r\|x_2 - x_1\| B.$$

Definition 5.5.8 (Open Covering) *Let X and Y be Banach spaces and let $F\colon X \to 2^Y$ be a closed multifunction. We say that F is an open covering with linear rate r at $(\bar{x}, \bar{y}) \in$ graph F provided that there exist neighborhoods W and V of $\bar{x}$ and $\bar{y}$ respectively such that, for any $x \in W$ and t sufficiently small,*

$$F(x) \cap V + tr B_Y \subset F(x + t B_X).$$

Now we can precisely state the relationship between metric regularity, the open covering property with linear rate and the pseudo-Lipschitz property.

Theorem 5.5.9 *Let X and Y be Banach spaces and let $F\colon X \to 2^Y$ be a closed multifunction. Then the following are equivalent:*

(i)   *F is an open covering with linear rate r at $(\bar{x}, \bar{y}) \in$ graph F;*
(ii)   *F is metrically regular with modulus $1/r$ at $(\bar{x}, \bar{y}) \in$ graph F;*
(iii) *F^{-1} is pseudo-Lipschitz with rank $1/r$ at $(\bar{y}, \bar{x}) \in$ graph F^{-1};*

Proof. (i) implies (ii). Let V, W and r be as in Definition 5.5.8 and let $y \in V$ and $x \in W$. Shrink V to start with if necessary, and choose a small positive number η so that

$$t = \frac{1}{r} d(F(x); y) + \eta$$

is sufficiently small that the inclusion in Definition 5.5.8 holds. Then

$$y \in F(x) \cap V + tr B_Y \subset F(x + t B_X)$$

and we can find a $u \in B_t(x)$ with $y \in F(u)$ or $u \in F^{-1}(y)$. It follows that

$$d(F^{-1}(y);x) \le \|u - x\| < t = \frac{1}{r}d(F(x);y) + \eta.$$

Letting $\eta \to 0$ we have

$$d(F^{-1}(y);x) \le \frac{1}{r}d(F(x);y).$$

(ii) implies (iii). Let V, W and $1/r$ be as in Definition 5.5.6. Note $F^{-1}: Y \to 2^X$ so that the roles of V and W switch as compared to Definition 5.5.7. Let y_1, y_2 lie in V. Choosing an arbitrary element $x \in F^{-1}(y_2) \cap W$ we have $y_2 \in F(x)$. Since F is metrically regular with modulus $1/r$, we have

$$d(F^{-1}(y_1);x) \le \frac{1}{r}d(F(x);y_1) \le \frac{1}{r}\|y_2 - y_1\|,$$

or

$$x \in F^{-1}(y_1) + \frac{1}{r}\|y_2 - y_1\|B_X.$$

Since $x \in F^{-1}(y_2) \cap W$ is arbitrary we arrive at

$$F^{-1}(y_2) \cap W \subset F^{-1}(y_1) + \frac{1}{r}\|y_2 - y_1\|B_X.$$

(iii) implies (i). Let

$$y_1 \in F(x) \cap V + trB_Y$$

be an arbitrary element of the set. Note that when t is sufficiently small $y_1 \in V$. Choose $y_2 \in F(x) \cap V$ with $\|y_1 - y_2\| < tr$. Since F^{-1} is pseudo-Lipschitz with rank $1/r$ we have

$$x \in F^{-1}(y_2) \cap W \subset F^{-1}(y_1) + \frac{1}{r}\|y_1 - y_2\|B_X.$$

Therefore, there exists $x' \in F^{-1}(y_1)$ with $\|x' - x\| \le \|y_1 - y_2\|/r < t$. That is to say

$$y_1 \in F(x') \subset F(x + tB_X).$$

Since $y_1 \in F(x) \cap V + trB_Y$ is arbitrary we have

$$F(x) \cap V + trB_Y \subset F(x + tB_X).$$

•

Using the implicit multifunction theorem of Theorem 5.5.5 and the equivalence relationship in Theorem 5.5.9 we have the following sufficient conditions to be an open mapping, or for metric regularity and pseudo-Lipschitz property.

Theorem 5.5.10 *Let X and Y be Fréchet smooth Banach spaces. Let U be an open set in $X \times Y$ and let $F: U \to 2^Y$ be a closed valued multifunction satisfying:*

(i)   there exists $(\bar{x}, \bar{y}) \in U$ such that $\bar{y} \in F(\bar{x})$,
(ii) F is upper semicontinuous in U,
(iii) there exists $r > 0$ such that for any $(x, y) \in U$, $y \notin F(x)$

$$r \leq \liminf_{\eta \to 0} \left\{ \|x^*\| \,\middle|\, x^* \in D_F^* F(x'; y')(y^*), \|y^*\| = 1, \right.$$

$$\left. \text{with } x' \in B_\eta(x), y' \in \mathrm{pr}_\eta(F(x'); y) \right\}.$$

Then

(a) (Open Covering) F is an open covering with linear rate r at $(\bar{x}, \bar{y})$;
(b) (Metric Regularity) F is metrically regular at $(\bar{x}, \bar{y})$ with modulus $1/r$;
(c) (pseudo-Lipschitz) F^{-1} is pseudo-Lipschitz with rank $1/r$ at $(\bar{y}, \bar{x})$.

Proof.   Exercise 5.5.3.   ●

The methods we have used so far in dealing with properties of multifunctions largely rely on distance functions. An alternative approach is to study a smooth function restricted to a closed set. We illustrate this method by revisiting metric regularity. Let X and Y be Banach spaces, let S be a closed subset of X and let $f \colon X \to Y$ be a C^1 mapping. We say that f is *metrically regular on S* at $\bar{x} \in S$ with modulus r provided that there exists a neighborhood W of $\bar{x}$ such that for any $x, z \in W \cap S$,

$$d(S \cap f^{-1}(f(x)); z) \leq r\|f(z) - f(x)\|.$$

It is not hard to check (Exercise 5.5.7) that this definition is consistent with Definition 5.5.6 if we identify f and S with the multifunction

$$F_S(x) := \begin{cases} \{f(x)\} & x \in S, \\ \emptyset & \text{otherwise.} \end{cases} \tag{5.5.9}$$

On the other hand a closed multifunction $F \colon X \to 2^Y$ is metrically regular at $(\bar{x}, \bar{y}) \in \mathrm{graph}\, F$ if and only if the C^1 mapping $f \colon X \times Y \to Y$ defined by $f(x, y) = y$ is metrically regular on $\mathrm{graph}\, F$ at $(\bar{x}, \bar{y})$ (Exercise 5.5.8). Now, it is easy to deduce a sufficient condition for a C^1 function metrically regular on a set from Theorem 5.5.10. To do so we need the following proposition on representing the coderivative of the multifunction defined by (5.5.9) in terms of the derivative of f and normal cone of S. The proof follows directly from the definitions and is left as an exercise.

Proposition 5.5.11 *Let X and Y be Banach spaces, let S be a closed subset of X and let $f \colon X \to Y$ be a C^1 mapping. Define the multifunction $F_S \colon X \to 2^Y$ by (5.5.9). Suppose $x \in S$ and $x^* \in D_F^* F_S(x, f(x))(y^*)$. Then*

$$x^* - (f'(x))^* y^* \in N_F(S; x).$$

Proof. Exercise 5.5.9.   ●

Combining Theorem 5.5.10 and Proposition 5.5.11 we obtain the following sufficient condition for the regularity of a function on a closed set.

Corollary 5.5.12 *Let* X *and* Y *be Fréchet smooth Banach spaces, let* S *be a closed subset of* X *and let* $f: X \to Y$ *be a* C^1 *mapping. Suppose that for any* x *in a neighborhood of* $\bar{x} \in S$, *any* $x^* \in X^*$ *and* $y^* \in Y^*$ *satisfying* $x^* - (f'(x))^* y^* \in N_F(S; x)$ *we have*

$$\|x^*\| \geq r\|y^*\|. \tag{5.5.10}$$

Then f *is metrically regular at* $\bar{x}$ *with modulus* $1/r$.

Proof. Exercise 5.5.10.   ●

Conversely using Corollary 5.5.12 and Exercise 5.5.7 one can also recover a sufficient condition for a multifunction to be metrically regular. We leave the verification of this fact as Exercise 5.5.11. Similarly, one can also study the open covering for multifunctions by way of studying open covering for functions on a closed sets (Exercise 5.5.12).

5.5.5 Commentary and Exercises

Implicit multifunction theorems discussed here are generalizations of implicit function theorems that have fundamental importance in analysis. They are closely related to many other concepts in the analysis of multifunctions and nonsmooth functions, such as solvability [91], open covering [103, 104, 264], metric regularity [149, 227, 30], inverse function theorems [83, 264], pseudo-Lipschitzness [5] and stability [227, 228, 253]. The relationship between these properties were discussed in [73, 219, 199]. Our exposition here, which emphasizes characterizing properties of multifunctions by their related distance functions, largely follows [174]. Since the distance function to a set can be viewed as an infimum function, an important tool in doing so is the representation of the subdifferential of an infimum [173]. This type of result has also been studied (under the name subdifferential for marginal functions) in different generality earlier (see e.g., [48, 151, 247, 174] and Exercise 5.5.5). Theorem 5.5.2 is taken from Ledyaev and Treiman [173] (see also Ledyaev [172] for an alternative form) which refines earlier results in that it asserts that the subdifferential of f at x is approximated by the subdifferential of f_a at some nearby points z for *all* a such that $f_a(x)$ is close enough to $f(x)$. This is a significant improvement and leads to new, interesting applications (see Exercises 5.5.5 and 5.5.6). The equivalence between open covering and metric regularity can be found in [103] and the connection between these and the pseudo-Lipschitzian property for F^{-1} was derived in [73] and is studied carefully in [199, 205]. Sufficient conditions for open covering with linear rate and regularity using other generalized derivative constructions have also been studied by Ioffe [149, 142] and Warga [264]. Studying open covering and metric regularity for C^1 functions on closed sets and deducing corresponding results for multifunctions is the prevailing method in early research on these subjects. They are convenient for deriving sufficient conditions in terms of various tangent cones (see Exercises 5.5.13 and [8] for additional references). Metric regularity is very useful in dealing with constrained minimization problems. Details can be found in the guided exercises below (Exercises 5.5.14 and 5.5.15).

Exercise 5.5.1 Prove Corollary 5.5.3.

Exercise 5.5.2 Prove Theorem 5.5.5.

Exercise 5.5.3 Prove Theorem 5.5.10.

Exercise 5.5.4 Deduce the following weaker result in [174] from Theorem 5.5.2.

Theorem 5.5.13 *Let X be a Fréchet smooth Banach space, let $f_a \colon X \to \mathbb{R} \cup \{+\infty\}, a \in A$ be a family of lsc function and define $f(x) := \inf_{a \in A} f_a(x)$. Suppose that $x^* \in \partial_F f(x)$. Then for any $\varepsilon > 0$, there exist $(a, z) \in A \times B_\varepsilon(x)$ satisfying $f_a(z) < \underline{f(x)} + \varepsilon$ and $z^* \in \partial_F f_a(z)$ such that*

$$\|z^* - x^*\| < \varepsilon.$$

***Exercise 5.5.5** Stegall's variational principle allows a linear perturbation. Here is a reflexive Banach space version. More general versions of this result and their applications will be discussed in Section 6.3.

Theorem 5.5.14 (Stegall's Variational Principle) *Let X be a reflexive Banach space, let $f \colon X \to \mathbb{R} \cup \{+\infty\}$ be a lsc function bounded from below and let A be a closed subset of X. Then for any $\varepsilon > 0$, there exists $x^* \in X^*$ with $\|x^*\| < \varepsilon$ such that*

$$x \to f(x) + \langle x^*, x \rangle$$

attains a minimum on A.

Deduce the Stegall's variational principle [243] in a reflexive Banach space from Theorem 5.5.2. Hint:

(i) Define $g_a(x^*) := f(a) + \langle x^*, a \rangle, a \in A$ and define $g \colon X^* \to \mathbb{R}$ by

$$g(x^*) := \inf_{a \in A} g_a(x^*).$$

Show that g is upper semicontinuous, concave and finite on X^* and therefore continuous on X^*.

(ii) Apply Theorem 5.5.2 to conclude that for any $\varepsilon > 0$, there exists $x^* \in \varepsilon B_{X^*}$ such that $\partial_F g(x^*) \neq \emptyset$.

(iii) Suppose that $x \in \partial_F g(x^*)$. Apply Theorem 5.5.2 to conclude that for any minimizing sequence $a_i \in A$ (i.e., $g_{a_i}(x^*) \to g(x^*)$), there exist sequences (x_i^*) and (x_i) with $x_i \in \partial_F g_{a_i}(x_i^*)$, such that $x_i \to x$.

(iv) Observing that $x_i = a_i$ we conclude x is the limit of every minimizing sequence, and therefore $x \in A$ and

$$f(a) + \langle x^*, a \rangle$$

attains a minimum at x over A.

*Exercise 5.5.6 Deduce the following result on the closest point to a subset in a Hilbert space from Theorem 5.5.2.

Theorem 5.5.15 (Closest Point: Hilbert Space Case) *Let H be a Hilbert space and let A be a closed subset of X. Then, there exists a dense subset D of H such that for any $x \in D$, there is a unique closest point from A.*

Hint: First show that

$$d_A^2(x) := \inf_{a \in A} \|x - a\|^2$$

is a continuous function, and therefore

$$D := \mathrm{dom}\partial_F d_A^2$$

is a dense subset of H. For any $x \in D$ and $x^* \in \partial_F d_A^2(x)$ apply Theorem 5.5.2 to show that for any minimizing sequence $\{a_n\} \in A$ of $d_A^2(x)$, $a_n \to x - x^*/2$, and therefore $x - x^*/2$ is the unique closest point in A to x. A similar argument works in a reflexive Banach space.

Exercise 5.5.7 Let X and Y be Banach spaces, let S be a closed subset of X and let $f: X \to Y$ be a C^1 mapping. Show that f is metrically regular on S at $\bar{x} \in S$ with modulus r if and only if the multifunction F_S defined by (5.5.9) is metrically regular at $(\bar{x}, f(\bar{x}))$ with modulus r.

Exercise 5.5.8 Let X and Y be Banach spaces and let $F: X \to 2^Y$ be a closed multifunction. Show that F is metrically regular at $(\bar{x}, \bar{y}) \in \mathrm{graph}\, F$ if and only if the C^1 mapping $f: X \times Y \to Y$ defined by $f(x, y) = y$ is metrically regular on $\mathrm{graph}\, F$ at $(\bar{x}, \bar{y})$.

Exercise 5.5.9 Prove Proposition 5.5.11.

Exercise 5.5.10 Prove Proposition 5.5.12.

Exercise 5.5.11 Use Corollary 5.5.12 to derive a sufficient condition for a multifunction to be metrically regular.

Exercise 5.5.12 Define open covering with linear rate for a C^1 function on a closed set and use it to study open covering with linear rate for multifunctions.

*Exercise 5.5.13 (Tangential Metric Regularity Condition) Let X and Y be finite dimensional Banach spaces. Let $f\colon X \to Y$ be a C^1 function and let S be a closed subset of X. Suppose that $\bar{x} \in S$ and there exists a neighborhood W of $\bar{x}$ such that for all $x \in W \cap S$,

$$f'(x)T_B(S; x) = Y. \tag{5.5.11}$$

Show that f is metrically regular on S at $\bar{x}$. Hint: Use a category argument and the compactness of S to show that (5.5.11) implies there exists $r > 0$ such that $rB_Y \subset f'(x)[T_B(S; x) \cap B_X]$ for all x in a neighborhood of $\bar{x}$, which in turn implies (5.5.10).

*Exercise 5.5.14 (Transversality) Let X and Y be finite dimensional Banach spaces. Let $f\colon X \to Y$ be a C^1 function and let S and R be closed subsets of X and Y, respectively. Suppose that $\bar{x} \in S$ and there exists a neighborhood W of $\bar{x}$ such that for all $x \in W \cap S$, we have $f(x) \in R$ and

$$f'(x)\,[T_B(S; x)] - T_B(R; f(x)) = Y. \tag{5.5.12}$$

(i) Define function $g\colon X \times Y \to Y$ by $g(x, y) = f(x) - y$. Prove g is metrically regular on $S \times R$ at $(\bar{x}, f(\bar{x}))$.

(ii) Apply the exact penalization in Example 3.0.3 to deduce the existence of a constant k such that

$$d_{(S \times R) \cap g^{-1}(g(\bar{x}, f(\bar{x})))}(z, y) \le k \left\{ \|f(z) - y\| + d_S(z) + d_R(y) \right\}$$

holds for all points (z, y) in a neighborhood of $(\bar{x}, f(\bar{x}))$.

(iii) Deduce the inequality

$$d_{(S \cap f^{-1}(R))}(z) \le k \left\{ d_S(z) + d_R(f(z)) \right\}$$

holds for all points z in a neighborhood of $\bar{x}$.

(iv) Deduce the inclusions

$$T_C(S; \bar{x}) \cap (f'(x))^{-1} T_C(R; f(\bar{x})) \subset T_C(S \cap f^{-1}(R); \bar{x})$$

and

$$T_B(S; \bar{x}) \cap (f'(x))^{-1} T_C(R; f(\bar{x})) \subset T_B(S \cap f^{-1}(R); \bar{x}).$$

(v) Suppose f is the identity map, so $T_C(S; \bar{x}) - T_C(R; \bar{x}) = Y$. If either R or S is tangentially regular at $\bar{x}$, prove

$$T_B(R \cap S; \bar{x}) = T_B(R; \bar{x}) \cap T_B(S; \bar{x}).$$

(vi) (Guignard) By taking the polar and applying the Krein–Rutman polar cone calculus of Theorem 4.4.5 and condition (5.5.12) again, deduce

$$N_C(S \cap f^{-1}(R); \bar{x}) \subset N_C(S; \bar{x}) + [f'(x)]^* N_C(R; f(\bar{x})).$$

(vii) If C_1 and C_2 are convex subsets of X satisfying $0 \in \mathrm{core}(C_1 - C_2)$, and the point $\bar{x}$ lies in $C_1 \cap C_2$, use part (iv) to prove

$$T_C(C_1 \cap C_2; \bar{x}) = T_C(C_1; \bar{x}) \cap T_C(C_2; \bar{x}).$$

Reference: [56, p. 158].

*__Exercise 5.5.15__ (Guignard Optimality Conditions) Let X and Y be finite dimensional Banach spaces and let S and R be closed subsets of X and Y, respectively. Suppose that $\bar{x}$ is a local minimizer for the optimization problem

$$\text{minimize} \quad f(x), \quad x \in S$$

$$\text{subject to} \quad g(x) \in R,$$

where $f \colon X \to \mathbb{R}$ and $g \colon X \to Y$ are C^1 functions with G satisfying the following transversality condition: for all x in a neighborhood of $\bar{x}$,

$$g'(x)\,[T_B(S; x)] - T_B(R; g(x)) = Y.$$

Use Exercise 5.5.14 to prove the optimality condition

$$0 \in f'(\bar{x}) + [g'(\bar{x})]^* \, N_C(R; g(\bar{x})) + N_C(S; \bar{x}).$$

Reference: [131].

6

Variational Principles in Nonlinear Functional Analysis

Calculus of variations provides the stimulation for the development of both nonlinear functional analysis and variational techniques. The development of nonlinear functional analysis and variational techniques have gone hand in hand ever since. We have seen that modern variational principles grew out of results in Banach space geometry. In this chapter we collect several interesting applications to showcase different approaches and to highlight the diversity of results.

6.1 Subdifferential and Asplund Spaces

6.1.1 Asplund Spaces

Recall that a subset S of a Banach space X is a G_δ set provided that it is the intersection of a countable number of open sets.

Definition 6.1.1 *Let X be a Banach space. We say that X is an* Asplund *space if every continuous convex function defined on a nonempty open convex subset C of X is Fréchet differentiable on a dense G_δ subset of C.*

The following lemmas provide convenient ways of checking whether a space is Asplund.

Lemma 6.1.2 *Let X be a Banach space and let C be an open convex subset of X. Suppose that f is a convex and continuous function on C. Then f is Fréchet differentiable at $x \in C$ if and only if for any $\varepsilon > 0$ there exists $\delta > 0$ such that for all $\|y\| = 1$ and $t \in (0, \delta)$,*

$$f(x + ty) + f(x - ty) - 2f(x) < t\varepsilon. \qquad (6.1.1)$$

Proof. The necessity is easy and left as Exercise 6.1.1. We prove the sufficiency. Choose $x^* \in \partial f(x)$. It follows that for all y and all sufficiently small $t > 0$ such that $x \pm ty \in C$,

$$\langle x^*, ty \rangle \le f(x + ty) - f(x) \tag{6.1.2}$$

and

$$-\langle x^*, ty \rangle \le f(x - ty) - f(x). \tag{6.1.3}$$

By hypothesis, for any $\varepsilon > 0$ there exists $\delta > 0$ such that (6.1.1) holds for any $t \in (0, \delta)$ and any y with $\|y\| = 1$, that is,

$$f(x + ty) - f(x) - \langle x^*, ty \rangle \le t\varepsilon + f(x) - f(x - ty) - \langle x^*, ty \rangle.$$

Inequality (6.1.2) shows that the left side is greater or equal to 0 while (6.1.3) shows the right side is less than or equal to $t\varepsilon$, which completes the proof. $\bullet$

A similar result holds for the Gâteaux differentiability (Exercise 6.1.2).

Lemma 6.1.3 *Let X be a Banach space, let C be an open subset of X and let f be a continuous and convex function on C. Then the set G of points in C where f is Fréchet differentiable is a G_δ set.*

Proof. For each i, define

$$G_i := \left\{ x \in C \ \middle| \ \sup_{\|y\|=1} \frac{f(x + \delta y) + f(x - \delta y) - 2f(x)}{\delta} < \frac{1}{i}, \ \text{for some } \delta > 0 \right\}.$$

By Exercise 4.2.9, for any fixed x and y the functions $t \to (f(x \pm ty) - f(x))/t$ are decreasing as $t \to 0+$, hence from Lemma 6.1.2 we can conclude that $G = \bigcap_{i=1}^{\infty} G_i$. It remains to prove that each G_i is open. Suppose that $x \in G_i$. Since f is locally Lipschitz, there exist $\delta_1 > 0$ and $L > 0$ such that f satisfies the Lipschitz condition with Lipschitz constant L on $B_{\delta_1}(x)$. Since $x \in G_i$ there exist $\delta > 0$ and $r > 0$ such that for all y with $\|y\| = 1$ we have $x \pm \delta y \in C$ and

$$\frac{f(x + \delta y) + f(x - \delta y) - 2f(x)}{\delta} \le r < \frac{1}{i}.$$

Choose $\delta_2 \in (0, \delta_1)$ small enough so that $B_{2\delta_2}(x) \subset C$ and $r + 4M\delta_2/\delta < 1/i$. We show that $B_{\delta_2}(x) \subset G_i$. In fact, for any $z \in B_{\delta_2}(x)$

$$\frac{f(z + \delta y) + f(z - \delta y) - 2f(z)}{\delta} \le \frac{f(x + \delta y) + f(x - \delta y) - 2f(x)}{\delta}$$

$$+ \frac{|f(z) - f(x)|}{2\delta} + \frac{|f(z + \delta y) - f(x + \delta y)|}{\delta}$$

$$+ \frac{|f(z - \delta y) - f(x - \delta y)|}{\delta}$$

$$\le r + \frac{4M\|z - x\|}{\delta} \le r + \frac{4M\delta_2}{\delta} < \frac{1}{i}.$$

6.1.2 A Sum Rule Characterization

It turns out that Asplund spaces are natural places to use Fréchet subdifferentials in the sense that they are characterized by having a natural Fréchet subdifferential calculus. Here we need to note that an Asplund space is not necessarily Fréchet smooth so that in such a space the Fréchet subdifferential and the viscosity Fréchet subdifferential may differ. In this chapter we use the Fréchet subdifferential in Definition 3.1.1. We will establish the equivalence of the Asplund space property and the Fréchet subdifferential (approximate) calculus by showing first the equivalence of the Asplund property with a simple Lipschitz local approximate Fréchet subdifferential sum rule and then establish the equivalence of other calculus rules with this Lipschitz sum rule. The following proposition provides the easy direction which essentially says that if an approximate local sum rule holds in X then X is an Asplund space.

Definition 6.1.4 (Simple Lipschitz Local Approximate Fréchet Subdifferential Sum Rule) *Let X be a Banach space. We say that X has a simple Lipschitz local approximate Fréchet subdifferential sum rule if, for any lsc function $f_1 \colon X \to \mathbb{R} \cup \{+\infty\}$, any $\varepsilon > 0$ and any Lipschitz function $f_2 \colon X \to \mathbb{R}$ with $\bar{x} \in \mathrm{argmin}(f_1 + f_2)$, there exist x_n and $x_n^* \in \partial_F f_n(x_n)$, $n = 1, 2$ satisfying $(x_n, f_n(x_n)) \in B_\varepsilon((\bar{x}, f_n(\bar{x})))$, and*

$$\|x_1^* + x_2^*\| < \varepsilon.$$

Proposition 6.1.5 *Let X be a Banach space. Suppose that X has a simple Lipschitz local approximate Fréchet subdifferential sum rule. Then X is Asplund.*

Proof. Let f be a continuous convex function on an open convex subset C of X. For any $x \in C$, f is locally Lipschitz around x by Theorem 4.1.3. Applying the simple Lipschitz local approximate Fréchet subdifferential sum rule to $f_1 = \iota_{\{x\}}$ and $f_2 = -f$, we conclude that for any $\varepsilon > 0$, $-f$ is Fréchet subdifferentiable at some point in $B_\varepsilon(x)$. Since x is arbitrary, $-f$ is densely Fréchet subdifferentiable in C. On the other hand, Theorem 4.2.7 and Exercise 4.2.10 imply that for any $x \in C$, $\partial f(x) = \partial_F f(x) \neq \emptyset$. Thus, f is densely Fréchet differentiable on C (Exercise 6.1.3) and therefore Fréchet differentiable on a dense G_δ set by Lemma 6.1.3.   ●

The converse of Proposition 6.1.5 is more involved. The proof relies on the separable reduction technique. One of the pillars of this method is the following result in Banach space geometry.

Theorem 6.1.6 *Let X be a separable Asplund space. Then X has an equivalent Fréchet smooth norm.*

Since the proof of this theorem relates more to Banach space geometry than to variational techniques, we will not present it here. Interested readers can find related discussions in [99, 101, 221].

The key of the separable reduction method is to show that for any Banach space X, one can construct a separable subspace in such a way that any Lipschitz local approximate Fréchet subdifferential sum rule in this subspace can be lifted to the whole space. This is achieved through the primary space characterization of the sum rule. We build up this result starting with convex functions.

Lemma 6.1.7 *Let X be a Banach space and let $f\colon X \to \mathbb{R} \cup \{+\infty\}$ be a proper convex function. Then, $\partial f(0) \neq \emptyset$ if and only if there exists $c \geq 0$ such that for all $h \in X$,*

$$f(h) \geq f(0) - c\|h\|.$$

Proof. Exercise 6.1.4. ●

Using Lemma 6.1.7 we can derive the following more general result.

Lemma 6.1.8 *Let X be a Banach space and let $f_n\colon X \to \mathbb{R} \cup \{+\infty\}, n = 1, \ldots, N$ be convex functions not identically equal to $+\infty$. Then,*

$$0 \in \sum_{n=1}^{N} \partial f_n(0)$$

if and only if there exists $c \geq 0$ such that for all $h_n \in X, n = 1, \ldots, N$,

$$\sum_{n=1}^{N} f_n(h_n) \geq \sum_{n=1}^{N} f_n(0) - c \sum_{n=1}^{N} \|h_n - h_N\|. \qquad (6.1.4)$$

Proof. The necessity is easy and left as Exercise 6.1.5. We prove sufficiency and consider the nontrivial case of $N > 1$. Define $f\colon X^{N-1} \to \mathbb{R} \cup \{+\infty\}$ by

$$f(x_1, \ldots, x_{N-1}) := \inf\left\{ \sum_{n=1}^{N-1} f_n(x_n + h) + f_N(h) \,\middle|\, h \in X \right\},$$

where $(x_1, \ldots, x_{N-1}) \in X^{N-1}$.

Clearly, f is a convex function. Moreover, it follows from inequality (6.1.4) that $f(0, \ldots, 0) = \sum_{n=1}^{N} f_n(0)$ and that

$$f(x_1, \ldots, x_{N-1}) \geq f(0, \ldots, 0) - c \sum_{n=1}^{N-1} \|x_n\|$$

for all $(x_1, \ldots, x_{N-1}) \in X^{N-1}$. Applying Lemma 6.1.7 to f on X^{N-1} we have that there exist $x_n^* \in X^*, n = 1, \ldots, N-1$ such that

$$f(x_1, \ldots, x_{N-1}) \geq f(0, \ldots, 0) + \sum_{n=1}^{N-1} \langle x_n^*, x_n \rangle,$$

that is,

$$\sum_{n=1}^{N-1} f_n(x_n + h) + f_N(h) \geq \sum_{n=1}^{N} f_n(0) + \sum_{n=1}^{N-1} \langle x_n^*, x_n \rangle \qquad (6.1.5)$$

for all $x_1, \ldots, x_{N-1}, h \in X$. For any $n \in \{1, \ldots, N-1\}$, setting $h = 0$ and $x_m = 0$ for all $m \in \{1, \ldots, N-1\} \backslash \{n\}$ in (6.1.5) we get $x_n^* \in \partial f_n(0)$. On the other hand, putting $x_1 = \cdots = x_{N-1} = -h$ in (6.1.5) we deduce that $-x_1^* - \cdots - x_{N-1}^* \in \partial f_n(0)$. Thus, we have shown that $0 \in \sum_{n=1}^{N} \partial f_n(0)$. $\bullet$

We now turn to a more general result in which the convex functions in Lemma 6.1.8 are replaced by general functions and the convex subdifferentials by the Fréchet subdifferential. This is the main lemma summarizing the technical part of the separable reduction technique.

Lemma 6.1.9 *Let X be a Banach space and let $f_n \colon X \to \mathbb{R} \cup \{+\infty\}, n = 1, \ldots, N$ be functions not identically equal to $+\infty$. Suppose that $x_n \in \operatorname{dom} f_n$. Then*

$$0 \in \sum_{n=1}^{N} \partial_F f_n(x_n) \qquad (6.1.6)$$

if and only if there exists $c \geq 0$ and sequences $\delta_{ni}, n = 1, \ldots, N, i = 1, 2, \ldots,$ of positive numbers such that

$$\sum_{n=1}^{N} \sum_{i=1}^{m_n} \beta_{ni} \sum_{j=1}^{k_{ni}} \alpha_{nij} [f_n(x_n + h_{nij}) + \frac{1}{i} \|h_{nij}\|] \qquad (6.1.7)$$

$$\geq \sum_{n=1}^{N} f_n(x_n) - c \sum_{n=1}^{N-1} \Big\| \sum_{i=1}^{m_n} \beta_{ni} \sum_{j=1}^{k_{ni}} \alpha_{nij} h_{nij} - \sum_{i=1}^{m_N} \beta_{Ni} \sum_{j=1}^{k_{Ni}} \alpha_{Nij} h_{Nij} \Big\|.$$

whenever $h_{nij} \in \delta_{ni} B_X$, $\alpha_{nij} \geq 0$, $j = 1, \ldots, k_{ni}$, $\sum_{j=1}^{k_{ni}} \alpha_{nij} = 1$, $k_{ni} = 1, 2, \ldots,$ $\beta_{ni} \geq 0$, $i = 1, \ldots, m_n$, and $\sum_{i=1}^{m_n} \beta_{ni} = 1$, $m_n = 1, 2, \ldots,$ $n = 1, \ldots, N$.

Proof. Necessity. Let $x_n^* \in \partial_F f_n(x_n)$ be such that $\sum_{n=1}^{N} x_n^* = 0$. For $n = 1, \ldots, N$ find sequences $\delta_{ni}, n = 1, \ldots, N, i = 1, 2, \ldots,$ of positive numbers such that, for any $\|h\| \leq \delta_{ni}$,

$$f_n(x_n + h) - f_n(x_n) \geq \langle x_n^*, h \rangle - \frac{1}{i} \|h\|.$$

Then, for any $h_{nij} \in \delta_{ni} B_X$, $\alpha_{nij} \geq 0$, $j = 1, \ldots, k_{ni}$, $\sum_{j=1}^{k_{ni}} \alpha_{nij} = 1$, $k_{ni} = 1, 2, \ldots,$ $\beta_{ni} \geq 0$, $i = 1, \ldots, m_n$, $\sum_{i=1}^{m_n} \beta_{ni} = 1$, $m_n = 1, 2, \ldots,$ $n - 1, \ldots, N$, we have

$$\sum_{n=1}^{N}\sum_{i=1}^{m_n}\beta_{ni}\sum_{j=1}^{k_{ni}}\alpha_{nij}[f_n(x_n+h_{nij})+\frac{1}{i}\|h_{nij}\|]$$

$$\geq \sum_{n=1}^{N}\sum_{i=1}^{m_n}\beta_{ni}\sum_{j=1}^{k_{ni}}\alpha_{nij}[f_n(x_n)+\langle x_n^*, h_{nij}\rangle]$$

$$= \sum_{n=1}^{N}f_n(x_n)+\sum_{n=1}^{N-1}\Big\langle x_n^*, \sum_{i=1}^{m_n}\beta_{ni}\sum_{j=1}^{k_{ni}}\alpha_{nij}h_{nij}-\sum_{i=1}^{m_N}\beta_{Ni}\sum_{j=1}^{k_{Ni}}\alpha_{Nij}h_{Nij}\Big\rangle$$

$$\geq \sum_{n=1}^{N}f_n(x_n)-c\sum_{n=1}^{N-1}\Big\|\sum_{i=1}^{m_n}\beta_{ni}\sum_{j=1}^{k_{ni}}\alpha_{nij}h_{nij}-\sum_{i=1}^{m_N}\beta_{Ni}\sum_{j=1}^{k_{Ni}}\alpha_{Nij}h_{Nij}\Big\|,$$

where $c := \max\{\|x_1^*\|, \ldots, \|x_{N-1}^*\|\}$.

Sufficiency: For $n = 1, \ldots, N$ and $i = 1, 2, \ldots$, we define functions $\phi_{ni}\colon X \to \mathbb{R} \cup \{+\infty\}$ by

$$\phi_{ni}(h) := \inf\Big\{\sum_{j=1}^{k}\alpha_j[f_n(x_n+h_j)+\frac{1}{i}\|h_j\|] \,\Big|\, h_j \in \delta_{ni}B_X,$$

$$\alpha_j \geq 0, j = 1, \ldots, k, \sum_{j=1}^{k}\alpha_j = 1, \sum_{j=1}^{k}\alpha_j h_j = h, k = 1, 2, \ldots\Big\},$$

if $\|h\| \leq \delta_{ni}$, and $\phi_{ni}(h) = \infty$ otherwise. It is not hard to check that ϕ_{ni} are proper convex functions (Exercise 6.1.6). It follows from (6.1.7) that

$$\phi_{ni}(0) \leq f_n(x_n) \tag{6.1.8}$$

and

$$\sum_{n=1}^{N}\sum_{i=1}^{m_n}\beta_{ni}\phi_{ni}(h_{ni}) \geq \sum_{n=1}^{N}f_n(x_n) \tag{6.1.9}$$

$$-c\sum_{n=1}^{N-1}\Big\|\sum_{i=1}^{m_n}\beta_{ni}h_{ni}-\sum_{i=1}^{m_N}\beta_{Ni}h_{Ni}\Big\|.$$

for all $h_{ni} \in X$, $i = 1, \ldots, m_n$, $m_n = 1, 2, \ldots$, $\beta_{ni} \geq 0$, $\sum_{i=1}^{m_n}\beta_{ni} = 1$, $n = 1, \ldots, N$. Further, for $n = 1, \ldots, N$ define $\phi_n\colon X \to \mathbb{R} \cup \{+\infty\}$ by

$$\phi_n(h) := \inf\Big\{\sum_{i=1}^{m}\beta_i\phi_{ni}(h_i) \,\Big|\, h_i \in X, \beta_i \geq 0, i = 1, \ldots, m,$$

$$\sum_{i=1}^{m}\beta_i = 1, \sum_{i=1}^{m}\beta_i h_i = h, m = 1, 2, \ldots\Big\}.$$

Again ϕ_n is convex (Exercise 6.1.6). It follows from inequality (6.1.9) that

$$\sum_{n=1}^{N} \phi_n(0) \geq \sum_{n=1}^{N} f_n(x_n).$$

Combining this with inequality (6.1.8) and (6.1.9) we have $\phi_n(0) = f_n(x_n)$ and

$$\sum_{n=1}^{N} \phi_n(h_n) \geq \sum_{n=1}^{N} \phi_n(0) - c \sum_{n=1}^{N-1} \|h_n - h_N\|$$

for all $h_n \in X, n = 1, \ldots, N$. By Lemma 6.1.7 this implies that

$$0 \in \sum_{n=1}^{N} \partial \phi_n(0).$$

Now to show (6.1.6), it suffices to show that $\partial \phi_n(0) \subset \partial_F f_n(x_n)$. So let $x_n^* \in \partial \phi_n(0)$. Then for $\|h\| \leq \delta_{ni}$, by the definition of ϕ_{ni} and ϕ_n we have

$$f_n(x_n + h) + \frac{1}{i}\|h\| \geq \phi_{ni}(h) \geq \phi_n(h)$$
$$\geq \phi_n(0) + \langle x_n^*, h \rangle = f_n(x_n) + \langle x_n^*, h \rangle.$$

Hence, for all i,

$$\liminf_{\|h\| \to 0} \frac{f_n(x_n + h) - f_n(x_n) - \langle x_n^*, h \rangle}{\|h\|} \geq -\frac{1}{i}.$$

Therefore, $x_n^* \in \partial_F f_n(x_n)$. ●

Now we are ready to prove the key result in our separable reduction argument.

Theorem 6.1.10 *Let X be a Banach space, let Y_0 be a separable subspace of X and let $f_n \colon X \to \mathbb{R} \cup \{+\infty\}$, $n = 1, \ldots, N$ be functions locally bounded from below. Then there exists a separable subspace Y of X, containing Y_0 such that $x_n \in Y, n = 1, \ldots, N$ and*

$$0 \in \sum_{n=1}^{N} \partial_F(f_n|_Y)(x_n),$$

implies

$$0 \in \sum_{n=1}^{N} \partial_F f_n(x_n).$$

Proof. Given $x = (x_1, \ldots, x_N) \in X^N$, let $\eta(x)$ denote the supremum of all $\eta > 0$ such that f_n is bounded from below on $B_\eta(x_n)$ for each $n = 1, \ldots, N$. Define

$$A_{mk} := \Big\{ \alpha = \{\alpha_{nij} : n = 1, \dots, N, i = 1, \dots, m, j = 1, \dots, k\} \Big|$$

$$\text{all } \alpha_{nij} \ge 0 \text{ are rational and } \sum_{j=1}^{k} \alpha_{nij} = 1$$

$$\text{for } n = 1, \dots, N, i = 1, \dots, m \Big\},$$

and

$$B_m := \Big\{ \beta = \{\beta_{ni} : n = 1, \dots, N, i = 1, \dots, m\} \Big|$$

$$\text{all } \beta_{ni} \ge 0 \text{ are rational and } \sum_{i=1}^{m} \beta_{ni} = 1$$

$$\text{for } n = 1, \dots, N \Big\}.$$

Clearly, A_{mk} and B_m are countable. For any $x = (x_1, \dots, x_N) \in X^N$, sequences $\eta_{ni} \in (0, \eta(x)), n = 1, \dots, N, i = 1, 2, \dots, m, k, l \in \mathcal{N}, \alpha \in A_{mk}$, $\beta \in B_m$ and $r = (r_1, \dots, r_N) \in (0, +\infty)^N$, we can find

$$h_{nij}(x, r, \alpha, \beta, l, \eta_{ni}) \in \eta_{ni} B_X,$$

such that

$$\Big\| \sum_{i=1}^{m} \beta_{ni} \sum_{j=1}^{k} \alpha_{nij} h_{nij}(x, r, \alpha, \beta, l, \eta_{ni})$$

$$- \sum_{i=1}^{m} \beta_{Ni} \sum_{j=1}^{k} \alpha_{Nij} h_{Nij}(x, r, \alpha, \beta, l, \eta_{ni}) \Big\| < r_n, \quad (6.1.10)$$

for $n = 1, \dots, N$, and

$$\sum_{n=1}^{N} \sum_{i=1}^{m} \beta_{ni} \sum_{j=1}^{k} \alpha_{nij} \Big[f_n(x_n + h_{nij}) + \frac{1}{i} \|h_{nij}\| \Big]$$

$$\ge \sum_{n=1}^{N} \sum_{i=1}^{m} \beta_{ni} \sum_{j=1}^{k} \alpha_{nij} \Big[f_n(x_n + h_{nij}(x, r, \alpha, \beta, l, \eta_{ni}))$$

$$+ \frac{1}{i} \|h_{nij}(x, r, \alpha, \beta, l, \eta_{ni})\| \Big] - \frac{1}{l} \quad (6.1.11)$$

whenever $h_{nij} \in \eta_{ni} B_X$ and

$$\Big\| \sum_{i=1}^{m} \beta_{ni} \sum_{j=1}^{k} \alpha_{nij} h_{nij} - \sum_{i=1}^{m} \beta_{Ni} \sum_{j=1}^{k} \alpha_{Nij} h_{Nij} \Big\| < r_n, n = 1, \dots, N.$$

Now we construct the space Y by induction. We build separable subspaces $Y_1, Y_2, \dots$ of X and dense countable subsets C_l of Y_l, $l = 0, 1, \dots$, as follows. Let C_0 be any dense countable subset of Y_0. If we have constructed $C_0, Y_0, \dots, C_l, Y_l$ for some $l \ge 0$, let Y_{l+1} be the closed linear span of the set

$$C_l \bigcup \Big\{ h_{nij}(x, r, \alpha, \beta, l, \eta_{ni}) \ \Big| \ x \in C_l^N, \alpha \in A_{mk}, \beta \in B_m,$$

$$r = (r_1, \ldots, r_N), r_n > 0 \text{ rational}, \eta_{ni} \in (0, \eta(x)) \text{ rational},$$

$$i = 1, \ldots, m, j = 1, \ldots, k, n = 1, \ldots, N \Big\}.$$

Since the above set is countable, the subspace Y_{l+1} is separable. Let C_{l+1} be some dense countable subset of Y_{l+1}. Finally, we define Y as the closure of $\bigcup_{l=1}^{\infty} Y_l$. Thus Y is a separable subspace of X.

Let $x = (x_1, \ldots, x_N)$ with $x_n \in Y$, $n = 1, \ldots, N$ be such that

$$0 \in \sum_{n=1}^{N} \partial_F(f_n|_Y)(x_n).$$

We need to verify inequality (6.1.7) in Lemma 6.1.9. Clearly it suffices to consider all the α_{nij} and β_{ni} rational. By Lemma 6.1.9 there are sequences $\delta_{ni} > 0$ and $c \geq 0$ such that (6.1.7) holds on replacing $h_{nij} \in X$ by $h_{nij} \in Y$. Take rational sequences $\eta_{ni} \in (0, \min(\eta(x), \delta_{ni})/2)$. Let $m, k \in \mathcal{N}$, $\alpha \in A_{mk}$, $\beta \in B_m$ and $h_{nij} \in \eta_{ni} B_X$ be fixed. We proceed to show that (6.1.7) holds. Let γ be an arbitrary positive number and let $r = (r_1, \ldots, r_N)$ be such that all $r_n > 0$ are rational and for $n = 1, \ldots, N$,

$$\gamma > r_n - \Big\| \sum_{i=1}^{m} \beta_{ni} \sum_{j=1}^{k} \alpha_{nij} h_{nij} - \sum_{i=1}^{m} \beta_{Ni} \sum_{j=1}^{k} \alpha_{Nij} h_{Nij} \Big\|. \qquad (6.1.12)$$

As $\bigcup_{l=1}^{\infty} Y_l$ is dense in Y and C_l is dense in Y_l, there exists an l and $y = (y_1, \ldots, y_N) \in C_l^N$ such that

$$\|x_n - y_n\| < \min(\gamma, \eta(x)/2), \|x_n - y_n + h_{nij}\| < \eta_{ni}/2 \qquad (6.1.13)$$

and

$$\Big\| \sum_{i=1}^{m} \beta_{ni} \sum_{j=1}^{k} \alpha_{nij}(h_{nij} - h_{Nij}) + x_n - y_n - x_N + y_N \Big\| < r_n \qquad (6.1.14)$$

for all n, i. Denote $\tilde{h}_{nij} = x_n - y_n + h_{nij}$. Then by inequalities (6.1.13) and (6.1.14) we have

$$\|\tilde{h}_{nij}\| < \eta_{ni},$$

$$\Big\| \sum_{i=1}^{m} \beta_{ni} \sum_{j=1}^{k} \alpha_{nij}(\tilde{h}_{nij} - \tilde{h}_{Nij}) \Big\| < r_n$$

and $\|y_n + h - x_n\| < \eta(x)$ whenever $h \in \frac{1}{2}\eta(x)B_X$. Thus, $\eta(y) \geq \frac{1}{2}\eta(x)$ and so $h_{nij}(y, r, \alpha, \beta, l, \eta_{ni})$ is well defined since $\eta_{ni} < \frac{1}{2}\eta(x)$. Now we can estimate

$$\mathcal{T} := \sum_{n=1}^{N} \sum_{i=1}^{m} \beta_{ni} \sum_{j=1}^{k} \alpha_{nij} \left[f_n(x_n + h_{nij}) + \frac{1}{i} \|h_{nij}\| \right]$$

$$\geq \sum_{n=1}^{N} \sum_{i=1}^{m} \beta_{ni} \sum_{j=1}^{k} \alpha_{nij} \left[f_n(y_n + \tilde{h}_{nij}) + \frac{1}{i} \|\tilde{h}_{nij}\| \right] - \frac{\gamma N}{i} \quad \text{by (6.1.13)}$$

$$\geq \sum_{n=1}^{N} \sum_{i=1}^{m} \beta_{ni} \sum_{j=1}^{k} \alpha_{nij} \left[f_n(y_n + h_{nij}(y, r, \alpha, \beta, l, \eta_{ni})) \right.$$

$$\left. + \frac{1}{i} \|h_{nij}(y, r, \alpha, \beta, l, \eta_{ni})\| \right] - \frac{1}{l} - \frac{\gamma N}{i} \quad \text{by (6.1.11)}$$

$$\geq \sum_{n=1}^{N} \sum_{i=1}^{m} \beta_{ni} \sum_{j=1}^{k} \alpha_{nij} \left[f_n(x_n + y_n - x_n + h_{nij}(y, r, \alpha, \beta, l, \eta_{ni})) \right.$$

$$\left. + \frac{1}{i} \|y_n - x_n + h_{nij}(y, r, \alpha, \beta, l, \eta_{ni})\| \right] - \frac{\gamma N}{i} - \frac{1}{l} - \frac{\gamma N}{i} \quad \text{by (6.1.13).}$$

Since $\|y_n - x_n + h_{nij}(y, r, \alpha, \beta, l, \eta_{ni})\| < \eta_{ni}$ (by (6.1.13) and $\|h_{nij}(y, r, \alpha, \beta, l, \eta_{ni})\| < \eta_{ni}/2$) and (6.1.7) holds in Y, we have

$$\mathcal{T} \geq \sum_{n=1}^{N} f_n(x_n) - c \sum_{n=1}^{N} \left\| \sum_{i=1}^{m} \beta_{ni} \sum_{j=1}^{k} (\alpha_{nij} h_{nij}(y, r, \alpha, \beta, l, \eta_{ni}) \right.$$

$$\left. -\alpha_{Nij} h_{Nij}(y, r, \alpha, \beta, l, \eta_{Ni})) + y_n - x_n - y_N + x_N \right\| - \frac{2N\gamma}{i} - \frac{1}{l}$$

$$> \sum_{n=1}^{N} f_n(x_n) - c \sum_{n=1}^{N} r_n - 2cN\gamma - \frac{2N\gamma}{i} - \frac{1}{l} \quad \text{by(6.1.10), (6.1.13)}$$

$$> \sum_{n=1}^{N} f_n(x_n) - c \sum_{n=1}^{N} \left\| \sum_{i=1}^{m} \beta_{ni} \sum_{j=1}^{k} (\alpha_{nij} h_{nij} - \alpha_{Nij} h_{Nij}) \right\|$$

$$-cN\gamma - 2cN\gamma - \frac{2N\gamma}{i} - \frac{1}{l} \quad \text{by(6.1.12).}$$

Since γ could be taken arbitrarily small and l arbitrarily large we have verified inequality (6.1.7). Thus, by Lemma 6.1.9,

$$0 \in \sum_{n=1}^{N} \partial_F f_n(x_n).$$

Combining Proposition 6.1.5, Theorem 6.1.6 and Theorem 6.1.10 we can now conclude that the simple Lipschitz local approximate Fréchet subdifferential sum rule characterizes an Asplund space.

Theorem 6.1.11 *Let X be a Banach space. Then X is Asplund if and only if X has a simple Lipschitz local approximate Fréchet subdifferential sum rule.*

Proof. Exercise 6.1.7.   ●

6.1.3 Subdifferential Characterizations

We now show that the Lipschitz local approximate sum rule and other sub-differential calculus rules are equivalent in the sense that in any Banach space the fact that one of them holds implies that all the others are also valid. Consequently, all the calculus rules discussed in this section characterize Asplund spaces. First we precisely define the subdifferential calculus rules.

Definition 6.1.12 *We say that* X *has a Fréchet subdifferential nonlocal approximate sum rulei if, for lower semicontinuous functions* $f_1, \ldots, f_N \colon X \to \mathbb{R} \cup \{+\infty\}$ *bounded below with* $\bigwedge[f_1, \ldots, f_N](X) < \infty$ *and any* $\varepsilon > 0$, *there exist* $x_n, n = 1, \ldots, N$ *and* $x_n^* \in \partial_F f_n(x_n)$ *with*

$$\mathrm{diam}(x_1, \ldots, x_N) \times \max(1, \|x_1^*\|, \ldots, \|x_N^*\|) < \varepsilon, \qquad (6.1.15)$$

and

$$\sum_{n=1}^{N} f_n(x_n) < \bigwedge[f_1, \ldots, f_N](X) + \varepsilon \qquad (6.1.16)$$

such that

$$\left\| \sum_{n=1}^{N} x_n^* \right\| < \varepsilon.$$

Definition 6.1.13 *We say that* X *has a Fréchet subdifferential multidirectional mean value theorem if for any nonempty, closed and convex subset* S *of* X, *any element* $x \in X$ *and any lower semicontinuous function* $f \colon X \to \mathbb{R} \cup \{+\infty\}$ *bounded below on a neighborhood of* $[x, S]$ *and*

$$r < \bigwedge[f](S) - f(x),$$

given $\varepsilon > 0$, *there exist* $z \in B_\varepsilon([x, S])$ *and* $z^* \in \partial_F f(z)$ *such that*

$$f(z) < \bigwedge[f]([x, S]) + |r| + \varepsilon.$$

and

$$r < \langle z^*, y - x \rangle + \varepsilon \|y - x\| \text{ for all } y \in S.$$

Definition 6.1.14 *We say that* X *has a Fréchet subdifferential local approximate sum rule if, for lower semicontinuous functions* $f_1, \ldots, f_N \colon X \to \mathbb{R} \cup \{+\infty\}$ *and* $\bar{x} \in \bigcap_{n=1}^{N} \mathrm{dom} f_n$ *satisfying*

$$\sum_{n=1}^{N} f_n(\bar{x}) \leq \bigwedge[f_1, \ldots, f_N](B_h(\bar{x})) \qquad (6.1.17)$$

for some $h > 0$, *and for any* $\varepsilon > 0$, *there exist* $x_n, n = 1, \ldots, N$ *with* $(x_n, f_n(x_n)) \in B_\varepsilon((\bar{x}, f_n(\bar{x})))$ *and* $x_n^* \in \partial_F f_n(x_n)$ *such that*

$$\left\| \sum_{n=1}^N x_n^* \right\| < \varepsilon.$$

Definition 6.1.15 *We say that* X *has a* Fréchet subdifferential Lipschitz local approximate sum rule *if condition (6.1.17) in Definition 6.1.14 is replaced by the requirement that all but one of the functions are locally Lipschitz at* $\bar{x}$.

The two function version of this Fréchet subdifferential Lipschitz local approximate sum rule is called a simple Fréchet subdifferential Lipschitz local approximate sum rule (see Definition 6.1.4).

Definition 6.1.16 *We say that* X *has a* Fréchet subdifferential extremal principle *if for every local extremal point* $\bar{x}$ *of an extremal system* $(S_1, \ldots, S_N)$ *at* $(\bar{p}_1, \ldots, \bar{p}_N)$ *and any* $\varepsilon > 0$, *there exist* $p_n \in B_\varepsilon(\bar{p}_n), x_n \in B_\varepsilon(\bar{x}), n = 1, \ldots, N$ *and* $x_n^* \in N_F(S_n(p_n), x_n) + \varepsilon B_{X^*}, n = 1, \ldots, N$ *such that* $\|x_1^*\| + \|x_2^*\| + \cdots + \|x_N^*\| \geq 1$ *and*

$$x_1^* + x_2^* + \cdots + x_N^* = 0.$$

Definition 6.1.17 *We say that* X *has a* simple Fréchet subdifferential extremal principle *if a Fréchet subdifferential extremal principle holds for two multifunctions.*

Now we can state and prove our main equivalence result.

Theorem 6.1.18 *Let* X *be a Banach space. Then the following are equivalent:*

(i)   *for any positive integer* N, X^N *has a simple Fréchet subdifferential Lipschitz approximate sum rule;*

(ii)   *for any positive integer* N, X^N *has a Fréchet subdifferential nonlocal approximate sum rule;*

(iii) *for any positive integer* N, X^N *has a Fréchet subdifferential multidirectional mean value inequality;*

(iv)   *for any positive integer* N, X^N *has a Fréchet subdifferential local approximate sum rule;*

(v)   *for any positive integer* N, X^N *has a Fréchet subdifferential Lipschitz local approximate sum rule;*

(vi)   *for any positive integer* N, X^N *has a Fréchet subdifferential extremal principle;*

(vii) *for any positive integer* N, X^N *has a simple Fréchet subdifferential extremal principle.*

Proof. We outline a cyclic proof and leave some of the details as exercises.
(i) $\Rightarrow$ (ii): Consider the extended-valued function f on X^N defined by

$$f(y) = f(y_1, \ldots, y_N) := \sum_{n=1}^{N} f_n(y_n)$$

and Lipschitz function $s_1(y) := s_1(y_1, \ldots, y_N)$ as in Lemma 3.2.2. For any real number $r > 0$, define
$$w_r := f + rs_1$$

and $M_r := \inf w_r$. Then we can show that (M_r) is increasing with respect to r and is bounded above by $\bigwedge[f_1, \ldots, f_N](X)$. Thus, (M_r) converges, say to M. For each r, applying the Ekeland variational principle of Theorem 2.1.2 to function w_r, we obtain a point $u^r = (u_1^r, \ldots, u_N^r)$ such that

$$w_r(y) + \frac{\varepsilon}{2}\|y - u^r\|_{X^N} = f(y) + rs_1(y) + \frac{\varepsilon}{2}\|y - u^r\|_{X^N}$$

attains a minimum at $y = u^r$ and

$$w_r(u^r) < \inf w_r + 1/r < M + 1/r. \tag{6.1.18}$$

Note that $rs_1 + (\varepsilon/2)\|\cdot\|$ is Lipschitz. By the simple Lipschitz approximate sum rule (i) there exist $x^r, z^r \in B_{\varepsilon/r}(u^r)$ with $f(x^r) \leq f(u^r) + \varepsilon/r$ such that

$$0 \in \partial_F f(x^r) + \partial_F \left(rs_1 + \frac{\varepsilon}{2}\|\cdot - u^r\|_{X^N}\right)(z^r) + \frac{\varepsilon}{2N} B_{X^{N*}}. \tag{6.1.19}$$

Let $\xi^r = (\xi_1^r, \ldots, \xi_N^r) \in \partial_F f(x^r)$ and

$$\zeta^r = (\zeta_1^r, \ldots, \zeta_N^r) \in \partial_F \left(rs_1 + \frac{\varepsilon}{2}\|\cdot - u^r\|_{X^N}\right)(z^r) \tag{6.1.20}$$

satisfy

$$0 \in \xi^r + \zeta^r + \frac{\varepsilon}{2N} B_{X^{N*}}. \tag{6.1.21}$$

Then $\xi_n^r \in \partial_F f(x_n^r)$. Moreover, similar to Lemma 3.2.2 we can show that (6.1.20) implies that

$$\|\zeta_1^r + \cdots + \zeta_N^r\| < \frac{\varepsilon}{2}.$$

Thus,

$$\|\xi_1^r + \cdots + \xi_N^r\| < \varepsilon.$$

The rest of the proof is similar to that of Theorem 3.2.3.

(ii) $\Rightarrow$ (iii): The proof of Theorem 3.6.1 still works here.

(iii) $\Rightarrow$ (iv): Let $y = (y_1, \ldots, y_N) \in X^N$, $f(y) := f_1(y_1) + f_2(y_2) + \cdots + f_N(y_N): X^N \to \mathbb{R} \cup \{+\infty\}$. Choose h small enough such that $\sum_{n=1}^{N} f_n(\bar{x}) \leq$

$\bigwedge[f_1,\ldots,f_N](B_h(\bar{x}))$ and choose $S := \{(y_1, y_2, \ldots, y_N) \mid y_n \in B_h(\bar{x}), n = 1,\ldots,N\}$. Then

$$\bigwedge[f](S) - f(\tilde{x}) \geq 0,$$

where $\tilde{x} = (\bar{x}, \bar{x}, \ldots, \bar{x})$. Applying the multidirectional mean value theorem (iii) to f, $\tilde{x}$ and S yields (iv).

(iv) $\Rightarrow$ (v): Obvious.

(v) $\Rightarrow$ (vi): Let U be a neighborhood of $\bar{x}$ as in the definition of an extremal point. Without loss of generality we may assume that $U = B_r(\bar{x})$. Let $\varepsilon' > 0$ be a positive number to be determined later and let $p_1, p_2, \ldots, p_N$ be as in the definition of the extremal point with $\varepsilon = \varepsilon'$. Let s_1 be as in Lemma 3.2.2 and define $f_1(y_1, y_2, \ldots, y_N) := s_1(y_1, y_2, \ldots, y_N)$ and

$$f_2(y_1, y_2, \ldots, y_N) := \sum_{n=1}^{N} \iota_{S_n(p_n)}(y_n).$$

Choose $y_n' \in S_n(p_n), n = 1, \ldots, N$ such that $\|y_n' - y_m'\| < d(S_n(p_n); \bar{x}) + d(S_m(p_m); \bar{x}) + \varepsilon' < 3\varepsilon'$. Then $f_1 + f_2 > 0$ and

$$(f_1 + f_2)(y_1', y_2', \ldots, y_N') < 4N^2\varepsilon'.$$

By Ekeland's variational principle of Theorem 2.1.2 there exist $x_n' \in B_{2N\sqrt{\varepsilon'}}(y_n'), n = 1, \ldots, N$, such that, for

$$f_3(y_1, \ldots, y_N) := 2N\sqrt{\varepsilon'} \sum_{n=1}^{N} \|y_n - x_n'\|,$$

$f_1 + f_2 + f_3$ attains a minimum at $(x_1', x_2', \ldots, x_N')$. It remains to apply the Lipschitz local approximate sum rule of (v) with a sufficiently small ε'.

(vi) $\Rightarrow$ (vii): Obvious.

(vii) $\Rightarrow$ (i): Let f_1 be a locally Lipschitz function with a Lipschitz constant L and let f_2 be a lower semicontinuous function. Assume, without loss of generality, that $f_1 + f_2$ attains a minimum at 0 and $f_1(0) = f_2(0) = 0$. Define

$$S_1 := \{(x, \mu) \in X \times \mathbb{R} \mid f_1(x) \leq \mu\}$$

and

$$S_2 := \{(x, \mu) \in X \times \mathbb{R} \mid f_2(x) \leq -\mu\}.$$

Then it is easy to check that $(0, 0)$ is an extremal point for sets (S_1, S_2). For any $\varepsilon > 0$, let $\varepsilon' = \min\{\varepsilon/4(1 + L)^2, 1/4(1 + L)\}$. By the simple extremal principle (vii), there exist $(x_n, \mu_n) \in S_n \cap \varepsilon B_{X \times \mathbb{R}}$, and $(\xi_n, -\lambda_n) \in X^* \times \mathbb{R}$, $n = 1, 2$ such that

$$(\xi_n, -\lambda_n) \in N_F(S_n; (x_n, \mu_n)),$$

$$\|(\xi_n, -\lambda_n)\| \geq 1 - \varepsilon' \geq 1/2, n = 1, 2 \tag{6.1.22}$$

and

$$|\lambda_1 + \lambda_2| < \varepsilon', \quad \|\xi_1 + \xi_2\| < \varepsilon'.$$

Since f_1 is Lipschitz with Lipschitz constant L, for any $h \in X$ with $\|h\| = 1$ we have $\iota_{S_1}(x_1 + th, \mu_1 + tL) = 0$ for $t > 0$ sufficiently small. It follows from $(\xi_1, -\lambda_1) \in N_F(S_1; (x_1, \mu_1))$ that

$$\liminf_{t \to o} \frac{\iota_{S_1}(x_1 + th, \mu_1 + tL) - \iota_{S_1}(x_1, \mu_1) - t\langle \xi_1, h \rangle + t\lambda_1 L}{t\|(h, L)\|}$$

$$= \liminf_{t \to o} \frac{-t\langle \xi_1, h \rangle + t\lambda_1 L}{t\|(h, L)\|} = \frac{-\langle \xi_1, h \rangle + \lambda_1 L}{\|(h, L)\|} \geq 0.$$

That is to say

$$\langle \xi_1, h \rangle - \lambda_1 L \leq 0, \quad \text{for all } \|h\| = 1. \tag{6.1.23}$$

Combining (6.1.22) and (6.1.23) we get $\lambda_1 > \frac{1}{2(1+L)} > 0$ and $\lambda_2 < \varepsilon' - \lambda_1 < -\frac{1}{4(1+L)} < 0$ which forces $\mu_1 = f_1(x_1)$ and $\mu_2 = f_2(x_2)$. By Theorem 3.1.8 we have $x_1^* := \xi_1/\lambda_1 \in \partial_F f_1(x_1)$ and $x_2^* := -\xi_2/\lambda_2 \in \partial_F f_2(x_2)$. Then, observing that $\|x_1^*\| \leq L$, we obtain

$$\|x_1^* + x_2^*\| = \left\| \frac{\xi_1}{\lambda_1} - \frac{\xi_2}{\lambda_2} \right\| = \left\| \frac{\xi_1(\lambda_1 + \lambda_2)}{\lambda_1 \lambda_2} - \frac{\xi_1 + \xi_2}{\lambda_2} \right\|$$

$$\leq \|x_1^*\| \times \left| \frac{\lambda_1 + \lambda_2}{\lambda_2} \right| + \frac{\|\xi_1 + \xi_2\|}{|\lambda_2|}$$

$$\leq 4L(1+L)\varepsilon' + 4(1+L)\varepsilon' = 4(1+L)^2\varepsilon' < \varepsilon.$$

$\bullet$

6.1.4 Commentary and Exercises

The concept of an Asplund space was introduced by Asplund in [4] (where it is called a *strong differentiability* space). This class of Banach space plays an important role in Banach space geometry (see [99, 221]). The Lipschitz approximate sum rule characterization of Asplund spaces is derived in [110, 113, 111]. The method of separable reduction can be traced back to [222] and elsewhere. The equivalence Theorem 6.1.18 is proved in Zhu [273] (for more general smooth subdifferentials) developing earlier partial results in [86, 113, 146, 206, 272]. This result shows that all these basic subdifferential rules are different facets of the variational principle in conjunction with a decoupling method. In [211] it is also shown that a variant of Theorem 3.1.10 characterize Asplund spaces.

The fact that all the Fréchet subdifferential calculus rules characterize Asplund spaces further emphasizes the importance of this class of spaces in the subdifferential theory, and Mordukhovich and Shao's paper [208] provides a systematic accounting of this material. Early indications of this are to be found in [61, 62]. It is also shown in [150, 170] that the equivalence of (i) through (iv) holds for *abstract subdifferentials* that satisfy a few very plausible axioms. In those papers one can also find additional equivalent subdifferential rules.

Exercise 6.1.1 Let X be a Banach space, let C be an open convex subset of X and let f be a convex and continuous function on C. Suppose that f is Fréchet differentiable at $x \in C$. Prove that for any $\varepsilon > 0$, there exists $\delta > 0$ such that for all $\|y\| = 1$ and $t \in (0, \delta)$,

$$f(x + ty) + f(x - ty) - 2f(x) < t\varepsilon.$$

Exercise 6.1.2 Let X be a Banach space, let C be an open convex subset of X and let f be a convex and continuous function on C. Prove that f is Gâteaux differentiable at $x \in C$ if and only if for each $y \in X$,

$$\lim_{t \to 0+} \frac{f(x + ty) + f(x - ty) - 2f(x)}{t} = 0.$$

Exercise 6.1.3 Let X be a Banach space and let $f \colon X \to \mathbb{R}$. Prove that f is Fréchet differentiable at x if and only if both $\partial_F f(x)$ and $\partial_F (-f)(x)$ are nonempty.

Exercise 6.1.4 Prove Lemma 6.1.7. Hint: The necessity follows directly from the definition of the convex subdifferential. To prove the sufficiency observe that $x \to f(x) + c\|x\|$ attains a minimum at $x = 0$, and therefore $0 \in \partial(f + c\| \cdot \|)(0)$. It remains to apply the subdifferential sum rule of Theorem 4.3.3.

Exercise 6.1.5 Prove the necessity in Lemma 6.1.8.

Exercise 6.1.6 Prove that the functions ϕ_{ni} and ϕ_n defined in the proof of Lemma 6.1.9 are convex.

Exercise 6.1.7 Prove Theorem 6.1.11.

Exercise 6.1.8 Show that the net (M_r) defined in (i) $\Rightarrow$ (ii) in the proof of Theorem 6.1.18 is bounded by $\bigwedge [f_1, \ldots, f_N](X)$.

Exercise 6.1.9 Let $s_p(y_1, \ldots, y_N), p \geq 1$ be as in Lemma 3.2.2 and let $g(y_1, \ldots, y_N)$ be a Lipschitzian function with constant L. Show that

$$(x_1^*, \ldots, x_N^*) \in \partial_F (s_1 + g)(x_1, \ldots, x_N)$$

implies that

$$\left\| \sum_{n=1}^{N} x_n^* \right\| \leq L.$$

Exercise 6.1.10 Provide the details for (iii) $\Rightarrow$ (iv) in the proof of Theorem 6.1.18.

Exercise 6.1.11 Supply the details for (v) $\Rightarrow$ (vi) in the proof of Theorem 6.1.18.

6.2 Nonconvex Separation Theorems and Welfare Economies

The convex separation theorem is one of the fundamental results in analysis. Here we discuss a nonconvex version of this result and its applications in welfare economy and optimization. We also show that it is equivalent to the extremal principle and therefore to all the other subdifferential calculus rules discussed in the previous section.

6.2.1 Nonconvex Separation Theorems

We start with the definition of a separable point.

Definition 6.2.1 (Separable Points for Multifunctions) *Let X be a Banach space and let $M_n, n = 1, \ldots, N$ be metric spaces. Consider multifunctions $S_n \colon M_n \to 2^X$, $n = 1, \ldots, N$. We say that $(\bar{x}_1, \ldots, \bar{x}_N)$ is locally separable from $(S_1, S_2, \ldots, S_N)$ at $(\bar{p}_1, \bar{p}_2, \ldots, \bar{p}_N)$ provided that $(\bar{x}_1, \ldots, \bar{x}_N) \in S_1(\bar{p}_1) \times S_2(\bar{p}_2) \times \cdots \times S_N(\bar{p}_N)$ and there is a positive number h such that for any $\varepsilon > 0$, there exists*

$$(p_1, p_2, \ldots, p_N) \in B_\varepsilon((\bar{p}_1, \bar{p}_2, \ldots, \bar{p}_N)) \backslash \{(\bar{p}_1, \bar{p}_2, \ldots, \bar{p}_N)\},$$

satisfying

$$d\Big(\sum_{n=1}^{N} \big(S_n(p_n) \cap B_h(\bar{x}_n) \big); \sum_{n=1}^{N} \bar{x}_n \Big) > 0.$$

Next we discuss an infinitesimal necessary condition for the separable points.

Theorem 6.2.2 (Separation Theorem for Multifunctions) *Let X be an Asplund space and let $M_n, n = 1, \ldots, N$ be metric spaces. Consider multifunctions $S_n \colon M_n \to 2^X$, $n = 1, \ldots, N$. Suppose that $(\bar{x}_1, \ldots, \bar{x}_N)$ is locally separable from $(S_1, S_2, \ldots, S_N)$ at $(\bar{p}_1, \bar{p}_2, \ldots, \bar{p}_N)$. Then for any $\varepsilon > 0$, there exist $p_n \in B_\varepsilon(\bar{p}_n), x_n \in B_\varepsilon(\bar{x}) \cap S_n(p_n), n = 1, \ldots, N$ and $x^* \in X^*$ with $\|x^*\| = 1$ such that*

$$x^* \in \bigcap_{n=1}^{N} \big(N_F(S_n(p_n); x_n) + \varepsilon B_{X^*} \big).$$

Proof. Let $h > 0$, $\varepsilon' := \varepsilon^2/16N^2(2N+1)$ and $p_1, p_2, \ldots, p_N$ be as in Definition 6.2.1 for $\varepsilon = \varepsilon'$. Define, $g(y) = g(y_1, y_2, \ldots, y_N) := \left\| \sum_{n=1}^N (y_n - \bar{x}_n) \right\|$,

$$f_1(y) = f_1(y_1, y_2, \ldots, y_N) := g(y_1, y_2, \ldots, y_N) + \sum_{n=1}^N i_{S_n(p_n) \cap B_\delta(\bar{x}_n)}(y_n),$$

and

$$f_2(y) = f_2(y_1, y_2, \ldots, y_N) = \sum_{n=1}^N \|y_n - \bar{x}_n\|^2.$$

Choose $y_n' \in S_n(p_n), n = 1, \ldots, N$ such that $\|y_n' - \bar{x}_n\| < d(S_n(p_n); \bar{x}_n) + \varepsilon' < 2\varepsilon'$. Then

$$\liminf_{\eta \to 0}\{f_1(u) + f_2(v) : \|u - v\| \leq \eta\} \leq (f_1 + f_2)(y_1', y_2', \ldots, y_N')$$
$$< N\varepsilon' + N(\varepsilon')^2 < 2N\varepsilon'$$

Applying the nonlocal sum rule of Theorem 3.2.3 we have that there exist $x = (x_1, \ldots, x_N)$ and $z = (z_1, \ldots, z_N)$, $(x_1^*, \ldots, x_N^*) \in \partial_F f_1(x)$, such that

$$\|x - z\| < \varepsilon', \tag{6.2.1}$$

$$f_1(x) + f_2(z) < \liminf_{\eta \to 0}\{f_1(u) + f_2(v) : \|u - v\| \leq \eta\} + \varepsilon'$$
$$< (2N+1)\varepsilon', \tag{6.2.2}$$

and

$$\|f_2'(z) + (x_1^*, \ldots, x_N^*)\| < \varepsilon'. \tag{6.2.3}$$

Note that (6.2.2) implies that $x_n \in S_n(p_n)$ and

$$f_2(z) = \sum_{n=1}^N \|z_n - \bar{x}\|^2 < (2N+1)\varepsilon'. \tag{6.2.4}$$

Consequently,

$$\|f_2'(x)\| < 2N\sqrt{(2N+1)\varepsilon'} = \varepsilon/2, \tag{6.2.5}$$

and therefore

$$\|(x_1^*, \ldots, x_N^*)\| < \varepsilon. \tag{6.2.6}$$

By the choice of $(p_1, \ldots, p_N)$ we have $g(x) > 0$. It follows that g is Fréchet differentiable at x. Suppose that $g'(x) = (\xi_1, \ldots, \xi_N)$ then for any $h_n \in X, n = 1, \ldots, N$ we have

$$\sum_{n=1}^{N} \langle \xi_n, h_n \rangle$$

$$= \lim_{t \to 0} \frac{g(x_1 + th_1, \ldots, x_N + th_N) - g(x_1, \ldots, x_N)}{t}$$

$$= \lim_{t \to 0} \frac{\| \sum_{n=1}^{N} (x_n - \bar{x}_n) + t \sum_{n=1}^{N} h_n \| - \| \sum_{n=1}^{N} (x_n - \bar{x}_n) \|}{t}. \quad (6.2.7)$$

Now, let h be an arbitrary element in X. Setting $h_n = -h_m = h$ and $h_k = 0$ for $k \neq n, m$ in (6.2.7) we have $\langle \xi_n - \xi_m, h \rangle = 0$. Thus, all the components of $g'(x)$ are the same and we can write $g'(x) = (-x^*, -x^*, \ldots, -x^*)$. Next, setting $h = \sum_{n=1}^{N} (x_n - \bar{x}_n)$ for all $n = 1, \ldots, N$ in (6.2.7) yields

$$N \Big\langle -x^*, \sum_{n=1}^{N} (x_n - \bar{x}_n) \Big\rangle = N \Big\| \sum_{n=1}^{N} (x_n - \bar{x}_n) \Big\|$$

so that $\|p^*\| = 1$. Finally,

$$(x_1^*, \ldots, x_N^*) \in g'(x) + N_F(S_1(p_1); x_1) \times \cdots \times N_F(S_N(p_N); x_N). \quad (6.2.8)$$

Combining (6.2.6) and (6.2.8) we have, for $n = 1, \ldots, N$,

$$x^* \in N_F(S_n(p_n); x_n) + cB_{X^*}.$$

$\bullet$

As an easy corollary we can derive the following separation theorem for nonconvex sets in [49].

Corollary 6.2.3 (Separation Theorem for Sets) *Let X be a Fréchet-smooth Banach space and let $S_n, n = 1, \ldots, N$ be N closed subsets of X. Suppose that $(\bar{x}_1, \ldots, \bar{x}_N) \in S_1 \times \cdots \times S_N$ satisfies*

$$\sum_{n=1}^{N} \bar{x}_n \in \mathrm{bd}\Big(\sum_{n=1}^{N} S_n \Big).$$

Then for any $\varepsilon > 0$, there exist $x_n \in B_\varepsilon(\bar{x}) \cap S_n, n = 1, \ldots, N$ and $x^* \in X^*$ with $\|x^*\| = 1$ such that

$$x^* \in \bigcap_{n=1}^{N} \big(N_F(S_n; x_n) + \varepsilon B_{X^*} \big).$$

Proof. Let $M_n = X$ and define $S_n(p_n) = S_n + p_n - \bar{x}_n$. Then one can check that $(\bar{x}_1, \ldots, \bar{x}_N)$ is locally separable from $(S_1, \ldots, S_N)$ at $(\bar{x}_1, \ldots, \bar{x}_N)$. Applying Theorem 6.2.2 there exists $p_n \in B_{\varepsilon/2}(\bar{x}_n)$,

$$y_n \in S_n(p_n) \cap B_{\varepsilon/2}(\bar{x}_n) = (S_n + p_n - \bar{x}_n) \cap B_{\varepsilon/2}(\bar{x}_n)$$

and $x^* \in X^*$ with $\|x^*\| = 1$ such that

$$x^* \in \bigcap_{n=1}^{N} \left(N_F(S_n(p_n); y_n) + \varepsilon B_{X^*} \right).$$

Define $x_n = y_n - p_n + \bar{x}_n$. Then $x_n \in B_\varepsilon(\bar{x}_n) \cap S_n$ and $N_F(S_n(p_n); y_n) = N_F(S_n; x_n)$ so that

$$x^* \in \bigcap_{n=1}^{N} \left(N_F(S_n; x_n) + \varepsilon B_{X^*} \right).$$

●

6.2.2 Relationships With Subdifferential Calculus

Next we show that the nonconvex separation theorem is also equivalent to the subdifferential calculus rules discussed in subsection 6.1.3. Thus, it provides yet another characterization of Asplund space.

Definition 6.2.4 (Separation Property for Multifunctions) *Let X be a Banach space. We say that X has the Fréchet subdifferential nonconvex separation property if for any closed multifunctions S_n from metric spaces M_n to X, $n = 1, \ldots, N$, for any local separable point $(\bar{x}_1, \ldots, \bar{x}_N)$ of $(S_1, S_2, \ldots, S_N)$ at $(\bar{p}_1, \bar{p}_2, \ldots, \bar{p}_N)$ and any $\varepsilon > 0$, there exist $p_n \in B_\varepsilon(\bar{p}_n), x_n \in B_\varepsilon(\bar{x}) \cap S_n(p_n), n = 1, \ldots, N$ and $x^* \in X^*$ with $\|x^*\| = 1$ such that*

$$x^* \in \bigcap_{n=1}^{N} \left(N_F(S_n(p_n); x_n) + \varepsilon B_{X^*} \right).$$

By carefully examining the proof of Theorem 6.2.2 we can see that the assumption that X is an Asplund space is used only to ensure the Fréchet subdifferential nonlocal approximate sum rule is valid in X. In other words, what we actually prove in Theorem 6.2.2 is that, if X^N has a nonlocal approximate sum rule for each positive integer N, then X^N has a nonconvex separation property for each positive integer N. Since the nonlocal approximate sum rule is one of the equivalent subdifferential calculus rules discussed in Subsection 6.1.3, to prove the equivalence of the nonconvex separation property to all the subdifferential calculus in that subsection, we need only deduce one of the calculus rules there from it. Our next theorem shows that the simple extremal principle can easily be deduced from the nonconvex separation property.

Theorem 6.2.5 *Let X be a Banach space. Suppose that X has a nonconvex separation property. Then X has a simple extremal principle.*

Proof. Let $\bar{x}$ be an extremal point of the extremal system (S_1, S_2) at $(\bar{p}_1, \bar{p}_2)$ where $\bar{p}_n, n = 1, 2$ are elements of metric spaces $M_n, n = 1, 2$, respectively. Then $(\bar{x}, -\bar{x})$ is a separable point of the multifunctions $(S_1, -S_2)$ at $(\bar{p}_1, \bar{p}_2)$ (Exercise 6.2.2). Since X has a nonconvex separation property, there exist $U = B_r(\bar{x})$, for any $\varepsilon > 0$, $p_n \in B_{\varepsilon/2}(\bar{p}_n), n = 1, 2$, $y_1 \in S_1(p_1) \cap B_{\varepsilon/2}(\bar{x})$, $y_2 \in -S_2(p_2) \cap B_{\varepsilon/2}(-\bar{x})$ and $x^* \in X^*$ with $\|x^*\| = 1$ such that

$$x^* \in N_F(S_1(p_1); y_1) + \varepsilon B_{X^*}$$
$$x^* \in N_F(-S_2(p_2); y_2) + \varepsilon B_{X^*}.$$

Note that $N_F(-S_2(p_2); y_2) = -N_F(S(p_2); -y_2)$. Setting $x_1 = y_1$, $x_2 = -y_2$ and $x_1^* = -x_2^* = x^*$ we see that X has a Fréchet extremal principle. ●

Corollary 6.2.6 *Let X be a Banach space. Then X is Asplund if and only if for any N, X^N has the Fréchet subdifferential nonconvex separation property.*

6.2.3 Welfare Economies

Now we consider a welfare economy model that can be analyzed conveniently with the separation theorem.

Let X represent a commodity space. We consider an economy with N consumers and M suppliers. For consumer $n \in \{1, \ldots, N\}$ we use $C_n \subset X$ representing its consumption set, i.e., the set of commodities potentially needed by consumer n. Similarly, for $m \in \{1, \ldots, M\}$ we use $S_m \subset X$ representing its supply set, i.e., commodities that can be produced by supplier m. Now we consider a welfare economy model $\mathcal{E} = (C_1, \ldots, C_N, S_1, \ldots, S_M, W)$. Here $W \subset X$ is a net demand constraint set representing constraints related to the initial inventory of commodities in the economy. Denote $x = (x_1, \ldots, x_N) \in C_1 \times \cdots \times C_N$ and $y = (y_1, \ldots, y_M) \in S_1 \times \cdots \times S_M$. Then (x, y) is a possible state of economy $\mathcal{E}$.

Definition 6.2.7 (Feasible Allocation) *We say that $(x, y) \in \Pi_{n=1}^N C_n \times \Pi_{m=1}^M S_m$ is a feasible allocation of $\mathcal{E}$ if*

$$\sum_{n=1}^N x_n - \sum_{m=1}^M y_m \in W. \tag{6.2.9}$$

To understand the role of W let us consider two special cases:

(i) $W = \{w\}$ where w is usually referred to as the initial aggregate endowment, representing the initial total available commodity. In this case, the feasibility condition (6.2.9) reduces to the classical market clear condition. It requires an ideal status of the economy in which all the inventory and products are exactly consumed without any waste.

(ii) The commodity space X has a natural positive cone X_+ and $W = w - X_+$. Then the feasibility condition (6.2.9) implies that commodities can be freely disposed.

A general set W allows us to flexibly handle more complicated constraints that allows partial disposal of commodities or/and to consider virtual consumption/production through various contracts such as commodity futures and options. With such a general net demand constraint set W, many feasible allocations of $\mathcal{E}$ are possible.

The goal of a welfare economy is to best satisfy the preference of the consumers which is described as follows: the preference of consumer n is characterized by a multifunction $P_n: C_1 \times \cdots \times C_N \to 2^{C_n}$. At each point $x \in C_1 \times \cdots \times C_N$, $P_n(x)$ consists of elements in C_n preferred to x_n by consumer n. What is the "best" solution in terms of satisfying consumer's preference? There are many possible options. Here we discuss optimality in the sense of weak Pareto.

Definition 6.2.8 (Weak Pareto Local Optimal Allocation) *We say that $(\bar{x}, \bar{y})$ is a* local weak Pareto optimal allocation *if there exists $r > 0$ such that for every feasible allocation $(x, y) \in B_r(\bar{x}, \bar{y})$, one has $x_n \notin P_n(\bar{x}) \cap B_r(\bar{x})$ for some $n \in \{1, \dots, N\}$.*

Our next theorem provides a necessary optimality condition for a weak Pareto local optimal allocation.

Theorem 6.2.9 (Existence of Approximate Shadow Price) *Let*

$$\mathcal{E} = (C_1, \dots, C_N, S_1, \dots, S_M, W)$$

be an economy. We assume that all the sets $C_1, \dots, C_N, S_1, \dots, S_M, W$ are closed. Let $(\bar{x}, \bar{y})$ be a local weak Pareto optimal allocation of economy $\mathcal{E}$. Suppose that $P(x) = \Pi_{n=1}^N P_n(x)$ satisfies the following conditions: there exists $h > 0$ such that for any $\varepsilon > 0$ there exists $x \in B_\varepsilon(\bar{x})$ satisfying

$$\sum_{n=1}^N \overline{P_n(x)} \cap B_h(\bar{x}_n) - \sum_{m=1}^M S_m \cap B_h(\bar{y}_m) - W \cap B_h(\bar{w})$$

$$\subset \sum_{n=1}^N P_n(\bar{x}) \cap B_h(\bar{x}_n) - \sum_{m=1}^M S_m \cap B_h(\bar{y}_m) - W \cap B_h(\bar{w}). \quad (6.2.10)$$

Then for any $\varepsilon > 0$, there exist $(x, y, w) \in B_\varepsilon((\bar{x}, \bar{y}, \bar{w}))$ and a vector $x^* \in X^*$ with $\|x^*\| = 1$ satisfying

$$-x^* \in N_F(\overline{P_n(x)}; x_n) + \varepsilon B_{X^*}, \quad n = 1, \dots, N,$$
$$x^* \in N_F(S_m; y_m) + \varepsilon B_{X^*}, \qquad m = 1, \dots, M$$

and

$$x^* \in N_F(W; w) + \varepsilon B.$$

Proof. Let $M_n = C_n, n = 1, \ldots, N$ and $M_n = \mathbb{R}, n = N+1, \ldots, N+M+1$. Denote $\bar{w} = \sum_{n=1}^{N} \bar{x}_n - \sum_{m=1}^{M} \bar{y}_m$. Then $(\bar{x}, \bar{y}, \bar{w})$ is locally separable from $(P_1, \ldots, P_N, S_1, \ldots, S_M, W)$ at $(\bar{x}_1, \ldots, \bar{x}_N, 0, \ldots, 0, 0)$. In fact, assume to the contrary that for any $h > 0$, there exists $\varepsilon > 0$ such that for any $x \in B_\varepsilon(\bar{x})$,

$$ d\left(\sum_{n=1}^{N} P_n(x) \cap B_h(\bar{x}) - \sum_{m=1}^{M} S_m \cap B_h(\bar{x}) - W \cap B_h(\bar{x}); 0 \right) = 0. $$

Now selecting h and $x \in B_\varepsilon(\bar{x})$ satisfying condition (6.2.10) leads to

$$ 0 \in \sum_{n=1}^{N} \overline{P_n(x)} \cap B_h(\bar{x}) - \sum_{m=1}^{M} S_m \cap B_h(\bar{x}) - W \cap B_h(\bar{x}) $$

$$ \subset \sum_{n=1}^{N} P_n(\bar{x}) \cap B_h(\bar{x}) - \sum_{m=1}^{M} S_m \cap B_h(\bar{x}) - W \cap B_h(\bar{x}), $$

which contradicts $(\bar{x}, \bar{y}, \bar{w})$ being a local weak Pareto optimal allocation. Now the conclusion directly follows from the separation theorem for sets.   ●

In finite dimensional spaces we can derive a limiting form of this result under additional assumptions. Note that any finite dimensional Banach space is Fréchet smooth. We say a multifunction $F : X \to 2^Y$ is *normal upper semicontinuous* at $(x, y) \in$ graph F provided that for any $(x_i, y_i) \to (x, y)$ and $N_F(F(x_i); y_i) \ni \xi_i \to \xi$ we have $\xi \in N_L(F(x); y)$.

Theorem 6.2.10 (Existence of Shadow Price) *Let*

$$ \mathcal{E} = (C_1, \ldots, C_N, S_1, \ldots, S_M, W) $$

be an economy. We assume that all the sets $C_1, \ldots, C_N, S_1, \ldots, S_M, W$ are closed and the commodity space X is finite dimensional. Let $(\bar{x}, \bar{y})$ be a weak Pareto local optimal allocation of economy $\mathcal{E}$. Suppose that $P(x) = \Pi_{n=1}^{N} P_n(x)$ satisfies the condition (6.2.10) and in addition we assume that $P_n, n = 1, \ldots, N$ are normal upper semicontinuous at $(\bar{x}, \bar{x}_n)$. Then there exists a unit $x^* \in X^*$ satisfying

$$ -x^* \in N_L(\overline{P_n(\bar{x})}; \bar{x}_n), \quad n = 1, \ldots, N, $$
$$ x^* \in N_L(S_m; \bar{y}_m), \qquad m = 1, \ldots, M $$

and

$$ x^* \in N_L(W; \bar{w}). $$

Proof. Take limits in Theorem 6.2.9.   ●

This result can be explained economically as follows. At the Pareto optimal point there exists a shadow price x^* such that the weak Pareto optimal allocation can be achieved when each consumer minimizes his or her cost in purchasing commodities and each producer maximizes his or her revenue in arranging their production. We also see that the initial inventory plays a role of a producer.

6.2.4 Commentary and Exercises

The convex separation theorem has long been an important tool in studying the welfare economy. A nonconvex separation theorem and its application in deriving the existence of the shadow price in a welfare economy involving nonconvex consumption and production sets was initiated by Borwein and Jofré in [49]. Further refinement can be found in [154, 153]. Mordukhovich showed that one can also use extremal principles to derive similar results [203]. The separation theorem for multifunctions in Theorem 6.2.2 derived in [278] models the generalized extremal principles in [210]. The qualification condition (6.2.10) weakens earlier similar conditions in [26, 154, 153, 159, 203]. The relationship between the nonconvex separation theorem and the extremal principle is discussed in [49, 278].

Exercise 6.2.1 Show that the conclusion of Theorem 6.2.2 can also be expressed as

$$x^* \in \bigcap_{n=1}^{N} \left(\partial_F d(S_n(p_n); x_n) + \varepsilon B_{X^*} \right).$$

where $c \le \|x^*\| \le 1$ for some $c > 0$

Exercise 6.2.2 Show that in the proof of Theorem 6.2.5, $(\bar{x}, -\bar{x})$ is a separable point for the multifunctions $(S_1, -S_2)$ at $(\bar{p}_1, \bar{p}_2)$.

6.3 Stegall Variational Principles

We have seen several times in a variational principle the situation where the better the property of the underlying space, the more we can hope for in the perturbation function. The Stegall variational principle asserts that in a Banach space with the *Radon–Nikodym property* (RNP) as defined below, one can actually make the perturbation function linear.

6.3.1 The Radon–Nikodym Property

Definition 6.3.1 (Slice) *Let X be a Banach space and let A be a nonempty subset of X. For $\alpha > 0$ and $x^* \in X^*$, we call*

$$S(x^*, A, \alpha) := \{x \in A \mid \langle x^*, x \rangle > \sigma_A(x^*) - \alpha\}$$

a slice of A. If A is a nonempty subset of X^*, we define a weak-star slice analogously, with the functional coming from X rather than from X^{**}.

Definition 6.3.2 (Dentable Set) *Let X be a Banach space and let A be a subset of X (X^*). We say that A is (weak-star) dentable provided for any $\varepsilon > 0$ there exists $x^* \in X^*$ $(x \in X)$ and $\alpha > 0$ such that*

$$\text{diam } S(x^*, A, \alpha) < \varepsilon \ (\text{diam } S(x, A, \alpha) < \varepsilon).$$

For a closed subset an alternate characterization of the dentable property is given in Exercise 6.3.1. Now we can define the Radon–Nikodym property.

Definition 6.3.3 (Radon–Nikodym Property) *Let X be a Banach space and let A be a subset of X. We say A has the Radon–Nikodym property (RNP) if every nonempty bounded subset of A is dentable.*

We note that a dual space has RNP if and only if the predual is Asplund, in particular, reflexive spaces and separable dual spaces such as ℓ_1 have the RNP (see [221]).

6.3.2 The Stegall Variational Principle

To state the Stegall variational principle we recall the concept of a *strong minimum*.

Definition 6.3.4 *Let X be a Banach space and let S be a nonempty subset of X. Suppose that f is a function on S bounded from below. We say that $x \in S$ is a strong minimum of f on S if $f(x) = \inf_S f$ and for any sequence (x_i) in S, $\|x - x_i\| \to 0$ whenever $f(x_i) \to f(x)$.*

We can define a more general slice related to a function f that is bounded above on S by

$$S(f, S, \alpha) := \{x \in S \mid f(x) > \sup_S f - \alpha\}.$$

Then a necessary and sufficient condition for a function f to attain a strong minimum on a closed set S is diam $S(-f, S, \alpha) \to 0$ as $\alpha \to 0+$ (Exercise 6.3.2).

Theorem 6.3.5 (Stegall Variational Principle) *Let X be a Banach space and let $C \subset X$ be a nonempty closed and bounded convex set with the Radon–Nikodym property and let f be a lsc function on C bounded from below. Then for any $\varepsilon > 0$ there exists $x^* \in X^*$ such that $\|x^*\| < \varepsilon$ and $f + x^*$ attains a strong minimum on C.*

The following variant due to M. Fabián is often convenient in applications.

Corollary 6.3.6 *Let X be a Banach space with the Radon–Nikodym property and let f be a lsc function on X. Suppose that there exists $a > 0$ and $b \in \mathbb{R}$ such that*

$$f(x) > a\|x\| + b, \quad x \in X.$$

Then for any $\varepsilon > 0$ there exists $x^* \in X^*$ such that $\|x^*\| < \varepsilon$ and $f + x^*$ attains a strong minimum on X.

Proof. Since we can replace f by $f - b$, we may assume that $b = 0$. Note that if $x^* \in X^*$ with $\|x^*\| < a/2$, then for any $x \in X$,

$$f(x) + \langle x^*, x \rangle \geq a\|x\| - \frac{a}{2}\|x\| = \frac{a}{2}\|x\|. \tag{6.3.1}$$

Let $r = (2/a)[f(0) + 1]$ and apply Theorem 6.3.5 to f restricted to the ball rB, with $0 < \varepsilon < a/2$. Thus, there exists $x^* \in X^*$, $\|x^*\| < \varepsilon$, and a point $\bar{x} \in rB$ such that $f + x^*$ attains a strong minimum at $\bar{x}$ in rB. We need only show that $\bar{x}$ is a strong minimum of $f + x^*$ in X. If $x \in X$ is such that $f(x) + \langle x^*, x \rangle \leq f(\bar{x}) + \langle x^*, \bar{x} \rangle = \inf_{rB}(f + x^*) \leq f(0)$, then from (6.3.1) we conclude that $\|x\| \leq (2/a)f(0) < r$, so $x = \bar{x}$. Similarly, if (x_i) is a sequence in X and $f(x_i) + \langle x^*, x_i \rangle \to f(\bar{x}) + \langle x^*, \bar{x} \rangle$, then eventually $f(x_i) + \langle x^*, x_i \rangle < f(0) + 1$, so from (6.3.1) it follows that $x_i \in rB$, and therefore $x_i \to \bar{x}$.   ●

We break the proof of the Stegall's variational principle into several lemmas. Our first lemma provides a recipe for constructing nondentable sets which will be used in the proof of the Stegall's variational principle by contradiction.

Lemma 6.3.7 *Let X be a Banach space and let (A_i) be a sequence of (eventually) nonempty subsets of X. Suppose that there exist constants $\varepsilon > 0$ and $\lambda > 0$ such that for all $i = 1, 2, \ldots$, $x \in \mathrm{conv}(A_i)$ and $y \in X$,*

$$d(\mathrm{conv}(A_{i+1} \backslash B_\varepsilon(y)); x) \leq \lambda/2^i.$$

Then the set

$$A = \bigcap_{i=1}^{\infty} \overline{\bigcup_{j \geq i} \mathrm{conv}(A_j)}$$

is nonempty and not dentable.

Proof. By Exercise 6.3.1 it suffices to show that $x \in \overline{\mathrm{conv}(A \backslash B_{\varepsilon/2}(x))}$ for each $x \in A$. First we show that A is nonempty and that for each $i \geq 1$,

$$\mathrm{conv}(A_i) \subset B_{4\lambda/2^i}(A). \tag{6.3.2}$$

To this end, fix i sufficiently large that A_i is nonempty and suppose that $x_0 \in \mathrm{conv}(A_i)$. By hypothesis, there exists a point $x_1 \in \mathrm{conv}(A_{i+1})$ such that $\|x_0 - x_1\| \leq 2\lambda/2^i$. Similarly there exists a point $x_2 \in \mathrm{conv}(A_{i+2})$ such that $\|x_1 - x_2\| \leq 2\lambda/2^{i+1}$. Continuing by induction, we obtain $x_k \in \mathrm{conv}(A_{i+k})$, $k = 0, 1, \ldots$, such that $\sum_{k=0}^{\infty} \|x_k - x_{k+1}\| \leq 4\lambda/2^i < \infty$. This implies that the series $\sum_{k=1}^{\infty}(x_k - x_{k+1})$ converges to an element $y \in X$, of norm at most $4\lambda/2^i$. By writing y as a limit of the collapsing partial sums we get $y = x_0 - z$ where $z = \lim_{k \to \infty} x_k$. It follows that $z \in A$ (so A is nonempty) and $x_0 \in B_{4\lambda/2^i}(A)$, which proves (6.3.2). Suppose that $x \in A$. Fix j such that $4\lambda/2^j < \varepsilon/2$. Since we have $x \in \overline{\bigcup_{k \geq i} \mathrm{conv}(A_k)}$ for all i, for each $i \geq j$ there exists $k \geq i$ and $y_i \in \mathrm{conv}(A_k)$ such that $\|x - y_i\| \leq \lambda/2^i$. By hypothesis,

$$d(\text{conv}(A_{k+1}\backslash B_\varepsilon(x)); y_i) \le \lambda/2^k < 2\lambda/2^i,$$

so there exists $z_i \in \text{conv}(A_{k+1}\backslash B_\varepsilon(x))$ such that $\|y_i - z_i\| < 2\lambda/2^i$. We can write z_i as a finite convex combination

$$z_i = \sum \lambda_n u_n, \quad u_n \in A_{k+1}\backslash B_\varepsilon(x).$$

By (6.3.2), for each n we have $u_n \in B_{4\lambda/2^k}(A) \subset B_{4\lambda/2^i}(A)$ so there exist $v_n \in A$ such that $\|u_n - v_n\| \le 4\lambda/2^i$. Let $w_i = \sum \lambda_n v_n$; it follows that $\|z_i - w_i\| \le 4\lambda/2^i$ and hence $\|x - w_i\| \le 7\lambda/2^i$. Since $\|u_n - x\| \ge \varepsilon$ for each n, we have

$$\|v_n - x\| \ge \varepsilon - 4\lambda/2^i \ge \varepsilon - 4\lambda/2^j > \varepsilon/2,$$

that is, $w_i \in \text{conv}(A\backslash B_{\varepsilon/2}(x))$. It follows that $x \in \overline{\text{conv}(A\backslash B_{\varepsilon/2}(x))}$ as required. ●

The next two lemmas will allow us to reduce the proof of Theorem 6.3.5 to simply showing that there exist arbitrary small perturbations $f + x^*$ of f that define small slices. The first, whose proof is left as an exercise, follows directly from the definition.

Lemma 6.3.8 *Let X be a Banach space and let S be a nonempty subset of the unit ball of X. Suppose that f is bounded from above on S. Then for any $\alpha > 0$, we have $S(f + x^*, S, \beta) \subset S(f, S, \alpha)$ provided $\|x^*\| < \alpha/2$ and $0 < \beta < \alpha - 2\|x^*\|$.*

Proof. Exercise 6.3.4. ●

Lemma 6.3.9 *Let X be a Banach space and let S be a nonempty bounded closed subset of X. Suppose that for every lsc function f bounded from below on S and every $\varepsilon > 0$, there exists $x^* \in X^*$, $\|x^*\| < \varepsilon$, and $\alpha > 0$ such that diam $S(-(f+x^*), S, \alpha) \le 2\varepsilon$. Then for any lsc function f bounded from below on S and any $\varepsilon > 0$ there exists $x^* \in X^*$ such that $\|x^*\| < \varepsilon$ and $f + x^*$ attains a strong minimum on S.*

Proof. We assume without loss of generality that S is contained in the unit ball of X and that $0 < \varepsilon < 1$. By hypothesis, there exist $x_1^* \in X^*$, $\|x_1^*\| < \varepsilon/2$, and $0 < \alpha_1 < 1$ such that diam $S_1 < \varepsilon$, where $S_1 = S(-(f + x_1^*), S, \alpha_1)$. By applying the hypothesis to $f + x_1^*$ and $\varepsilon_1 = \alpha_1\varepsilon/2^2$, we obtain $x_2^* \in X^*$, $\|x_2^*\| < \varepsilon_1$, and $0 < \alpha_2 < \alpha_1 < 1$ such that diam $S_2 < 2\varepsilon_1$, where $S_2 = S(-(f + x_1^* + x_2^*), S, \alpha_2)$. Continuing by induction with $\alpha_0 = 1$, we obtain sequences

$$\varepsilon_i > 0, \quad 0 < \alpha_i < 1, \quad x_i^* \in X^*, \quad S_i = S(-(f + \sum_{k=1}^{i} x_k^*), S, \alpha_i)$$

such that

$$\text{diam}\, S_i \leq 2\varepsilon_{i-1}, \quad \|x_i^*\| < \varepsilon_{i-1}, \quad \varepsilon_i = \varepsilon\alpha_i/2^{i+1} \text{ and } \alpha_i < \alpha_{i-1}.$$

Obviously, the series $\sum_{i=1}^{\infty} x_i^*$ converges to a point $x^* \in X^*$ of norm at most $\varepsilon \sum_{i=1}^{\infty} \alpha_i 2^{-(i+1)} < \varepsilon$. Note that diam $S_i \to 0$, since $\varepsilon_i \to 0$. We claim $f + x^*$ attains a strong minimum. By Exercise 6.3.2 it suffices to show that diam $S(-(f + x^*), S, \alpha) \to 0$ as $\alpha \to 0+$, and for this it suffices to prove that for all i there exists $\alpha > 0$ such that $S(-(f + x^*), S, \alpha) \subset S_i$. This follows from Lemma 6.3.8, using the fact that $f + x^* = f + \sum_{k=1}^{i} x_k^* + w_i^*$ where

$$\|w_i^*\| \leq \sum_{k=i+1}^{\infty} \varepsilon\alpha_{k-1}/2^k < \alpha_i/2;$$

we need only choose α such that $0 < \alpha < \alpha_i - 2\|w_i^*\|$.   ●

Now we can prove the Stegall variational principle of Theorem 6.3.5.

Proof of Theorem 6.3.5. By utilizing the previous lemma, we need only show that given any $\varepsilon > 0$ there exist $x^* \in X^*$, $\|x^*\| < \varepsilon$, and $\alpha > 0$ such that diam $S(-(f + x^*), C, \alpha) \leq 2\varepsilon$. Proceeding by contradiction, suppose that for every $\|x^*\| < \varepsilon$ and each $\alpha > 0$, we have diam $S(-(f + x^*), C, \alpha) > 2\varepsilon$. For each i let

$$A_i = \bigcup\{S(-(f + x^*), C, 1/4^i) \mid \|x^*\| \leq \varepsilon - 2^{-i}\}.$$

For all sufficiently large i, we have $\varepsilon - 2^{-i} > 0$, so that A_i is nonempty. Let $\lambda = 5/2$; we will show that for this choice of λ, the sequence (A_i) satisfies the hypothesis of Lemma 6.3.7, the conclusion of which will contradict the hypothesis of Theorem 6.3.5 (that C has the RNP). Restating the main hypothesis of Lemma 6.3.7, we want to show that for any $y \in X$,

$$\text{conv}(A_i) \subset \text{conv}(A_{i+1} \backslash B_\varepsilon(y)) + (\lambda/2^i)B. \tag{6.3.3}$$

Since the set on the right hand side is convex, it suffices to prove that it contains A_i. Suppose then that $x \in A_i$, but for some $y \in X$ it is not in the right hand side of (6.3.3) (which has nonempty interior). By the separation theorem of Theorem 4.3.8 there exists $y^* \in X^*$, $\|y^*\| = 1$, such that

$$\langle y^*, x \rangle \leq \inf\{\langle y^*, u \rangle \mid u \in A_{i+1} \backslash B_\varepsilon(y)\} - \lambda/2^i. \tag{6.3.4}$$

Since $x \in A_i$, there exists $x^* \in X^*$ with $\|x^*\| < \varepsilon - 2^{-i}$ such that x is in the slice $S(-(f + x^*), C, 1/4^i)$. Write

$$z^* = x^* + y^*/2^{i+1};$$

it follows that $\|z^*\| \leq \varepsilon - 2^{-i} + 2^{-(i+1)} = \varepsilon - 2^{-(i+1)}$ so

$$S(-(f + z^*), C, 1/4^{i+1}) \subset A_{i+1}.$$

Since diam $S(-(f + z^*), C, 1/4^{i+1}) > 2\varepsilon$, this slice is not contained in $B_\varepsilon(y)$. This implies that there exists $z \in C \backslash B_\varepsilon(y)$ such that

$$f(z) + \langle z^*, z \rangle < \inf_C (f + z^*) + 1/4^{i+1}. \tag{6.3.5}$$

Thus, $z \in A_{i+1} \backslash B_\varepsilon(y)$ and hence, from (6.3.4), we conclude that

$$\langle y^*, x \rangle \le \langle y^*, z \rangle - \lambda/2^i. \tag{6.3.6}$$

We will prove the inclusion (6.3.3) by showing that this cannot be true. Note first that since $x \in S(-(f + x^*), C, 1/4^i)$ and $z \in C$, we necessarily have

$$f(x) + \langle x^*, x \rangle < \inf_C(f + x^*) + 1/4^i \le f(z) + \langle x^*, z \rangle + 1/4^i. \tag{6.3.7}$$

Similarly, from (6.3.5) we have

$$\begin{aligned}
f(z) + \langle z^*, z \rangle &= f(z) + \langle x^*, z \rangle + \langle y^*, z \rangle/2^{i+1} \\
&< \inf_C(f + x^* + y^*/2^{i+1}) + 1/4^{i+1} \\
&\le f(x) + \langle x^*, x \rangle + \langle y^*, x \rangle/2^{i+1} + 1/4^{i+1}. \tag{6.3.8}
\end{aligned}$$

Using (6.3.7) and (6.3.8) we obtain

$$f(z) + \langle x^*, z \rangle + \langle y^*, z \rangle/2^{i+1} < f(z) + \langle x^*, z \rangle + 1/4^i + \langle y^*, x \rangle/2^{i+1} + 1/4^{i+1}$$

or

$$\langle y^*, z - x \rangle/2^{i+1} < 1/4^i + 1/4^{i+1} = 5/4^{i+1} = 5/2^{2i+2}.$$

Equivalently, $\langle y^*, z - x \rangle < (5/2)(1/2^i) = \lambda/2^i$, which contradicts (6.3.6) and completes the proof. ●

Next we discuss two attractive applications of the Stegall variational principle.

6.3.3 Representation of Radon–Nikodym Sets

Let us first recall the concept of strongly exposed points of a convex set.

Definition 6.3.10 *Let X be a Banach space and let C be a closed convex subset of X. We say that $x \in C$ is strongly exposed by $x^* \in X^*$ if for any sequence (x_i) in C, $\langle x^*, x_i \rangle \to \sigma_C(x^*)$ implies $\|x_i - x\| \to 0$. A functional x^* with the property above is called a strongly exposing functional.*

A strongly exposed point and its exposing functional can be characterized by using the slice (Exercise 6.3.3).

Theorem 6.3.11 *Let X be a Banach space and let $C \subset X$ be a nonempty bounded closed convex set with the Radon–Nikodym property. Then C is the closed convex hull of its strongly exposed points. Moreover, the functionals which strongly expose points of C constitute a dense G_δ subset of X^*.*

Proof. We first prove the second assertion. To this end, define

$$G_i = \{x^* \in X^* \mid \text{diam } S(x^*, C, \alpha) < 1/i \text{ for some } \alpha > 0\},$$

and $G = \bigcap_{i=1}^{\infty} G_i$. If $y^* \in G_i$ (so that diam $S(y^*, C, \alpha) < 1/i$ for some α), then by Lemma 6.3.8, applied to $f = y^*$, the $\alpha/2$-neighborhood of y^* is contained in G_i, hence the latter is open. Theorem 6.3.5 (applied to any element $f \in X^*$) trivially implies that each G_i is dense in X^*. Obviously, if x^* strongly exposes C, then it defines slices of arbitrarily small diameter, hence is in G. Conversely, if $x^* \in G$, then it will define a nested sequence $(S(x^*, C, \alpha_i))$ of slices of C with diameters converging to 0. Their closures intersect in a point of C which is strongly exposed by x^*. To prove the first assertion of the theorem, let D be the closed convex hull of the strongly exposed points of C. If $D \neq C$, then by the separation theorem there exists $x^* \in X^*$ such that $\sigma_D(x^*) < \sigma_C(x^*)$. Since the support functions are norm continuous on X^*, there exists a functional in the dense set G for which the same inequality holds, contradicting the definition of D.   ●

Since strongly exposing functional can be used to define slices of arbitrarily small diameter, Theorem 6.3.11 leads to a nice characterization of a Banach space with the RNP.

Corollary 6.3.12 *A Banach space X has the RNP if and only if every bounded closed convex subset of X is the closed convex hull of its strongly exposed points.*

Proof. Exercise 6.3.5.   ●

6.3.4 Pitt's Theorem

Pitt's theorem is a classical result about compactness of bounded linear operators between sequence spaces. We give a brief proof of Pitt's theorem using the Stegall variational principle.

Theorem 6.3.13 (Pitt) *Suppose $1 \leq p < q < +\infty$. Then every bounded linear operator from ℓ_q into ℓ_p is compact.*

Proof. Let T be a bounded linear operator from ℓ_q into ℓ_p. Then the function

$$f(x) = \|x\|_q^q - \|Tx\|_p^p, \quad x \in \ell_q$$

is bounded below and $f(x) > \|x\|_q$ if $\|x\|_q$ is large enough. By Corollary 6.3.6 there is a point $x \in \ell_q$ and a functional $x^* \in \ell_q^*$ such that

$$f(x+h) - f(x) - \langle x^*, h \rangle \geq 0, \quad h \in \ell_q.$$

It follows that

$$f(x+h) - f(x-h) - 2f(x) \geq 0, \quad h \in \ell_q.$$

Thus, for all $h \in \ell_q$,

$$\|x+h\|_q^q + \|x-h\|_q^q - 2\|x\|_q^q \geq \|T(x+h)\|_p^p + \|T(x-h)\|_p^p - 2\|Tx\|_p^p.$$

Let (x_i) be a bounded sequence in ℓ_q. By passing to a subsequence if necessary we may assume that (x_i) converges weakly to some $y \in \ell_q$. We will show that $\|Tx_i - Ty\| \to 0$ as $i \to \infty$. Indeed, by substituting $h = t(x_i - y)$ in the last inequality, we get

$$\|x + t(x_i - y)\|_q^q + \|x - t(x_i - y)\|_q^q - 2\|x\|_q^q$$
$$\geq \|Tx + tT(x_i - y)\|_p^p + \|Tx - tT(x_i - y)\|_p^p - 2\|Tx\|_p^p$$

for all $i = 1, 2, \ldots$ and all $t > 0$. Using the fact that (Exercise 6.3.6) if $z \in \ell_r, r \geq 1$ and $w_i \to 0$ weakly in ℓ_r, then

$$\limsup_{i \to \infty} \|z + w_i\|_r^r = \|z\|_r^r + \limsup_{i \to \infty} \|w_i\|_r^r,$$

we have

$$\limsup_{i \to \infty} \|x \pm t(x_i - y)\|_q^q = \|x\|_q^q + t^q \limsup_{i \to \infty} \|x_i - y\|_q^q$$

and

$$\limsup_{i \to \infty} \|Tx \pm tT(x_i - y)\|_p^p = \|Tx\|_p^p + t^p \limsup_{i \to \infty} \|T(x_i - y)\|_p^p.$$

Thus,

$$t^q \limsup_{i \to \infty} \|x_i - y\|_q^q \geq t^p \limsup_{i \to \infty} \|T(x_i - y)\|_p^p,$$

for all $t > 0$, and therefore $\|T(x_i - y)\|_p \to 0$ as $i \to \infty$. ●

6.3.5 Commentary and Exercises

Stegall's variational principle appeared in [243]. Corollary 6.3.6 is in Fabián's paper [109]. Our exposition here including its application to the representation of the Radon–Nikodym sets largely follows that of [221]. The Pitt theorem is usually proved by using the theory of Schauder bases (see [115, p. 175] or [183, p. 54]). The variational proof here is due to Fabián and Zizler [112].

Exercise 6.3.1 Let X be a Banach space and let A be a closed subset of X. Show that A is dentable if and only if for every $\varepsilon >$ there exists $x \in A$ such that $x \notin \overline{\text{conv}(A \backslash B_\varepsilon(x))}$.

Exercise 6.3.2 Let X be a Banach space and let S be a nonempty closed subset of X. Suppose that f is a lsc function bounded from below on S. Show that f attains a strong minimum on S if and only if diam $S(-f, S, \alpha) \to 0$ as $\alpha \to 0+$.

Exercise 6.3.3 Let X be a Banach space and let C be a closed convex subset of X. Then $x \in C$ is strongly exposed with an strongly exposing functional $x^* \in X^*$ if and only if $x \in S(x^*, C, \alpha)$ for every $\alpha > 0$ and diam $S(x^*, C, \alpha) \to 0$ as $\alpha \to 0$.

Exercise 6.3.4 Prove Lemma 6.3.8.

Exercise 6.3.5 Prove Corollary 6.3.12.

Exercise 6.3.6 Let $z \in \ell_r, r \geq 1$ and let $w_i \to 0$ weakly in ℓ_r. Prove that

$$\limsup_{i \to \infty} \|z + w_i\|_r^r = \|z\|_r^r + \limsup_{i \to \infty} \|w_i\|_r^r.$$

Hint: First consider the case when z has only finitely many nonzero components (finitely supported). Then prove that general case by using the fact that finitely supported elements in ℓ_r are dense and $\| \cdot \|_r^r$ is Lipschitz on a bounded set.

Exercise 6.3.7 Prove that the dual of an Asplund space has the RNP by showing that points of differentiability of support functions produce denting points in the dual. The converse is more elaborate [221].

6.4 Mountain Pass Theorem

The variational techniques and applications we discussed so far are all based on generalizations of the Fermat principle that any minimum of a function is a critical point. It is well known that critical points also arise at minimax (saddle) points. Can variational principles be used in this situation? The answer is positive. As an example we discuss a variational proof of a mountain pass theorem. We adopt the view of considering the sup as a function in a problem involving inf sup and applying a variational principle to it.

6.4.1 The Mountain Pass Theorem

The mountain pass theorem is an important result in studying multiple solutions for nonlinear partial differential equations. The name comes from an interesting geometric picture. Consider a basin surrounded by mountains. To travel from a location in the basin to a city outside these surrounding mountains one must follow a mountain pass crossing the mountain ridge. Each such mountain pass will have a point with the highest elevation. Intuitively, one

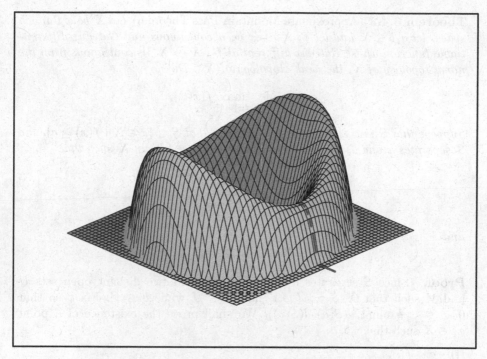

Fig. 6.1 A typical mountain pass.

can imagine that among all those possible mountain passes there must be *a pass with a least highest elevation.* This point of minimum highest elevation on this particular mountain pass must be a saddle point. In fact, along the mountain pass it has the highest elevation while it has the lowest elevation along the mountain ridge, as illustrated in Figure 6.1 One certainly expects it to be a critical point, and indeed it is, under appropriate conditions.

To state this result precisely we need to define what we mean by "mountain pass." Let X be a Banach space and let $C([0,1];X)$ be the Banach space of all continuous functions $x\colon [0,1] \to X$ with the supremum norm $\|x\|_\infty$. For $a, b \in X$ we define the set of *passes* from a to b by

$$\Gamma(a,b) := \{x \in C([0,1];X) \mid x(0) = a \text{ and } x(1) = b\}.$$

Our next definition helps to characterize the mountain ridge.

Definition 6.4.1 *Let X be a Banach space and let S be a closed subset of X. We say that S separates* two points *a and b in X provided that a and b belong to disjoint connected components of $X \setminus S$.*

Now we can state and prove the mountain pass theorem.

Theorem 6.4.2 (Approximate Mountain Pass Theorem) *Let X be a Banach space, let $a, b \in X$ and let $f \colon X \to \mathbb{R}$ be a continuous and Gâteaux differentiable function whose Gâteaux differential $f' \colon X \to X^*$ is continuous from the norm topology of X, the weak* topology of X^*. Define*

$$c = \inf_{x \in \Gamma(a,b)} \max_{t \in [0,1]} f(x(t)).$$

Suppose that S is a closed subset of X such that $S \subset \{x \in X \mid f(x) \geq c\}$ and S separates a and b. Then there exists a sequence (x_i) in X such that

$$\lim_{i \to \infty} d(S; x_i) = 0,$$

$$\lim_{i \to \infty} f(x_i) = c,$$

and

$$\lim_{i \to \infty} \|f'(x_i)\| = 0.$$

Proof. Since S separates a and b we can find two disjoint open sets U and V such that $X \backslash S = U \cup V$ and $a \in U$ while $b \in V$. Fix ε so that $0 < \varepsilon < \frac{1}{2} \min(1, d(S; a), d(S; b))$. We shall prove the existence of a point $x_\varepsilon \in X$ such that

(i) $c < f(x_\varepsilon) < c + \frac{5}{4}\varepsilon^2$,
(ii) $d(S; x_\varepsilon) < \frac{3}{2}\varepsilon$ and
(iii) $\|f'(x_\varepsilon)\| < \frac{3}{2}\varepsilon$.

To do so let $\bar{x} \in \Gamma(a, b)$ satisfy

$$\max\{f(\bar{x}(t)) \mid t \in [0,1]\} < c + \frac{\varepsilon^2}{4}. \tag{6.4.1}$$

Define

$$\alpha := \sup\{t \in [0,1] \mid \bar{x}(t) \in U \text{ and } d(S; \bar{x}(t)) \geq \varepsilon\},$$
$$\beta := \inf\{t \in [0,1] \mid \bar{x}(t) \in V \text{ and } d(S; \bar{x}(t)) \geq \varepsilon\},$$

so that necessarily $d(S; \bar{x}(t)) < \varepsilon$ whenever $t \in (\alpha, \beta)$. Clearly $\Gamma(\bar{x}(\alpha), \bar{x}(\beta))$ is a closed subset of $C([0,1]; X)$. Set $h(x) := \varepsilon \max(0, \varepsilon - d(S; x))$, and define a function $\varphi \colon \Gamma(\bar{x}(\alpha), \bar{x}(\beta)) \to \mathbb{R}$ by

$$\varphi(x) := \max\{f(x(t)) + h(x(t)) \mid t \in [0,1]\}.$$

Note that for any $x \in \Gamma(\bar{x}(\alpha), \bar{x}(\beta))$ we have that $x([0,1]) \cap S \neq \emptyset$, since $x(0) = \bar{x}(\alpha) \in U$, $x(1) = \bar{x}(\beta) \in V$ and $X \backslash S = U \cap V$. It follows that for any $x \in \Gamma(\bar{x}(\alpha), \bar{x}(\beta))$

$$\varphi(x) \geq \max\{f(x(t)) + h(x(t)) \mid t \in [0,1] \text{ and } x(t) \in S\} \geq c + \varepsilon^2$$

so that

$$\inf_{\Gamma(\bar{x}(\alpha), \bar{x}(\beta))} \varphi \geq c + \varepsilon^2. \tag{6.4.2}$$

On the other hand, let $z(t) := \bar{x}(\alpha + t(\beta - \alpha)) \in \Gamma(\bar{x}(\alpha), \bar{x}(\beta))$. We have

$$\varphi(z) \leq \max\{f(\bar{x}(t)) + h(\bar{x}(t)) \mid t \in [0,1]\} \leq \left(c + \frac{\varepsilon^2}{4}\right) + \varepsilon^2. \tag{6.4.3}$$

We can now apply the Ekeland variational principle of Theorem 2.1.2 to φ on $\Gamma(\bar{x}(\alpha), \bar{x}(\beta))$ to find a path $y \in \Gamma(\bar{x}(\alpha), \bar{x}(\beta))$ such that

$$\varphi(y) \leq \varphi(z), \tag{6.4.4}$$

$$\|y - z\| \leq \varepsilon/2. \tag{6.4.5}$$

and

$$\varphi(x) + \frac{\varepsilon}{2}\|x - y\| \geq \varphi(y) \quad \text{for all } x \in \Gamma(\bar{x}(\alpha), \bar{x}(\beta)). \tag{6.4.6}$$

Now let M be the subset of $[0,1]$ consisting of all points where $(f + h) \circ y$ attains its maximum on $[0,1]$. We prove first that there exists $\bar{t} \in M$ such that $\|f'(y(\bar{t})\| \leq \frac{3}{2}\varepsilon$. Indeed, first note that it follows from (6.4.6) that for any $\eta \in C([0,1]; X)$ with $\eta(0) = \eta(1) = 0$,

$$-\frac{\varepsilon}{2}\|\eta\| \leq \liminf_{s \to 0+} \frac{\varphi(y + s\eta) - \varphi(y)}{s}.$$

Using the definition of the Gâteaux differential of f and the fact that h has a Lipschitz constant ε, it follows that the last inequality is dominated by

$$\liminf_{s \to 0+} \frac{1}{s}\left[\max_{t \in [0,1]}\left((f + h)(y(t)) + s\langle f'(y(t)), \eta(t)\rangle\right) - \max_{t \in [0,1]}\left((f + h)(y(t))\right)\right] + \varepsilon\|\eta\|.$$

Hence

$$-\frac{3\varepsilon}{2}\|\eta\| \leq \liminf_{s \to 0+} \frac{m(k + sl) - m(k)}{s}, \tag{6.4.7}$$

where $k = (f+h) \circ y$, $l = \langle f'(y), \eta\rangle$ and m is the continuous convex function on $C([0,1]; X)$ defined by $m(x) = \max\{x(t) \mid t \in [0,1]\}$. Recall that the convex subdifferential of m has the following representation [147] $\partial m(x) = \{\mu \mid \mu$ is a Radon probability measure supported in $M(x)\}$ where $M(x) := \{t \in [0,1] \mid x(t) = m(x)\}$.

Thus, it follows from (6.4.7) and the max formula of Theorem 4.2.7 that

$$-\frac{3\varepsilon}{2}\|\eta\| \leq \liminf_{s \to 0+} \frac{m(k + sl) - m(k)}{s}$$
$$\leq \max\{\langle l, \mu\rangle \mid \mu \in \partial m(k)\}$$
$$= \max\{\int \langle f'(y), \eta\rangle \, d\mu \mid \mu \in \partial m(k)\}.$$

By a standard minimax theorem ([7, Theorem 6.2.7]) we have

$$-\frac{3\varepsilon}{2} = \inf_{\eta}\max_{\mu}\left\{\int \langle f'(y),\eta\rangle\, d\mu \;\Big|\; \mu\in\partial m(k), \|\eta\|\leq 1, \eta(0)=\eta(1)=0\right\}$$

$$= \max_{\mu}\inf_{\eta}\left\{\int \langle f'(y),\eta\rangle\, d\mu \;\Big|\; \mu\in\partial m(k), \|\eta\|\leq 1, \eta(0)=\eta(1)=0\right\}$$

$$= \max_{\mu}\left\{-\int \|f'(y)\|\, d\mu \;\Big|\; \mu\in\partial m(k)\right\}$$

$$\leq \min\left\{-\|f'(y(t))\| \mid t\in M(k)\right\}.$$

Combining (6.4.1) and (6.4.2) we can check that (Exercise 6.4.1)

$$M(k)\cap\{0,1\} = \emptyset. \tag{6.4.8}$$

Therefore, there exists $\bar{t}\in M(k) = M$ such that $\|f'(y(\bar{t}))\|\leq 3\varepsilon/2$. It remains to show that the point $x_\varepsilon = y(\bar{t})$ satisfies (i) and (ii). For (i) combining (6.4.2), (6.4.3) and (6.4.4) to get

$$c+\varepsilon^2 \leq \inf_{\Gamma(\bar{x}(\alpha),\bar{x}(\beta))}\varphi \leq f(y(\bar{t}))+h(y(\bar{t})) = \varphi(y)\leq\varphi(z)\leq c+\frac{5\varepsilon^2}{4}.$$

Since $0\leq h\leq\varepsilon^2$ we obtain $c\leq f(x_\varepsilon)\leq c+5\varepsilon^2/4$. For (ii) it is enough to notice that (6.4.8) implies $\bar{t}\in(0,1)$, hence $d(S;z(\bar{t})) = d(S;\bar{x}(\alpha+\bar{t}(\beta-\alpha)))\leq\varepsilon$. This combined with (6.4.5) gives that $d(S;x_\varepsilon) = d(S;y(\bar{t}))\leq 3\varepsilon/2$.   ●

6.4.2 The Palais–Smale Condition

Theorem 6.4.2 asserts the existence of a sequence (x_i) such that $(f'(x_i))$ converges to 0. To ensure the existence of an actual critical point for f we need the following Palais–Smale type condition.

Definition 6.4.3 (Palais–Smale Condition) *Let X be a Banach space, let S be a subset of X and let $f\colon X\to\mathbb{R}$ be a Gâteaux differentiable function. We say that f satisfies the* Palais–Smale condition *around S at level c if every sequence (x_i) in X verifying $\lim_{i\to\infty}f(x_i) = c$, $\lim_{i\to\infty}d(S;x_i) = 0$ and $\lim_{i\to\infty}\|f'(x_i)\| = 0$ has a convergent subsequence.*

Adding this condition to Theorem 6.4.2, we can ensure the existence of a critical point.

Theorem 6.4.4 (Mountain Pass Theorem) *Let X be a Banach space, let $a,b\in X$ and let $f\colon X\to\mathbb{R}$ be a continuous and Gâteaux differentiable function whoes Gâteaux differential $f'\colon X\to X^*$ is continuous from the norm topology of X to the weak* topology of X^*. Define*

$$c = \inf_{x\in\Gamma(a,b)}\max_{t\in[0,1]}f(x(t)).$$

Suppose that S is a closed subset of X such that $S \subset \{x \in X \mid f(x) \geq c\}$ and S separates a and b. Then there exists a point $\bar{x} \in S$ such that $f(\bar{x}) = c$ and $f'(\bar{x}) = 0$.

Proof. Exercise 6.4.2.   ●

6.4.3 Solutions of a Dirichlet Problem

The mountain pass theorem is a powerful tool for obtaining the existence and multiplicity results for solutions to many nonlinear partial differential equations. We illustrate by considering the Dirichlet problem for semilinear elliptic equations of the form

$$-\Delta x = F'(x) \quad y \in \Omega,$$
$$x = 0 \quad\quad y \in \partial\Omega. \tag{6.4.9}$$

Here F is a nonlinear function, Ω is an open subset of $\mathbb{R}^N$, $\partial\Omega$ signals the boundary of Ω and Δ is the Laplacian operator. The solution is in the sense of distribution. Let X be the Sobolev space $H_0^1(\Omega)$. Then solutions of (6.4.9) corresponding to critical points of the functional

$$f(x) := \frac{1}{2}\|x\|^2 - \int_\Omega F(x(y))\,dy.$$

As an example we consider the simple case of $F(x) = |x|^p$, where $2 < p < 2^* := (2N-2)/(N-2)$. Clearly, the Dirichlet problem (6.4.9) has the trivial solution $x(y) = 0$. We use the mountain pass theorem to show it has at least one nontrivial solution. The condition $2 < p < 2^* := (2N-2)/(N-2)$ is crucial. Under this condition $X = H_0^1(\Omega)$ is compactly imbedded in $L^p(\Omega)$, i.e., the imbedding $X \to L^p(\Omega)$ maps bounded closed subsets of X to compact subsets of $L^p(\Omega)$. Thus, by the Sobolev inequality, $f(x) \geq r$ for some $r > 0$ on the unit sphere S_X of X. Clearly, $f(0) = 0$. Moreover, fixing $x \neq 0$ we have,

$$f(tx) = \frac{1}{2}t^2\|x\|^2 - t^p \int_\Omega |x|^p\,dy \to -\infty$$

as $t \to +\infty$. Thus, there exists $b = tx$ such that $f(b) \leq 0$. This tells us that f satisfies the geometric conditions of Theorem 6.4.4 with $a = 0$ and $S = S_X$. To show that f has a critical point on S it remains to check the Palais–Smith condition. Let (x_i) be a sequence in X such that $f(x_i) \to c$ and $f'(x_i) \to 0$. First we derive a bound on (x_i) as follows. From $\langle f'(x_i), x_i \rangle \leq \varepsilon\|x_i\|$ we have that

$$p\int_\Omega |x_i|^p\,dy \leq \|x_i\|^2 + \varepsilon\|x_i\|,$$

while $f(x_i)$ is bounded implies that, for some constant k,

$$\|x_i\|^2 \le k + 2 \int_\Omega |x_i|^p \, dy,$$

and therefore

$$\|x_i\|^2 \le k + \frac{2}{p}\|x_i\|^2 + \frac{2}{p}\varepsilon\|x_i\|^2.$$

Since $p > 2$ it follows that $\|x_i\|$ is bounded. Then, taking a subsequence if necessary, we may assume that (x_i) converges weakly to $\bar{x}$ in X. Since X is compactly imbedded in $L^p(\Omega)$ we may assume that (x_i) converges strongly in $L^p(\Omega)$. Combining this with $f'(x_i) \to 0$ we have (x_i) converges strongly to $\bar{x}$ (Exercise 6.4.3) in X, and therefore $f'(\bar{x}) = 0$.

6.4.4 Commentary and Exercises

The idea of considering the existence of critical points other than minima was suggested by Morse [212] and Lusternik and Schirelman [184]. These variational methods are often referred to as calculus of variation in the large. The mountain pass theorem is due to Ambrosetti and Rabinowitz [2]. This theorem and its extensions can handle unbounded functions and have wide ranging applications in the existence of multiple solutions for nonlinear partial differential equations. The original proof of the mountain pass theorem in [2] relies on a deformation lemma. Our proof here using the Ekeland variational principle follows [125]. The Palais–Smale condition is in [215]. The survey papers [1, 224] provide insights and abundant applications of mountain pass theorems.

Exercise 6.4.1 Prove (6.4.8) in the proof of Theorem 6.4.2.

Exercise 6.4.2 Prove Theorem 6.4.4.

Exercise 6.4.3 Let $X = H_0^1(\Omega)$ and define $f \colon X \to \mathbb{R}$ by

$$f(x) := \frac{1}{2}\|x\|^2 - \int_\Omega F(x(y)) \, dy.$$

Suppose that (x_i) converges weakly to $\bar{x}$ in X and (x_i) converges strongly to $\bar{x}$ in $L^p(\Omega)$ where $2 < p < 2^*$ and $(\|f'(x_i)\|)$ converges to 0. Show that (x_i) converges strongly to $\bar{x}$ in X. Hint: $f'(x_i)(x_i - \bar{x}) \to 0$ implies that $\|x_i\|^2 \to \|\bar{x}\|^2$.

6.5 One-Perturbation Variational Principles

In a typical variational principle one is given a fixed real-valued lsc function f defined on a complete metric space X bounded from below and one seeks to find conditions under which there is at least one "nice" perturbation g such that the perturbed function $f + g$ attains a minimum.

Our focus here is in some sense to reverse this scheme and to investigate a situation in which we have one perturbation function, say g, such that for every function f from some large enough class of functions, $f + g$ always attains a minimum.

6.5.1 One-Perturbation Variational Principles

It is well-known, and due to Weierstrasse, that if a proper lsc function $g \colon X \to \mathbb{R} \cup \{+\infty\}$ defined on a Hausdorff topological space X has at least one non-empty compact sublevel set, then it is bounded from below and attains its minimum. Moreover, something stronger is true: the problem of minimizing g is *well-posed* in the sense that every minimizing net (or sequence in the metric case) has a subnet (subsequence) converging to a minimum .

Thus the following result holds:

Theorem 6.5.1 (One-Perturbation Variational Principle) *Let X be a Hausdorff topological space and let $\varphi \colon X \to \mathbb{R} \cup \{+\infty\}$ be a proper lsc function whose sublevel sets $\varphi^{-1}((-\infty, r]), r \subset \mathbb{R}$ are all compact. Then for any proper lsc function $f \colon X \to \mathbb{R} \cup \{+\infty\}$ bounded from below, the function $f + \varphi$ attains its minimum. In particular, if $\operatorname{dom}\varphi$ is relatively compact, the conclusion is true for any proper lsc function f (not necessarily bounded from below).*

Proof. Suppose $\operatorname{dom} f \cap \operatorname{dom}\varphi \neq \emptyset$ (if the latter set is empty there is nothing to prove). Then the function $f + \varphi$ is a proper lsc function. In particular, $(f+\varphi)^{-1}((-\infty, r])$ is non-empty (and closed) for some r. Now, if f is bounded from below then $f + \varphi$ is bounded below as well. But, $(f + \varphi)^{-1}((-\infty, r]) \subset \varphi^{-1}((-\infty, r - \inf f])$ and since the latter set is compact, the same is true for $(f + \varphi)^{-1}((-\infty, r])$. On the other hand, if the domain of φ is relatively compact, since $\operatorname{dom}(f+\varphi) \subset \operatorname{dom}\varphi$, then the domain of $f+\varphi$ is also relatively compact. Thus, again, $(f + \varphi)^{-1}((-\infty, r])$ is compact. The result then follows by the remark before the theorem. $\bullet$

In a metric space X, the conditions imposed on the unique perturbation φ in Theorem 6.5.1 are also necessary.

Theorem 6.5.2 *Let $\varphi \colon X \to \mathbb{R} \cup \{+\infty\}$ be a proper function on a metric space X. Suppose that for every bounded continuous function $f \colon X \to \mathbb{R}$, the function $f + \varphi$ attains its minimum. Then φ is a lsc function, bounded from below, whose sublevel sets are all compact.*

Proof. Obviously φ is bounded from below. Suppose it is not lower semi-continuous. Then its epigraph, $\operatorname{epi}\varphi$, is not closed in the product topology of $X \times \mathbb{R}$, i.e., there exists a sequence $((x_i, r_i))$ in $\operatorname{epi}\varphi$ such that $(x_i, r_i) \to (x_0, r) \notin \operatorname{epi}\varphi$. Fix such an x_0 and observe that, since φ is bounded from below, the sequence may be taken to have the form $((x_i, \varphi(x_i)))$. This shows that for the fixed x_0, the following set is non-empty:

$$R := \{r \in \mathbb{R} \mid \text{ there is } (x_i) \subset X \text{ such that } (x_i, \varphi(x_i)) \to (x_0, r) \notin \text{epi}\, \varphi\}.$$

Put $r_0 := \inf R$. Since R is bounded from below by $\inf \varphi$, the number r_0 is well-defined and obviously $r_0 < \varphi(x_0)$. Moreover, using a diagonal process, one easily sees that $r_0 \in R$.

Further, for any $i = 0, 1, 2, \ldots$, let V_i be a neighborhood of x_0 with the property:

$$\varphi(x) > r_0 - \frac{1}{2^{i+1}} \text{ for any } x \in V_i.$$

The existence of such a V_i is guaranteed by the definition of r_0. We may arrange without loss of generality that for any $i = 0, 1, 2, \ldots$, we have $\overline{V_{i+1}} \subset V_i$ and that the family $\{V_i\}_{i \in \mathbb{N}}$ provides a local base of neighborhoods for the point x_0.

Now, for any $i = 0, 1, 2, \ldots$, we define the following continuous function $h_i \colon X \to [0, 1]$:

$$h_i|_{X \setminus V_i} \equiv 1 \quad \text{and} \quad h_i(x_0) = 0.$$

The functions exist because X is a completely regular topological space. Choose also $\alpha \in \mathbb{R}_+$ so that $\alpha > r_0 - \inf \varphi$ and let

$$f(x) := \alpha h_0(x) + \sum_{i=1}^{\infty} \frac{1}{2^i} h_i(x), \quad x \in X.$$

The function f is well-defined, continuous, bounded and such that $f(x_0) = 0$. We show that for every $x \in X$ we have $f(x) + \varphi(x) > r_0$. Indeed, if $x = x_0$ we have

$$f(x_0) + \varphi(x_0) = \varphi(x_0) > r_0.$$

If $x \neq x_0$ we have two possibilities:

Case 1: $x \notin V_i$ for any $i = 0, 1, \ldots$. In this case $h_i(x) = 1$ for every i and consequently

$$f(x) + \varphi(x) = \alpha + 1 + \varphi(x) \geq \alpha + 1 + \inf \varphi > r_0.$$

Case 2: $x \in V_i$ for some n. Since $x \neq x_0$ and the family $\{V_i\}$ is a local base for x_0, we have $x \in V_k \setminus V_{k+1}$ for some $k \in \mathbb{N}$. Then, by the definition of f and the h_i's, we have

$$f(x) + \varphi(x) \geq \frac{1}{2^{k+1}} + \varphi(x) > r_0,$$

according to the definition of V_k. Thus

$$f(x) + \varphi(x) > r_0, \quad \text{for every } x \in X. \tag{6.5.1}$$

On the other hand, since $r_0 \in R$, there is $x_i \to x_0$ so that $\varphi(x_i) \to r_0$. Hence, since f is continuous, $f(x_i) + \varphi(x_i) \to f(x_0) + r_0 = r_0$, which together with (6.5.1) shows that $\inf(f + \varphi) = r_0$. But this infimum of the perturbation $f + \varphi$ is obviously not attained, according to the same inequality (6.5.1), which contradicts to the assumptions of the theorem. Therefore, the function φ must be lower semicontinuous.

Secondly, let us prove that every sublevel set of φ is compact. We may suppose, without loss of generality, that $\inf \varphi = 0$. Take some $r > 0$ and consider the sublevel set $X_r := \{x \in X \mid \varphi(x) \le r\}$. Observe that because φ is lsc, the set X_r is closed.

Suppose X_r is not compact. Then there is a sequence (x_i) in X_r which has no cluster point in X_r. We will, for the moment, restrict ourselves to the space X_r with the inherited metric topology. Using the above, one can see that there are sequences of sets (W_i), open in X_r, such that $x_i \in W_i$ for every i and $W_i \cap W_j = \emptyset$ for every $i \ne j$. Since φ is lsc we may assume that we have in addition:

$$\varphi(x) > \varphi(x_i) - \frac{1}{2i} \quad \text{for all } x \in W_i. \tag{6.5.2}$$

Further, for any $x \in X_r \setminus \bigcup_{i=1}^{\infty} W_i$ there is a relatively open neighborhood W_x so that $x_i \notin W_x$ for every i (indeed, otherwise x would be an accumulation point of the sequence (x_i)). Consider now the following open covering of X_r

$$\Gamma := \left\{ W_x : x \notin X_r \setminus \bigcup_{i=1}^{\infty} W_i \right\} \cup \{W_i : i = 1, 2, \dots\}.$$

Since X_r is a metric space this covering has an open refinement $\gamma = \{V_\alpha : \alpha \in \Lambda\}$ which is locally finite, i.e., for any $x \in X_r$ there is a neighborhood U of x so that U intersects at most finitely many members of γ. Let $\alpha_i \in \Lambda$ be such that $x_i \in V_{\alpha_i}$ for every $i = 1, 2, \dots$. Because of the definition of the covering Γ ($x_i \notin W_x$ for every i and $x \notin \cup W_i$) and because of the fact that the W_i are pair-wise disjoint, it follows that $V_{\alpha_i} \subset W_i$ for every n. In particular, V_{α_i} are pair-wise disjoint as well. Now, for every n there are continuous functions $h_i \colon X_r \to [-\varphi(x_i) - 1 + 1/i, 0]$:

$$h_i(x_i) = -\varphi(x_i) - 1 + \frac{1}{i} \quad \text{and} \quad h|_{X_r \setminus V_{\alpha_i}} \equiv 0.$$

Finally, let

$$h(x) := \sum_{i=1}^{\infty} h_i(x), \quad x \in X_r.$$

Because the open sets V_{α_i} are pair-wise disjoint, the function h is well-defined in X_r. Obviously h takes its values in the interval $[-r - 1, 0]$. Moreover, it is continuous when restricted to X_r. Indeed, if $x \in V_{\alpha_i}$ for some n, then it coincides with h_i on V_{α_i}, hence is continuous at x. If $x \notin V_{\alpha_i}$ for any i then due to the properties of the covering γ, there is an open neighborhood U of the point x so that U intersects at most finitely many V_{α_i}. Consequently in U the function h is a finite sum of continuous functions. Thus again, h is continuous at x.

Further, since X_r was closed in X, according to Dugundji's extension theorem, there is a continuous function $f\colon X \to [-r-1, 0]$ which is an extension of h. Let us consider the perturbation $f + \varphi$. We first show that

$$f(x) + \varphi(x) > -1 \quad \text{for all } x \in X. \tag{6.5.3}$$

Indeed, if $x \notin X_r$ we have $\varphi(x) > r$ and the above inequality is obvious having in mind the range of f. So, let $x \in X_r$. Then we again have two cases:

Case 1: $x \notin V_{\alpha_i}$ for any n. Then $f(x) = h(x) = 0$. Thus, $f(x) + \varphi(x) = \varphi(x) \geq 0$ and the inequality is obviously true;

Case 2: $x \in V_{\alpha_i}$ for some (unique!) i. Then $f(x) = h(x) = h_i(x)$, and using (6.5.2) we have

$$f(x) + \varphi(x) > \varphi(x_i) - \frac{1}{2i} + h_i(x) \geq \varphi(x_i) - \frac{1}{2i} - \varphi(x_i) - 1 + \frac{1}{i} = -1 + \frac{1}{2i} > -1.$$

Hence, (6.5.3) is also true in this case.

Finally, let us observe that $f(x_i) + \varphi(x_i) = \varphi(x_i) + h_i(x_i) = -1 + 1/n \to -1$ which, together with (6.5.3) yields that $\inf(f + \varphi) = -1$, and this infimum is not attained according to (6.5.3). This contradiction shows that X_r must be compact. The proof of the theorem is complete. ●

6.5.2 Perturbation Functions in Separable Banach Spaces

To apply Theorem 6.5.1, we need a collection of functions φ satisfying the hypotheses stated there and having enough additional regularity properties so that, combined with the existence of a minimizer of the perturbation, they have interesting consequences. The following result answers this need when X is separable.

Proposition 6.5.3 *Let X be a separable Banach space. Then there exists a convex lsc function $\varphi\colon X \to [0, +\infty]$ whose domain is relatively norm-compact and linearly dense in X. In addition φ possesses the following smoothness property:*

$$\lim_{t \to 0+} \sup_{h \in \mathrm{dom}\,\varphi} \frac{\varphi(x + th) + \varphi(x - th) - 2\varphi(x)}{t} = 0, \quad x \in \mathrm{dom}\,\varphi. \tag{6.5.4}$$

Proof. Let $\{x_i : i \in \mathbb{N}\}$ be a dense subset of B_X. Define $T\colon \ell_2 \to X$ (where ℓ_2 is the usual space of square summable sequences) by

$$T(\alpha) = \sum_{i=1}^{\infty} \alpha_i 2^{-i} x_i, \quad \alpha = (\alpha_i) \in \ell_2.$$

Clearly, T is a well-defined, linear, and bounded mapping. T is also compact, that is, $T(B_{\ell_2})$ is a norm-compact set. Indeed, take any sequence (α^k) of elements in B_{ℓ_2}. Since B_{ℓ_2} is weakly compact, there is a subsequence (we do not relabel) along which α^k converges weakly to $\alpha = (\alpha_1, \alpha_2, \dots)$: in particular, $\alpha \in B_{\ell_2}$ and for each $i \in \mathbb{N}$, $\alpha_i^k \to \alpha_i$ as $k \to \infty$. Let us show that along this same subsequence, $T(\alpha^k)$ norm-converges to $T(\alpha)$. Actually, all natural numbers k, K satisfy

$$\|T(\alpha^k) - T(\alpha)\| \le \Big\| \sum_{i=1}^{K} \frac{(\alpha_i^k - \alpha_i)x_i}{2^i} \Big\| + \Big\| \sum_{i=K+1}^{\infty} \frac{(\alpha_i^k - \alpha_i)x_i}{2^i} \Big\|$$

$$\le \Big\| \sum_{i=1}^{K} \frac{(\alpha_i^k - \alpha_i)x_i}{2^i} \Big\| + \Big[\sum_{i=K+1}^{\infty} (\alpha_i^k - \alpha_i)^2 \Big]^{\frac{1}{2}}$$

$$\times \Big[\sum_{i=K+1}^{\infty} 2^{-2i} \Big]^{\frac{1}{2}}$$

$$\le \Big\| \sum_{i=1}^{K} \frac{(\alpha_i^k - \alpha_i)x_i}{2^i} \Big\| + \Big[2 \sum_{i=1}^{\infty} (\alpha_i^k)^2 + 2 \sum_{i=1}^{\infty} \alpha_i^2 \Big]^{\frac{1}{2}}$$

$$\times \Big(\frac{4}{3 \times 2^{2(K+1)}} \Big)^{\frac{1}{2}},$$

and so

$$\limsup_{k \to \infty} \|T(\alpha^k) - T(\alpha)\| \le 0 + 2^{-K+1}/\sqrt{3}.$$

This holds for every $K \in \mathbb{N}$, so $\limsup_{k \to \infty} \|T(\alpha^k) - T(\alpha)\| = 0$, as required. (The compactness of T may also be established more abstractly by observing that since $T(B_{\ell_2})$ is weakly closed it suffices to show it is totally bounded.)

We have no guarantee that T is injective. Thus we introduce the quotient space $H = \ell_2/T^{-1}(0)$ of ℓ_2, and factorize $T = S \circ Q$, where $Q: \ell_2 \to H$ is the canonical map. It is not difficult to see, by using the parallelogram law, that H, with its quotient norm $\| \cdot \|_H$, is again a Hilbert space. The mapping $S: H \to X$ is then linear, bounded, compact, and injective. Also, the range of S, range(S), is dense in X, since range$(S) = $ range$(T) \supset \{x_i : i \in \mathbb{N}\}$.

Now we define a function with the desired properties as follows: for $x \in X$, let

$$\varphi(x) = \begin{cases} \tan\left(\|S^{-1}x\|_H^2\right) & \text{if } \|S^{-1}x\|_H^2 < \frac{\pi}{2}, \\ +\infty & \text{otherwise.} \end{cases}$$

Then $x \in \operatorname{dom}\varphi$ implies that $\|S^{-1}x\|_H^2 < \pi/2$, and hence $x = S(S^{-1}x) \in S\left(\sqrt{\frac{\pi}{2}}B_H\right)$, the last set being relatively norm-compact. Also, for every $i \in \mathbb{N}$, the standard basis vector e^i in ℓ_2 obeys $T(2^i te^i) = tx_i$ for all real t. This implies that $S^{-1}(tx_i)$ has small norm in H whenever t is sufficiently close to 0, and hence that $tx_i \in \operatorname{dom}\varphi$ whenever $|t|$ is sufficiently small. Thus $\operatorname{dom}\varphi$ is linearly dense in X.

The function φ is convex because $\tan(\cdot)$ is increasing and convex on $[0, \frac{\pi}{2})$ and $\|\cdot\|_H^2$ is convex on H.

Let us show that φ is lower semicontinuous. Fix any $p \in X$ and any sequence (p_i) in X, with $\|p_i - p\| \to 0$ as $i \to \infty$. We will show that

$$\varphi(p) \le \liminf_{i \to \infty} \varphi(p_i).$$

The inequality is obvious if the right side equals $+\infty$. Assume therefore that the right side is finite. Choose a subsequence (p_{i_j}) along which

$$\liminf_{n \to \infty} \varphi(p_i) = \lim_{j \to \infty} \varphi(p_{i_j}).$$

Then $\|S^{-1}(p_{i_j})\|_H^2 < \pi/2$ for all large i, and indeed

$$\lim_{j \to \infty} \|S^{-1}(p_{i_j})\|_H^2 < \frac{\pi}{2}.$$

By passing to a further subsequence if necessary we may assume that the bounded sequence $(S^{-1}(p_{i_j}))$ converges weakly to some $z \in \ell_2$ and that $\lim_{i \to \infty} \|S^{-1}(p_{i_j})\|_H$ exists: the weak lower semicontinuity of $\|\cdot\|_H$ then gives

$$\|z\|_H^2 \le \lim_{j \to \infty} \|S^{-1}(p_{i_j})\|_H^2 < \frac{\pi}{2}. \tag{6.5.5}$$

Now $p_{i_j} = S(S^{-1}p_{i_j})$ converges weakly to Sz, giving $Sz = p$, $z = S^{-1}p$. Using the continuity and monotonicity of $\tan(\cdot)$ on $[0, \pi/2)$ together with (6.5.5), we obtain the desired inequality:

$$\begin{aligned}
\varphi(p) = \tan\big(\|S^{-1}p\|_H^2\big) &= \tan\big(\|z\|_H^2\big) \\
&\le \lim_{j \to \infty} \tan\big(\|S^{-1}(p_{i_j})\|_H^2\big) \\
&= \lim_{j \to \infty} \varphi(p_{i_j}) = \liminf_{i \to \infty} \varphi(p_i).
\end{aligned}$$

Finally, take any $x, h \in \operatorname{dom}\varphi$. Then $x \pm th \in \operatorname{dom}\varphi$ for all $t > 0$ sufficiently small, so the parallelogram law gives

$$\|S^{-1}(x + th)\|_H^2 + \|S^{-1}(x - th)\|_H^2 - 2\|S^{-1}(x)\|_H^2$$
$$= 2t^2\|S^{-1}(h)\|_H^2 \tag{6.5.6}$$

(since S^{-1} is linear). Hence the convex function $t \mapsto \|S^{-1}(x + th)\|_H^2$ is differentiable at $t = 0$, and the same is true for $t \mapsto \varphi(x + th)$. This establishes the existence of all the directional derivatives

$$\varphi'(x; h) = \lim_{t \to 0} \frac{\varphi(x + th) - \varphi(x)}{t}, \quad x, h \in \operatorname{dom}\varphi.$$

In fact, since $h \in \operatorname{dom}\varphi$ if and only if $\|S^{-1}(h)\|_H^2 < \pi/2$, then (6.5.6) supports a type of uniform differentiability:

$$0 = \lim_{t \to 0+} \sup_{h \in \mathrm{dom}\,\varphi} \frac{\varphi(x + th) + \varphi(x - th) - 2\varphi(x)}{t}, \quad x \in \mathrm{dom}\,\varphi.$$

This completes the proof.   ●

Combining Theorem 6.5.1 and the above proposition leads to the main result of this section:

Theorem 6.5.4 *Let X be a separable Banach space. Then there exists a (proper) convex lsc function $\varphi \colon X \to [0, +\infty]$ whose domain $\mathrm{dom}\,\varphi$ is relatively norm-compact and linearly dense in X and which satisfies the smoothness property (6.5.4). In particular, for every proper lsc function $f \colon X \to \mathbb{R} \cup \{+\infty\}$, the perturbed function $f + \varphi$ attains its infimum.*

Proof. Exercise 6.5.3   ●

6.5.3 Generic Gâteaux Differentiability

In this section we see how the variational principle in the case of a separable space can be used to obtain some known differentiability results. First, we show the following well-known result:

Theorem 6.5.5 (Generic Gâteaux Differentiability) *Suppose X is a separable Banach space. Then every continuous convex function $f \colon X \to \mathbb{R}$ is Gâteaux differentiable at the points of a generic (that is a dense G_δ) subset of X.*

Proof. First, we will show that f is Gâteaux differentiable at the points of a dense subset of X. After translation, it suffices to show that every nonempty open set S of X with $0 \in S$, contains a point at which f is Gâteaux differentiable. Fix such an S and let φ be the function given by Theorem 6.5.4. We may suppose $\mathrm{dom}\,\varphi \subset S$. Then there is some $x \in \mathrm{dom}\,\varphi \subset S$ at which the function $-f + \varphi$ attains its minimum. In particular, for any $h \in \mathrm{dom}\,\varphi$ and $t > 0$ we have

$$-f(x \pm th) + \varphi(x \pm th) \geq -f(x) + \varphi(x).$$

Using this and the convexity of f we obtain

$$0 \leq f(x + th) + f(x - th) - 2f(x) \leq \varphi(x + th) + \varphi(x - th) - 2\varphi(x)$$

which together with the differentiability property (6.5.4) of φ shows that

$$\lim_{t \to 0+} \frac{f(x + th) + f(x - th) - 2f(x)}{t} = 0,$$

for every $h \in \operatorname{dom} \varphi$. Since f is locally Lipschitz and $\operatorname{dom} \varphi$ is linearly dense, in fact, the latter limit is 0 for any $h \in X$. Finally, the fact that f is convex yields its (linear) Gâteaux differentiability at x.

In order to show that the set of points of Gâteaux differentiability of f is *exactly* a G_δ-subset of X, let us observe that property (6.5.4) yields a stronger conclusion in the argument above: in fact, we obtain that X possesses a dense subset in which every x obeys the following stronger condition

$$\sup\left\{f(x+th) + f(x-th) - 2f(x);\ h \in \operatorname{dom}\varphi\right\} = o(t),\ t \to 0_+. \quad (6.5.7)$$

On the other hand, the set of all $x \in X$ satisfying (6.5.7) is always G_δ (Exercise 6.5.4). Therefore, f is Hadamard-like, as well as Gâteaux differentiable, on a dense G_δ-subset of X.   ●

The theorem above can be considerably extended.

Theorem 6.5.6 *Let Y be a Gâteaux differentiability space and X a separable Banach space. Then $Y \times X$ is a Gâteaux differentiability space.*

Proof. Let $f : Y \times X \to \mathbb{R}$ be a convex continuous function, and $S \subset Y \times X$ be a nonempty open set. Assume, for simplicity, that $2B_Y \times 2B_X \subset S$ and that f is bounded on S. Let $\varphi : X \to [0, +\infty]$ be the function provided by Theorem 6.5.4 with domain in B_X. Define $g : Y \to (-\infty, +\infty]$ by

$$g(y) = \begin{cases} \inf\{-f(y,x) + \varphi(x) \mid x \in X\} & \text{if } y \in 2B_Y, \\ +\infty & \text{if } y \in Y \setminus 2B_Y. \end{cases}$$

Then g is concave and continuous on $2B_Y$. As Y is a Gâteaux differentiability space, the function g is Gâteaux differentiable at some $y \in B_Y$. By Theorem 6.5.4, there is $x \in B_X$ so that $g(y) = -f(y,x) + \varphi(x)$. Thus, for every $k \in Y$ and every $h \in \operatorname{dom}\varphi$ we have for all $t > 0$ sufficiently small,

$$\begin{aligned}
f(y+tk, x+th) &+ f(y-tk, x-th) - 2f(y,x) \\
&\le -g(y+tk) + \varphi(x+th) \\
&\quad - g(y-tk) + \varphi(x-th) + 2g(y) - 2\varphi(x) \\
&= o(t) + o(t).
\end{aligned}$$

Finally, the local Lipschitzian property of f and the linear density of $\operatorname{dom}\varphi$ in X imply

$$f(y+tk, x+th) + f(y-tk, x-th) - 2f(y,x) = o(t), \quad t \to 0+$$

for every $k \in Y$ and every $h \in X$. Therefore f is Gâteaux differentiable at the point (y,x).   ●

6.5.4 Commentary and Exercises

The concept of a one-perturbation variational principle has been used by Tykhonov [252] in the case of convex functions f in a Hilbert space with the perturbation function φ to be the square of the norm. This is often referred to as Tykhonov regularization (so that the perturbation $f + \varphi$ always has a unique minimum). The general results in Theorems 6.5.1 and 6.5.2 are derived in [38]. Theorem 6.5.6 is a recent result from [78]. In the case of a normed space $(X, \|\cdot\|)$, allowing translations of the argument of the fixed perturbation φ, we can establish a localization of the minimum of the perturbation in Theorem 6.5.1 as in the classical variational principles discussed in Chapter 2 (Exercise 6.5.2).

Exercise 6.5.1 Prove that if $(X, \|\cdot\|)$ is a real normed space and φ is in addition convex, then the result in Theorem 6.5.1 remains true for every proper lsc convex f, provided only that either the sublevel sets of φ are only weakly compact or that $\operatorname{dom}\varphi$ is.

Exercise 6.5.2 Let X be a normed space. Suppose that X admits a function φ as in Theorem 6.5.1. Prove that for any proper lsc (bounded from below) function $f \colon X \to \mathbb{R} \cup \{+\infty\}$, for any $\bar{x} \in \operatorname{dom} f$ and each $\lambda > 0$, the function $f(x) + \varphi((\cdot - \bar{x})/\mu)$ (for a suitable $\mu > 0$), attains its minimum at some u with $\|u - \bar{x}\| \leq \lambda$.

Exercise 6.5.3 Prove Theorem 6.5.4.

Exercise 6.5.4 Let X be a separable Banach space. Prove that the set of all $x \in X$ satisfying (6.5.7) is always G_δ (possibly empty). Reference: [114, proof of Proposition 1.4.5].

7

Variational Techniques In the Presence of Symmetry

Symmetry is exploited in many physical and geometrical applications. The focus of this chapter is what happens when we apply variational methods to functions with additional symmetry. The mathematical characterization of symmetry is invariance under certain group actions. Typical examples are the spectral functions associated with a linear transformation, such as the maximum eigenvalue for a matrix. They are in general invariant with respect to the similarity transform. Another example is the distance function on a Riemannian manifold, which is invariant with respect to an isometric transform. It turns out that nonsmooth functions on smooth manifolds provide a convenient mathematical framework for such problems.

7.1 Nonsmooth Functions on Smooth Manifolds

7.1.1 Smooth Manifolds and Submanifolds

We start with a brief review of the smooth manifolds and related notation. In what follows k is either a nonnegative integer or ∞. Let Y be an N-dimensional C^k complex manifold with a C^k atlas $\{(U_a, \psi_a)\}_{a \in A}$. For each a, the N components $(x_a^1, \dots, x_a^N)$ of ψ_a form a *local coordinate system* on (U_a, ψ_a). A function $g \colon Y \to \mathbb{R}$ is C^k at $y \in Y$ if $y \in U_a$ and $g \circ \psi_a^{-1}$ is a C^k function in a neighborhood of $\psi_a(y)$. As usual C^0 represents the collection of continuous functions. It is well known that this definition is independent of the coordinate systems. If g is C^k at all $y \in Y$, we say g is C^k on Y. The collection of all C^k functions on Y is denoted by $C^k(Y)$. A map $v \colon C^1(Y) \to \mathbb{R}$ is called a *tangent vector* of Y at y provided that for any $f, g \in C^1(Y)$,

(i) $v(af + bg) = av(f) + bv(g)$ for all $a, b \in \mathbb{R}$ and
(ii) $v(f \cdot g) = v(f)g(y) + f(y)v(g)$.

The collection of all the tangent vectors of Y at y forms an (N-dimensional) vector space, called the *tangent space* of Y at y and denoted by $T_y(Y)$ or

$T(Y; y)$. The union $\bigcup_{y \in Y}(y, T_y(Y))$ forms a new space called the *tangent bundle* to Y, denoted by $T(Y)$. The dual space of $T_y(Y)$ is the *cotangent space* of Y at y, denoted by $T_y^*(Y)$ or $T^*(Y; y)$. Elements of $T_y^*(Y)$ are called *cotangent vectors*. The *cotangent bundle* to Y then is $T^*(Y) := \bigcup_{y \in Y}(y, T_y^*(Y))$. We will use π (resp., π^*) to denote the *canonical projection* on $T(Y)$ (resp., $T^*(Y)$) defined by $\pi(y, T_y(Y)) = y$ (resp., $\pi^*(y, T_y^*(Y)) = y$). A mapping $v \colon Y \to T(Y)$ is called a *vector field* provided that $\pi(v(y)) = y$. A vector field v is C^k at $y \in Y$ provided $v(g)$ is C^k for any $g \in C^k$. If a vector field v is C^k for all $y \in Y$ we say it is C^k on Y. The collection of all C^k vector fields on Y is denoted by $V^k(Y)$.

In particular, if (U, ψ) is a local coordinate neighborhood with $y \in U$ and $(x^1, \ldots, x^N)$ is the corresponding local coordinate system on U then $(\frac{\partial}{\partial x^n})_y$, $n = 1, \ldots, N$, defined by $(\frac{\partial}{\partial x^n})_y g = \frac{\partial g \circ \psi^{-1}}{\partial x^n}(\psi(y))$, is a *basis* of $T_y(Y)$. Let g be a C^1 function at y, the differential of g at y, $dg(y)$, is an element of $T_y^*(Y)$ defined by

$$dg(y)(v) = v(g) \quad \text{for all } v \in T_y(Y).$$

Let Y and Z be two C^k manifolds. Consider a map $\phi \colon Y \to Z$. Then for every function $g \in C^k(Z)$, ϕ induces a function $\phi^* g$ on Y defined by $\phi^* g = g \circ \phi$. A map $\phi \colon Y \to Z$ is called C^k at $y \in Y$ (on $S \subset Y$) provided that $\phi^* g$ is C^k for any $g \in C^k(Z)$. Let $\phi \colon Y \to Z$ be a C^1 map and let $y \in Y$ be a fixed element. For $v \in T_y(Y)$ and $g \in C^1(Z)$, define $(\phi_* v)(g) = v(\phi^* g)$. Then $\phi_* \colon T_y(Y) \to T_{\phi(y)}(Z)$ is a linear map. The dual map of ϕ_* is denoted by ϕ^*. It is a map from $T_{\phi(y)}^*(Z) \to T_y^*(Y)$ and has the property that for any $g \in C^1(Z)$, $\phi^* dg(\phi(y)) = d(\phi^* g)(y)$.

We often need to study submanifolds of a smooth manifold. Let Y be an N-dimensional C^k manifold. We say that $S \subset Y$ is an M-dimensional submanifold of Y provided that (i) S is a topological subspace of Y and (ii) for any $y \in S$ there exists a coordinate neighborhood (U, ψ) of y in Y with the following properties: (a) $\psi(y) = 0 \in \mathbb{R}^N$; (b) $\psi(U) = C_r^N := \{x \in \mathbb{R}^N \mid -r < x_n < r, n = 1, \ldots, N\}$; and (c) $\psi(U \cap S) = \{x \in C_r^N \mid x^{M+1} = \cdots = x^N = 0\}$. A coordinate neighborhood with the above property is called a *preferred neighborhood*.

Next we state two important theorems for identifying submanifolds whose proofs can be found, for example, in [27]. The first identifies submanifolds as fibers of smooth mappings and can be viewed as a generalization of the implicit function theorem on a smooth manifold.

Theorem 7.1.1 (Fiber Theorem) *Let Y and Z be C^1 manifolds and let $f \colon Y \to Z$ be a C^1 mapping. Suppose that Y is connected with dimension N and rank $(f_*)_y = M$ is a constant for all $y \in Y$. Then, for any $\bar{y} \in Y$, $S := f^{-1}(f(\bar{y})) \subset Y$ is a C^1 manifold with*

$$T_y(S) = \operatorname{Ker}\,(f_*)_y, \quad \text{for all } y \in S.$$

In particular, the dimension of S is $N - M$.

Our next result can be viewed as a generalization of the parametric representations of surfaces in $\mathbb{R}^3$.

Theorem 7.1.2 (Imbedding Theorem) *Let Y and Z be C^1 manifolds and let $f \colon Y \to Z$ be a C^1 mapping. Suppose that Y is compact and has dimension M, f is one-to-one and rank $(f_*)_y = M$ is a constant for all $y \in Y$. Then $S := f(Y) \subset Z$ is a C^1 manifold with*

$$T_{f(y)}(S) = f_* T_y(Y).$$

We end this subsection by recalling a convenient framework for studying many problems involving symmetry. A *Lie group* G is a group which is also a C^k smooth manifold such that

$$G \times G \ni (g, h) \to gh^{-1}$$

is a C^k map. A *Lie group action* of a Lie group G on a C^k manifold Y is a C^k map $G \times Y \to Y$ defined by $(g, y) \to g \cdot y$ satisfying

$$g \cdot (h \cdot y) = (gh) \cdot y, \quad e \cdot y = y,$$

for all $g, h \in G$ and $y \in Y$, where e is the unit element of G. The *orbit* of $y \in Y$ under a Lie group action of G is defined by

$$G \cdot y := \{g \cdot y \mid g \in G\},$$

and the *stabilizer* of $y \in Y$ with respect to G is defined as

$$\mathrm{Stab}(G; y) := \{g \in G \mid g \cdot y = y\}.$$

$\mathrm{Stab}(G; y)$ is a closed subgroup of G and is also a Lie group. The group action induces a diffeomorphism between $G/\mathrm{Stab}(G; y)$ and $G \cdot y$. It follows from the compact imbedding theorem of Theorem 7.1.2 that when G is compact the orbits of G are smooth, compact submanifolds of Y.

7.1.2 Subdifferentials on Smooth Manifolds

We now consider nonsmooth functions on a smooth manifold and their subdifferentials. We will consider both the Fréchet subdifferential and the limiting subdifferentials for which the limits of cotangent sequences are needed. Let $v_i^* \in T_{y_i}^*(Y)$, $i = 1, 2, \ldots$ be a sequence of cotangent vectors of Y and let $v^* \in T_y^*(Y)$. We say (v_i^*) converges to v^*, denoted by $\lim v_i^* = v^*$, provided that $y_i \to y$ and, for any $v \in V^1(Y)$, $\langle v_i^*, v(y_i) \rangle \to \langle v^*, v(y) \rangle$. Let (U, ψ) be a local coordinate neighborhood with $y \in U$. Since $y_i \to y$, we may assume without loss of generality that $y_i \in U$ for all i. Then $\lim v_i^* = v^*$ if and only if $\langle v_i^*, (\frac{\partial}{\partial x^n})_{y_i} \rangle \to \langle v^*, (\frac{\partial}{\partial x^n})_y \rangle$ for $n = 1, \ldots, N$. Another equivalent description is $(\psi^{-1})^*_{\psi(y_i)} v_i^* \to (\psi^{-1})^*_{\psi(y)} v^*$ (in the dual of $\mathbb{R}^N$).

Definition 7.1.3 (Subdifferentials) *Let Y be a C^1-manifold and let $f : Y \to$ $\mathbb{R} \cup \{+\infty\}$ be an extended-valued lsc function. We define the* Fréchet subdifferential *of f at $y \in \mathrm{dom}(f)$ by*

$$\partial_F f(y) := \{dg(y) \mid g \in C^1(Y) \text{ and } f - g \text{ attains a local minimum at } y\}.$$

We define the limiting subdifferential *and* singular subdifferential *of f at $y \in Y$ by*

$$\partial_L f(y) := \{\lim v_i^* \mid v_i^* \in \partial_F f(y_i), (y_i, f(y_i)) \to (y, f(y))\}$$

and

$$\partial^\infty f(y) := \{\lim t_i v_i^* \mid v_i^* \in \partial_F f(y_i), (y_i, f(y_i)) \to (y, f(y)) \text{ and } t_i \to 0^+\},$$

respectively.

Note that the Fréchet subdifferential of f depends only on the local behavior of the function f. Moreover, every locally C^1 function can be extended to a C^1 function on Y. Therefore, the support function g in the definition of the Fréchet subdifferential need only be C^1 in a neighborhood of y.

The elements of a Fréchet subdifferential are called Fréchet subgradients. As before, we define the Fréchet *superdifferential* by $\partial^F f(y) = -\partial_F(-f)(y)$ and its elements are called *supergradients*. An alternative definition of a supergradient $v^* \in \partial^F f(y)$ is: $v^* = dg(y)$ for some C^1 function g such that $f - g$ attains a local maximum at y on Y.

It follows directly from the definition that $\partial_F f(y) \subset \partial_L f(y)$ and $0 \in \partial^\infty f(y)$. Note that $\partial_F f(y)$ may be empty. However, if f attains a local minimum at y then $0 \in \partial_F f(y) \subset \partial_L f(y)$. These are the usual properties to be expected for a subdifferential.

As before, the geometric concept of the normal cones to a closed set can be naturally established by using the subdifferential for the corresponding indicator function.

Definition 7.1.4 (Normal Cones) *Let Y be a C^1-manifold and let S be a closed subset of Y with $s \in S$. We define the* Fréchet normal cone *of S at s by*

$$N_F(S; s) := \partial_F \iota_S(s).$$

We define the (limiting) normal cone *of S at s by*

$$N_L(S; s) := \partial_L \iota_S(s).$$

It is easy to see that a normal cone always contains 0 and at an interior point of a set it contains only 0 (Exercise 7.1.2). We will be interested mainly in nonzero normal cones. These are necessarily normal cones for the boundary points of the set by Exercise 7.1.2. The following necessary optimality condition for a constrained minimization problem is useful; its easy proof is left as an exercise.

Proposition 7.1.5 *Let Y be a C^1-manifold, let S be a closed subset of Y and let $f: Y \to \mathbb{R}$ be a C^1 function. Suppose that $\bar{y}$ is a solution of the constrained minimization problem*

$$\text{minimize} \quad f(y)$$

$$\text{subject to} \quad y \in S.$$

Then

$$0 \in df(\bar{y}) + N_F(S, \bar{y}).$$

Proof. Exercise 7.1.3   ●

Our next result relates the normal cone of a smooth submanifold to its tangent cone.

Proposition 7.1.6 (Normal Cone of a Submanifold) *Let Y be a C^1-manifold and let S be a C^1-submanifold of Y. Then, for any $s \in S$,*

$$N_F(S; s) = T_s(S)^{\perp} := \{v^* \in T_s^*(Y) : \langle v^*, v \rangle = 0 \ \text{for all } v \in T_s(S)\}.$$

In particular, $N_F(S; s)$ is a subspace of $T_s^*(Y)$.

Proof. Assume that the dimension of S is M. Then there exists a coordinate neighborhood (U, ψ) around s with a corresponding local coordinate system $(x^1, \ldots, x^N)$ such that $s = \psi^{-1}(0)$ and $S \cap U = \{\psi^{-1}(x^1, \ldots, x^M, 0, \ldots, 0) \mid (x^1, \ldots, x^M, 0, \ldots, 0) \in \psi(U)\}$. Then $T_s(S) = \text{span}((\frac{\partial}{\partial x^1})_s, \ldots, (\frac{\partial}{\partial x^M})_s)$ and $T_s(S)^{\perp} = \text{span}(dx^{M+1}(s), \ldots, dx^N(s))$. Let $v^* \in N_F(S; s)$. Then there exists $g \in C^1(Y)$ such that $g(s) = 0$, $dg(s) = v^*$ and for any $y \in S$, $g(y) \le 0$. Let $v \in T_s(S)$ and $\gamma: (-r, r) \to S$ be a C^1 curve such that $\gamma(0) = s$ and $\gamma'(0) = v$. Then, in particular, $g \circ \gamma(t) \le 0, t \in (-r, r)$ and $g \circ \gamma(0) = 0$. Thus, $d(g \circ \gamma)(0) = \langle v^*, v \rangle = 0$.x That is to say $N_F(S; s) \subset T_s(S)^{\perp}$.

On the other hand, let $v^* \in T_s(S)^{\perp} = \text{span}(dx^{M+1}(s), \ldots, dx^N(s))$. Then $v^* = \sum_{n=M+1}^N \alpha_n dx^n(s)$ where $\alpha_n \in \mathbb{R}, n = M + 1, \ldots, N$. Define $g: U \to R$ by

$$g(y) = \sum_{n=M+1}^N \alpha_n x^n(y).$$

Then, for any $y \in S \cap U$, $g(y) = 0$. Thus, $v^* = \sum_{n=M+1}^N \alpha_n dx^n(s) = dg(s) \in N_F(S; s)$, which establishes $T_s(S)^{\perp} \subset N_F(S; s)$.   ●

As an application we discuss a special case of the Courant–Fisher minimax theorem that characterizes the largest and the smallest eigenvalues of a symmetric matrix.

Proposition 7.1.7 *Let A be an $N \times N$ real symmetric matrix and let λ_1 and λ_N denote its largest and smallest eigenvalues, respectively. Then*

$$\lambda_1 = \max_{y \in S} \langle y, Ay \rangle \quad \text{and} \quad \lambda_N = \min_{y \in S} \langle y, Ay \rangle \qquad (7.1.1)$$

where $S = \{y \in \mathbb{R}^N \mid \|y\|^2 = 1\}$. More generally, the critical values and critical points of $\langle y, Ay \rangle$ on S exactly correspond to the eigenvalues and eigenvectors of A.

Proof. Note that S is a compact smooth manifold and by the fiber theorem, Theorem 7.1.1, for any $y \in S$, $T_y(S) = \{z \in \mathbb{R}^N \mid \langle y, z \rangle = 0\}$. Thus, by Proposition 7.1.6 $N_F(S; y) = \{\lambda y \mid \lambda \in \mathbb{R}\}$. Now consider the problem of minimizing $f(y) = \langle y, Ay \rangle$ subject to $y \in S$. Suppose that $\bar{y}$ is a solution. Then, by Proposition 7.1.5

$$0 \in df(\bar{y}) + N_F(S, \bar{y}),$$

that is, there exists $\lambda \in \mathbb{R}$ such that

$$A\bar{y} = \lambda\bar{y}.$$

Taking inner product with $\bar{y}$ on both sides of this equality we have

$$\lambda = \langle \bar{y}, A\bar{y} \rangle.$$

Note that the same conclusion also applies to $f(y) = -\langle y, Ay \rangle$. Thus, the critical values and critical points of $\langle y, Ay \rangle$ on S exactly correspond to the eigenvalues and eigenvectors of A. •

7.1.3 Smooth Chain Rules

When applying the variational technique to problems on a smooth manifold, it is often convenient to use the local coordinate system to project the problem into a Euclidean space. The following chain rule is an important tool for doing so.

Theorem 7.1.8 (Smooth Chain Rule) *Let Y and Z be smooth manifolds, let $g \colon Y \to Z$ be a C^1-mapping and let $f \colon Z \to \mathbb{R} \cup \{+\infty\}$ be a lsc function. Suppose that $z = g(y)$. Then*

$$g^* \partial_F f(z) \subset \partial_F (f \circ g)(y), \qquad (7.1.2)$$
$$g^* \partial_L f(z) \subset \partial_L (f \circ g)(y), \qquad (7.1.3)$$

and

$$g^* \partial^\infty f(z) \subset \partial^\infty (f \circ g)(y). \qquad (7.1.4)$$

Moreover, if g is a C^1-diffeomorphism then both sides of (7.1.2), (7.1.3) and (7.1.4) are equal.

Proof. Since (7.1.3) and (7.1.4) follow directly from (7.1.2) by taking limits, we prove (7.1.2). Let $z^* \in \partial_F f(z)$. Then there exists a C^1 function h such that $dh(z) = y^*$ and $f - h$ attains a local minimum at z. It follows that $f \circ g - h \circ g$ attains a local minimum at y. Observing that $h \circ g$ is a C^1 function on Y, we have

$$\partial_F(f \circ g)(y) \ni d(h \circ g)(y) = g^* dh(z) = g^* z^*.$$

Thus,

$$g^* \partial_F f(z) \subset \partial_F(f \circ g)(y).$$

When g is a diffeomorphism applying (7.1.2) to $f \circ g$ and g^{-1} yields the opposite inclusion. ●

Applying Theorem 7.1.8 to $g = \psi^{-1}$ for a local coordinate mapping ψ yields the following corollary.

Corollary 7.1.9 *Let Y be a C^1 manifold and let $f \colon Y \to \dot{\mathbb{R}} \cup \{+\infty\}$ be a lsc function. Suppose that (U, ψ) is a local coordinate neighborhood and $y \in U$. Then*

$$\partial_F f(y) = \psi^* \partial_F(f \circ \psi^{-1})(\psi(y)),$$
$$\partial_L f(y) = \psi^* \partial_L(f \circ \psi^{-1})(\psi(y)),$$

and

$$\partial^\infty f(y) = \psi^* \partial^\infty(f \circ \psi^{-1})(\psi(y)).$$

Proof. Exercise 7.1.5 ●

We illustrate the usage of Corollary 7.1.9 by proving the density of the domain of the Fréchet subdifferential.

Theorem 7.1.10 *Let Y be a C^1-manifold, let $f \colon Y \to \mathbb{R} \cup \{+\infty\}$ be a lsc function and S be a closed subset of Y. Then $\mathrm{dom}(f) \subset \overline{\mathrm{dom}(\partial_F f)} \subset \overline{\mathrm{dom}(\partial_L f)}$ and $\mathrm{bd}(S) \subset \overline{\{s : N_F(S; s) \neq \{0\}\}} \subset \{s : N(S; s) \neq \{0\}\}$.*

Proof. We need only to show that $\mathrm{dom}(f) \subset \overline{\mathrm{dom}(\partial_F f)}$. The rest follows easily. Let $y \in \mathrm{dom}(f)$ and W be an arbitrary neighborhood of y. Without loss of generality we may assume that there is a local coordinate neighborhood (U, ψ) such that $W \subset U$. Then $f \circ \psi^{-1} \colon \psi(U) \to \mathbb{R} \cup \{+\infty\}$ is a lower semicontinuous function and $\psi(y) \in \mathrm{dom}(f \circ \psi^{-1})$. Since $\psi(W)$ is a neighborhood of $\psi(y)$ by the density theorem of the Fréchet subdifferential in $\mathbb{R}^N$, there exists $x \in \psi(W)$ such that $\partial_F(f \circ \psi^{-1})(x) \neq \emptyset$. Then $z = \psi^{-1}(x) \in W$ and by Lemma 7.1.9

$$\partial_F f(z) = \psi_x^* \partial_F(f \circ \psi^{-1})(x) \neq \emptyset.$$

●

7.1.4 Commentary and Exercises

We refer the readers to Boothby [27] and Matsushima [192] for preliminaries on smooth manifolds. Materials in subsections 7.1.2 and 7.1.3 regarding nonsmooth analysis on smooth manifolds were developed by Ledyaev and Zhu [175, 176, 177] where one can also find more comprehensive discussion on the calculus of subdifferentials for nonsmooth functions on smooth manifolds and their applications in optimization and invariance theory on smooth manifolds. A slightly different approach to generalizing the derivative concept to nonsmooth functions on smooth manifolds was proposed by Chryssochoos and Vinter [79], motivated by applications to an interesting optimal control problem on a smooth manifold.

Exercise 7.1.1 Verify that $N_F(S; s)$ is a cone and $N_L(S; s) := \partial_L \iota_S(s) = \partial^\infty \iota_S(s)$.

Exercise 7.1.2 Show that $0 \in N_F(S; s) \subset N_L(S; s)$ for any $s \in S$ and $N_F(S; s) = N_L(S; s) = \{0\}$ for any $s \in \text{int } S$.

Exercise 7.1.3 Prove Proposition 7.1.5.

Exercise 7.1.4 (Rayleigh Quotient) Let A be a real symmetric matrix and define the *Rayleigh quotient* of A by

$$r_A(y) = \frac{\langle y, Ay \rangle}{\|y\|^2}, \quad y \neq 0.$$

Show that Proposition 7.1.7 can be rephrased as: the critical values and critical points of r_A on $\mathbb{R}^N \setminus \{0\}$ exactly correspond to the eigenvalues and eigenvectors of A.

Exercise 7.1.5 Prove Corollary 7.1.9.

Exercise 7.1.6 (Lagrange Multiplier Rule) Consider

$$\text{minimize} \quad f(x) \tag{7.1.5}$$
$$\text{subject to} \quad g(x) = 0,$$

where $f \colon \mathbb{R}^N \to \mathbb{R}$ and $g \colon \mathbb{R}^N \to \mathbb{R}^M$ are C^1 mappings. Suppose that $\bar{x}$ is a solution to the constrained minimization problem (7.1.5) and that $g'(\bar{x}) \colon \mathbb{R}^N \to \mathbb{R}^M$ is a surjective. Show that there exists $\lambda \in \mathbb{R}^M$ such that

$$f'(x) + \langle \lambda, g'(\bar{x}) \rangle = 0.$$

Hint: Note that problem (7.1.5) is equivalent to minimizing f over $S = g^{-1}(0) = g^{-1}(g(\bar{x}))$. Then use Propositions 7.1.5, 7.1.6 and Theorem 7.1.1.

7.2 Manifolds Of Matrices and Spectral Functions

7.2.1 Manifolds of Matrices

Let $E_{N \times M}$ be the Euclidean space of real $N \times M$ matrices. Square matrices are particularly interesting in many applications. We use $E(N) = E_{N \times N}$ to denote the space of N by N square matrices. For $A \in E_{N \times M}$ we denote its transpose by $A^\top$, an element of $E_{M \times N}$. When $A \in E(N)$ we say that A is symmetric if $A = A^\top$ and we say that A is skew symmetric if $A = -A^\top$. We denote the linear subspace of $N \times N$ symmetric matrices by $S(N)$ and the linear subspace of $N \times N$ skew symmetric matrices by $A(N)$. We define an inner product on $E_{N \times M}$ by

$$\langle A, B \rangle := \operatorname{tr}(A^\top B) = \sum_{n=1}^{N} \sum_{m=1}^{M} a_{nm} b_{nm},$$

and the norm of A is defined by $\|A\| := \sqrt{\langle A, A \rangle}$. We will often use the Lie bracket $[A, B] = AB - BA$ for $A, B \in E(N)$. The next proposition summarizes some easy yet useful facts of Lie brackets.

Proposition 7.2.1 *If $A, B \in S(N)$ then $[A, B] \in A(N)$ and if $A \in S(N)$ and $B \in A(N)$ then $[A, B] \in S(N)$.*

Proof. Exercise 7.2.1. ●

Example 7.2.2 (Manifold of Invertible Matrices) The set of invertible matrices $GL(N) := \{A \in E(N) \mid \det(A) \neq 0\}$ is a Lie group under matrix multiplication. Moreover, for any $A \in GL(N)$

$$T_A(GL(N)) = E(N).$$

More generally we have

Example 7.2.3 (Manifold of Full Rank Matrices) Let $N \geq M$. The set of full rank matrices $GL(N, M) := \{A \in E_{N \times M} \mid \operatorname{rank} A = M\}$ is a submanifold of $E_{N \times M}$. Moreover, for any $A \in GL(N, M)$

$$T_A(GL(N, M)) = E_{N \times M}.$$

Noticing that $GL(N)$ and $GL(N, M)$ are open subsets of $E(N)$ and $E_{N \times M}$, respectively, these facts are easy to check and we leave them as exercises.

Example 7.2.4 (Manifold of Orthogonal Matrices) The set of orthogonal matrices $O(N) := \{A \in GL(N) \mid A^\top A = I\}$, where I is the identity matrix, is a compact Lie group under matrix multiplication. Moreover, for any $A \in O(N)$,

$$T_A(O(N)) = \{B \in E(N) \mid A^\top B + B^\top A = 0\}, \qquad (7.2.1)$$

and

$$N_F(O(N); A) = T_A(O(N))^\perp = \{AX \in E(N) \mid X \in S(N)\}, \qquad (7.2.2)$$

To verify these results, let us consider the smooth function $f: GL(N) \to S(N)$ defined by $f(A) = A^\top A - I$. We can directly calculate that for any $A \in GL(N)$ and $B \in T_A(GL(N)) = E(N)$,

$$(f_*)_A(B) = \lim_{t \to 0} \frac{(A+tB)^\top(A+tB) - A^\top A}{t} = A^\top B + B^\top A.$$

Moreover, $(f_*)_A$ is a surjection onto $S(N)$ for any $A \in GL(N)$. In fact, as a linear subspace the tangent space of $S(N)$ at any point is $S(N)$ itself. Thus, range $(f_*)_A$ is a subspace of $S(N)$. Suppose that $C \in (\text{range}(f_*)_A)^\perp$, i.e., for any $B \in E(N)$,

$$\langle C, A^\top B + B^\top A \rangle = 0. \qquad (7.2.3)$$

By Exercise 7.2.4 we have

$$\begin{aligned}
\langle C, A^\top B + B^\top A \rangle &= \text{tr}(C^\top A^\top B + C^\top B^\top A) \\
&= \text{tr}(C^\top A^\top B) + \text{tr}(C^\top B^\top A) \\
&= \text{tr}((AC)^\top B) + \text{tr}(B^\top AC) \\
&= 2\,\text{tr}((AC)^\top B) = \langle AC, B \rangle.
\end{aligned}$$

Therefore, $AC = 0$. Since A is invertible we must have $C = 0$, which implies that $(\text{range}(f_*)_A)^\perp = \{0\}$ or $(f_*)_A$ is a surjection. Thus, by the fiber theorem (Theorem 7.1.1) we have $O(N) := f^{-1}(0)$ is a smooth submanifold of $GL(N)$ and therefore of $E(N)$. Observe that for any $A \in O(N)$, $\|A\| = 1$ so that $O(N)$ is compact. Moreover,

$$T_A(O(N)) = \ker(f_*)_A = \{B \in E(N) \mid A^\top B + B^\top A = 0\}.$$

In particular,

$$T_I(O(N)) = \{B \in E(N) \mid B + B^\top = 0\} = A(N). \qquad (7.2.4)$$

We now turn to the normal cone characterization (7.2.2). It follows from Proposition 7.1.6 that

$$N_F(O(N); A) = T_A(O(N))^\perp.$$

Let $X \in S(N)$. Using Exercise 7.2.4 we have, for any $B \in T_A(O(N))$,

$$\langle AX, B \rangle = \frac{1}{2}[\text{tr}((AX)^\top B) + \text{tr}(B^\top(AX))]$$
$$= \frac{1}{2}[\text{tr}(XA^\top B) + \text{tr}(XB^\top A)]$$
$$= \frac{1}{2}\langle X, A^\top B + B^\top A \rangle = 0.$$

Thus, $\{AX \in E(N) \mid X \in S(N)\} \subset T_A(O(N))^\perp$. On the other hand the dimensions of both subspaces are the same as that of $S(N)$, and therefore they must be the same.

The method used in discussing Example 7.2.4 can also be used to discuss the more general Stiefel manifolds. We state the results below and leave the proofs as guided exercises.

Example 7.2.5 (Stiefel Manifold) Define

$$\text{St}(N, M) := \{A \in GL(N, M) \mid A^\top A = I\}.$$

Then $\text{St}(N, M)$ is a C^∞ submanifold of $E_{N \times M}$, called the *Stiefel manifold*. Moreover, for any $A \in \text{St}(N, M)$,

$$T_A(\text{St}(N, M)) = \{B \in E_{N \times M} \mid A^\top B + B^\top A = 0\}, \qquad (7.2.5)$$

and

$$N_F(\text{St}(N, M); A) = T_A(\text{St}(N, M))^\perp$$
$$= \{AX \in E_{N \times M} \mid X \in S(M)\}. \qquad (7.2.6)$$

Having seen how to use the fiber theorem to identify submanifolds we turn to an example of using the parameterization theorem to identify a useful submanifold.

Example 7.2.6 Define a Lie group action of $O(N)$ on $S(N)$ by

$$U \cdot A = U^\top AU, U \in O(N), A \in S(N). \qquad (7.2.7)$$

Since $O(N)$ is a compact Lie group, for any $A \in S(N)$, the orbit $O(N) \cdot A$ is a compact smooth submanifold of $S(N)$. We show that

$$T_A(O(N) \cdot A) = \{[A, B] \mid B \in A(N)\} \qquad (7.2.8)$$

and

$$N_F(O(N) \cdot A; A) = \{B \in S(N) \mid [A, B] = 0\} \qquad (7.2.9)$$

To this end let us denote by $\phi\colon O(N)/\text{Stab}(O(N); A) \to O(N) \cdot A$ the diffeomorphism induced by the Lie group action $O(N) \times S(N) \ni (U, A) \to U^\top AU$. Then ϕ_* is an isomorphism between the corresponding tangent spaces

$$T_I(O(N)/\text{Stab}(O(N);A)) \text{ and } T_A(O(N) \cdot A).$$

On the other hand let $\pi \colon O(N) \to O(N)/\text{Stab}(O(N);A)$ be the quotient map. Then

$$\pi_* \colon T_I(O(N)) \to T_I(O(N)/\text{Stab}(O(N);A))$$

is onto. Now consider $\psi \colon O(N) \to O(N) \cdot A$ defined by $\psi(U) \to U^\top A U$. We have $\psi = \phi \circ \pi$. Thus, the chain rule gives

$$\psi_* T_I(O(N)) = T_A(O(N) \cdot A).$$

Direct calculation shows that for any $B = -B^\top$, $\psi_* B = AB - BA$. Combining this with (7.2.4) we arrive at (7.2.8). Finally, we can check that for $C \in A(N)$ and $A, B \in S(N)$ we have the identity

$$\langle [A, C], B \rangle = \langle C, [A, B] \rangle. \tag{7.2.10}$$

Now if a matrix $B \in S(N)$ commutes with A then any matrix C in $T_I(O(N)) = A(N)$ satisfies

$$\langle [A, C], B \rangle = 0.$$

Conversely, suppose that for all $C \in T_I(O(N)) = A(N)$

$$\langle [A, C], B \rangle = 0.$$

Choosing $C = [A, B]$ and using (7.2.10) we have $\|[A, B]\|^2 = 0$, that is $[A, B] = 0$.

7.2.2 Spectral and Eigenvalue Functions

For any $A \in S(N)$ we use $\lambda_1(A), \ldots, \lambda_N(A)$ to denote the N (including repeated) real eigenvalues of A in nonincreasing order. We call

$$\lambda(A) := (\lambda_1(A), \ldots, \lambda_N(A))$$

the *eigenvalue mapping*. Let $\mathbb{R}_\geq^N := \{(x_1, x_2, \ldots, x_N) \in \mathbb{R}^N \mid x_1 \geq x_2 \geq \cdots \geq x_N\}$. Then $\lambda \colon S(N) \to \mathbb{R}_\geq^N$. Let $P(N)$ be the set of N by N permutation matrices. We say that a function $f \colon \mathbb{R}^N \to \mathbb{R}^N$ is *permutation invariant* provided that $f(Py) = f(y)$ for any $P \in P(N)$ and any $y \in \mathbb{R}^N$. A *spectral function* is a function of the form $\phi := f \circ \lambda \colon S(N) \to \mathbb{R} \cup \{+\infty\}$ where f is permutation invariant. The concept of a spectral function encompasses many useful functions related to the eigenvalue mapping such as the *maximum eigenvalue* $\lambda_1(A) = \max\{\lambda_n(A), n = 1, \ldots, N\}$, the *spectral radius* $\sigma(A) = \max\{|\lambda_n(A)|, n = 1, \ldots, N\}$, the *determinant* $\det(A) = \Pi_{n=1}^N \lambda_n(A)$ and the *trace* $\text{tr}(A) = \sum_{n=1}^N \lambda_n(A)$. A spectral function inherits the symmetric property of a permutation invariant function f in the form of being *unitary invariant*. Recall that a *unitary mapping* $u \colon E(N) \to E(N)$ associated with an orthogonal matrix $U \in O(N)$ is defined by $u(A) = U \cdot A = U^\top A U$.

Lemma 7.2.7 (Unitary Invariance of Spectral Functions) *Let $\phi: S(N) \to \mathbb{R} \cup \{+\infty\}$ be a spectral function and let u be a unitary mapping associated with $U \in O(N)$. Then, for any $A \in S(N)$,*

$$(\phi \circ u)(A) = \phi(A).$$

Proof. Exercise 7.2.7.   ●

An important theme of this section is to relate the properties of a spectral function on $S(N)$ to that of the corresponding permutation function on $\mathbb{R}^N$. For this purpose the mapping diag: $\mathbb{R}^N \to S(N)$ that maps a vector in $x \in \mathbb{R}^N$ to a diagonal matrix diag $x \in S(N)$ with the components of x as the diagonal elements is useful. We can check that the dual mapping diag*: $S(N) \to \mathbb{R}^N$ is given by (Exercise 7.2.8)

$$\text{diag}^*(a_{nm}) = (a_{11}, \ldots, a_{NN}). \tag{7.2.11}$$

We now prove a useful formula for the sum of the k largest (or smallest) components of the eigenvalue mapping which generalizes the Courant–Fisher minimax theorem (Proposition 7.1.7). The proof of this result is a nice illustration of the variational argument in combination with the properties of manifolds of matrices.

Theorem 7.2.8 (Ky Fan Minimax Theorem) *Let $A \in S(N)$ and let $k \in \{1, \ldots, N\}$. Then*

$$\lambda_1(A) + \cdots + \lambda_k(A) = \max_{X \in St(N,k)} \text{tr}(X^\top AX) \tag{7.2.12}$$

and

$$\lambda_{N-k+1}(A) + \cdots + \lambda_N(A) = \min_{X \in St(N,k)} \text{tr}(X^\top AX). \tag{7.2.13}$$

Proof. We prove (7.2.13) and the proof for (7.2.12) is similar. Define $f_A: GL(N,k) \to \mathbb{R}$ by

$$f_A(B) = \frac{1}{2} \text{tr}(B^\top AB).$$

Then f is a smooth function with

$$df_A(B)(C) = \langle AB, C \rangle. \tag{7.2.14}$$

Consider the optimization problem of minimizing f on the smooth manifold $St(N,k) \subset GL(N,k)$. Since the Stiefel manifold $St(N,k)$ is a compact submanifold a minimizer must exist. Let B be such a minimizer. By Proposition 7.1.5, B must satisfy

$$0 \in df_A(B) + N_F(St(N,k); B).$$

Combining this necessary condition with (7.2.14) and the representation of the normal space of a Stiefel manifold in Example 7.2.5, we have that for some $X \in S(k)$,

$$AB = BX. \tag{7.2.15}$$

It is not hard to verify that any eigenvalue of X must also be an eigenvalue of A (Exercise 7.2.12). Thus,

$$2f_A(B) = \operatorname{tr}(B^\top AB) = \operatorname{tr}(B^\top BX) = \operatorname{tr}(X) = \lambda_{N-k+1}(A) + \cdots + \lambda_N(A).$$

The last equality is due to the fact that $f_A(B)$ is the minimum of f_A over $\operatorname{St}(N, k)$. $\qquad\bullet$

Two useful corollaries of the Ky Fan minimax theorem are given below. Their easy proofs are left as exercises.

Corollary 7.2.9 *Let $A \in S(N)$ and let $k \in \{2, \ldots, N\}$. Then*

$$\lambda_k(A) = \max_{B \in S(N,k)} f_A(B) - \max_{B \in S(N,k-1)} f_A(B).$$

Proof. Exercise 7.2.13. $\qquad\bullet$

Corollary 7.2.10 *Let $A, B \in S(N)$ and let $k \in \{1, \ldots, N\}$. Then*

$$\sum_{n=1}^{k} \lambda_k(A+B) \leq \sum_{n=1}^{k} \lambda_k(A) + \sum_{n=1}^{k} \lambda_k(B).$$

Proof. Exercise 7.2.14. $\qquad\bullet$

It follows from the representation in Corollary 7.2.9 that for any $k \in \{1, \ldots, N\}$, $\lambda_k \colon S(N) \to \mathbb{R}$ is a DC function defined on $S(N)$, and therefore is locally Lipschitz and directional differentiable (Exercise 7.2.15). It is well known that for such a function on a finite dimensional Banach space the directional derivatives provide a uniform approximation for the difference quotient. More precisely we have the following lemma.

Lemma 7.2.11 *Let X be a finite dimensional Banach space and let $f \colon X \to \mathbb{R}$ be a locally Lipschitz function. Suppose that for all $h \in X$, $f'(\bar{x}; h)$ exists. Then for any $\varepsilon > 0$, there exists $\delta > 0$ such that $\|h\| < \delta$ implies*

$$|f(\bar{x} + h) - f(\bar{x}) - f'(\bar{x}; h)| \leq \varepsilon \|h\|.$$

Proof. Suppose for contradiction that there is an $\varepsilon > 0$ and a sequence (h_i) with $\|h_i\| = t_i < 1/i$ such that, for $i = 1, 2, \ldots$,

$$|f(\bar{x} + h_i) - f(\bar{x}) - f'(\bar{x}; h_i)| \geq \varepsilon \|h_i\|.$$

Extracting a subsequence if necessary we may assume that $h_i/t_i \to d$ for some unit vector d. Let L be a local Lipschitz constant for f we have

$$\varepsilon t_i \leq |f(\bar{x} + h_i) - f(\bar{x}) - f'(\bar{x}; h_i)|$$
$$\leq |f(\bar{x} + h_i) - f(\bar{x} + t_i d)|$$
$$+ |f(\bar{x} + t_i d) - f(\bar{x}) - f'(\bar{x}; t_i d)| + |f'(\bar{x}; t_i d) - f'(\bar{x}; h_i)|.$$

It is not hard to verify that $h \to f'(\bar{x}; h)$ is Lipschitz with the same Lipschitz constant L (Exercise 7.2.17). Thus, we have

$$\varepsilon t_i \leq 2L \|h_i - t_i d\| + |f(\bar{x} + t_i d) - f(\bar{x}) - f'(\bar{x}; t_i d)|.$$

Dividing by $t_i > 0$ and taking limits when $i \to \infty$ we obtain the contradiction $\varepsilon < 0$.   ●

In particular, we have the following estimate for eigenvalue functions in terms of its directional derivatives.

Theorem 7.2.12 (Eigenvalue Derivative) *Let* $A \in S(N)$. *Then there exists* $h > 0$ *such that, for* $B \in S(N)$ *with* $\|B\| < h$,

$$\lambda(A + B) = \lambda(A) + \lambda'(A; B) + o(B).$$

Proof. Apply Lemma 7.2.11 to the components of $\lambda \colon S(N) \to \mathbb{R}^N$.   ●

7.2.3 The von Neumann–Theobald Inequality

The variational method can also be used to prove the von Neumann–Theobald inequality which we will need below. To do so we need the following combinatorial lemma.

Lemma 7.2.13 *Let* $x, y \in \mathbb{R}^N_{\geq}$ *and let* $P \in P(N)$. *Then*

(i) $\langle x, Py \rangle \leq \langle x, y \rangle$, *and*
(ii) $\langle x, Py \rangle = \langle x, y \rangle$ *if and only if there exists* $Q \in P(N)$ *such that* $Qx = x$ *and* $QPy = y$.

Proof. We prove (i) and leave (ii) as a guided exercise (Exercise 7.2.19). Denote $P_0 = P$ and apply the following algorithms: for $n = 1$ to $N - 1$ if $(P_{n-1}y)_n < (P_{n-1}y)_{\bar{n}} := \max\{(P_{n-1}y)_{n+1}, \ldots, (P_{n-1}y)_N\}$ then switch $(P_{n-1}y)_n$ and $(P_{n-1}y)_{\bar{n}}$ in $P_{n-1}y$. The resulting vector is clearly a permutation of y. We denote its permutation matrix with respect to y by P_n. We can check that $\langle x, P_n y \rangle$ is increasing with n (Exercise 7.2.18). Thus,

$$\langle x, Py \rangle \leq \langle x, P_N y \rangle.$$

Notice that $P_N y \in \mathbb{R}^N_{\geq}$, and therefore $P_N y = y$.   ●

Now we can prove the von Neumann–Theobald inequality. Recall that two matrices A and B in $S(N)$ have a *simultaneous ordered spectral decomposition* if there exists a orthogonal matrix $U \in O(N)$ such that $A = U^\top \operatorname{diag} \lambda(A) U$ and $B = U^\top \operatorname{diag} \lambda(B) U$.

Theorem 7.2.14 (von Neumann–Theobald Inequality) *Let* $A, B \in S(N)$. *Then*

$$\langle A, B \rangle \leq \langle \lambda(A), \lambda(B) \rangle.$$

Moreover, equality holds if and only if A and B have a simultaneous ordered spectral decomposition.

Proof. Consider the optimization problem

$$\alpha = \sup_{C \in O(N) \cdot A} \langle C, B \rangle. \tag{7.2.16}$$

Let $U \in O(N)$ satisfy $B = U \cdot \operatorname{diag} \lambda(B)$ and choosing $C = U \cdot \operatorname{diag} \lambda(A)$ shows that $\alpha \geq \langle \lambda(A), \lambda(B) \rangle$.

On the other hand, since the orbit $O(N) \cdot A$ is compact, problem (7.2.16) has an optimal solution, say $C = C_0$. By the stationary condition of Proposition 7.1.5 and the fact that $N_F(O(N) \cdot A; C_0)$ is a subspace (Proposition 7.1.6), we have

$$B \in N_F(O(N) \cdot A; C_0) = N_F(O(N) \cdot C_0; C_0) = N(O(N) \cdot C_0; C_0).$$

By the representation of $N_F(O(N) \cdot C_0; C_0)$ in Example 7.2.6 we have $[B, C_0] = 0$. Thus, there is a matrix $U \in O(N)$ that simultaneously diagonalizes B and C_0:

$$B = U \cdot \operatorname{diag}(P\lambda(B)) \text{ and } C_0 = U \cdot \operatorname{diag} \lambda(C_0), \tag{7.2.17}$$

where $P \in P(N)$ is a permutation matrix. It follows from Lemma 7.2.13 that

$$\alpha = \langle C_0, B \rangle = \langle \lambda(C_0), P\lambda(B) \rangle \leq \langle \lambda(C_0), \lambda(B) \rangle = \langle \lambda(A), \lambda(B) \rangle \leq \alpha,$$

whence $\alpha = \langle \lambda(A), \lambda(B) \rangle$.

We turn to the necessary and sufficient conditions for the equality. If $\langle A, B \rangle = \langle \lambda(A), \lambda(B) \rangle$ then A must be a solution of (7.2.16). Therefore, by the previous argument there exists an orthogonal matrix $U \in O(N)$ such that $A = U \cdot \operatorname{diag} \lambda(A)$ and $B = U \cdot \operatorname{diag}(P\lambda(B))$ for some $P \in P(N)$ satisfying

$$\langle \lambda(A), P\lambda(B) \rangle = \langle \lambda(A), \lambda(B) \rangle.$$

By (ii) of Lemma 7.2.13 there exists a permutation matrix $Q \in P(N)$ such that $Q\lambda(A) = \lambda(A)$ and $Q\lambda(B) = P\lambda(B)$. Then $QU \in O(N)$ provides a simultaneous ordered decomposition for A and B. Sufficiency is easy and is left as an exercise (Exercise 7.2.20).   ●

7.2.4 Subdifferentials of Spectral Functions

Spectral functions are often intrinsically nonsmooth. Thus, to analyze their properties we often need to study their subdifferentials. Our first result on the subdifferential of the spectral function follows directly from the smooth chain rule.

Lemma 7.2.15 (Chain Rule for Spectral Functions) *Let ϕ be a spectral function on $S(N)$ and let $U \in O(N)$. Then*

$$\partial\phi(U^\top AU) = U^\top \partial\phi(A)U,$$

where $\partial = \partial_F, \partial_L$ or ∂^∞.

Proof. For $U \in O(N)$, define a mapping $u \colon E(N) \to E(N)$ by $u(B) = U^\top BU$. We can check (Exercise 7.2.21) that u is a diffeomorphism and for any $A \in E(N)$, $T_A^*(E(N)) = E(N)$ and $(u^*)_A \colon E(N) \to E(N)$ is defined by

$$(u^*)_A B = UBU^\top. \tag{7.2.18}$$

It follows directly from the smooth chain rule of Theorem 7.1.8 that

$$(u^*)_A \partial\phi(u(A)) = \partial(\phi \circ u)(A),$$

where $\partial = \partial_F, \partial_L$ or ∂^∞. Since $\phi \circ u = \phi$, the conclusion follows from (7.2.18). ●

A similar result holds for a permutation invariant function on $\mathbb{R}^N$.

Lemma 7.2.16 (Chain Rule for Permutation Invariant Functions) *Let f be a permutation invariant function on $\mathbb{R}^N$ and let $P \in P(N)$. Then*

$$\partial f(Px) = P\partial f(x),$$

where $\partial = \partial_F, \partial_L$ or ∂^∞.

Proof. Exercise 7.2.22. ●

Next is a result on the commutativity of the subdifferential of spectral functions.

Lemma 7.2.17 (Commutativity of Subdifferential of Spectral Functions) *Let ϕ be a spectral function on $S(N)$ and let $A \in S(N)$. Suppose that $B \in \partial\phi(A)$. Then $AB^\top = B^\top A$, i.e., $[A, B] = 0$, where $\partial = \partial_F, \partial_L$ or ∂^∞.*

Proof. We need only to prove the case when $\partial = \partial_F$. The rest follows naturally from a limiting process. Observe that by the definition of the Fréchet subdifferential we have

$$\partial_F \phi(A) \subset N_F(\phi^{-1}(-\infty, \phi(A)); A).$$

Since ϕ is a constant on $O(N) \cdot A$, we have $O(N) \cdot A \subset \phi^{-1}(-\infty, \phi(A))$. Thus,

$$\partial_F \phi(A) \subset N_F(O(N) \cdot A; A).$$

The lemma follows from the representation of $N(O(N) \cdot A; A)$ in Example 7.2.6.   ●

Theorem 7.2.18 *Let ϕ be a spectral function on $S(N)$. Suppose that $B \in \partial\phi(A)$. Then, for some $P \in P(N)$,*

$$\operatorname{diag} P\lambda(B) \in \partial\phi(\operatorname{diag} \lambda(A)), \tag{7.2.19}$$

where $\partial = \partial_F, \partial_L$ or ∂^∞.

Proof. By Lemma 7.2.17 $AB^\top = B^\top A$. Then there exists a $U \in O(N)$ that diagonalizes both A and B simultaneously. That is to say $\operatorname{diag} P\lambda(B) = U^\top BU$ and $\operatorname{diag} \lambda(A) = U^\top AU$ for some permutation matrix $P \in P(N)$. The conclusion then follows from Lemma 7.2.15.   ●

Next we show that $\partial\phi(\operatorname{diag} \lambda(A)) = \operatorname{diag} \partial f(\lambda(A))$ so that the subdifferential of a spectral function $\phi = f \circ \lambda$ on $S(N)$ can be represented in terms of its corresponding permutation invariant function f defined on $\mathbb{R}^N$. For this purpose we need several auxiliary results.

Lemma 7.2.19 *Let $w \in \mathbb{R}_\geq^N$. Then the function $A \to \langle w, \lambda \rangle(A) := \langle w, \lambda(A) \rangle$ is convex and, for any vector $x \in \mathbb{R}_\geq^N$,*

$$\operatorname{diag} w \in \partial\langle w, \lambda \rangle(\operatorname{diag} x). \tag{7.2.20}$$

Proof. We leave the convexity of the function $A \to \langle w, \lambda \rangle(A)$ as a guided exercise (Exercise 7.2.23). It follows from the von Neumann–Theobald inequality that for any $C \in S(N)$,

$$\langle \operatorname{diag} w, C \rangle \leq \langle w, \lambda \rangle(C).$$

Thus, for any $x \in \mathbb{R}_\geq^N$,

$$\langle \operatorname{diag} w, C - \operatorname{diag} x \rangle \leq \langle w, \lambda \rangle(C) - \langle w, \lambda \rangle(\operatorname{diag} x).$$

This verifies (7.2.20).   ●

Lemma 7.2.20 *Let $v \in \mathbb{R}^N$ and $x \in \mathbb{R}_\geq^N$. Suppose that $\operatorname{Stab}(P(N); x)$ is a subgroup of $\operatorname{Stab}(P(N); v)$. Then*

$$\langle v, \lambda \rangle'(\operatorname{diag} x) = \operatorname{diag} v.$$

Proof. Exercise 7.2.24.   ●

Theorem 7.2.21 *Let* $x \in \mathbb{R}^N_{\geq}$ *and* $A \in S(N)$. *Then*

$$\text{diag}^* A \in \text{conv}\{P\lambda'(\text{diag}\, x; A) \mid P \in \text{Stab}(P(N); x)\}. \qquad (7.2.21)$$

Proof. Partition the set $\{1, \ldots, N\}$ into consecutive blocks of integers $S_1, \ldots, S_K$ with $N_1, \ldots, N_K$ elements, respectively, so that $x_n = x_m$ if and only if n and m belong to the same block. Correspondingly, write any vector y in the form of

$$y = (y^1, \ldots, y^K), \text{ where } y^k \in \mathbb{R}^{N_k}, k = 1, \ldots, K.$$

Notice that the stabilizer $\text{Stab}(P(N); x)$ consists of matrices of permutations fixing each block S_k.

Now suppose relation (7.2.21) fails. By the convex separation theorem there exists some vector $y \in \mathbb{R}^N$ satisfying

$$\langle y, \text{diag}^* A \rangle > \langle y, P\lambda'(\text{diag}\, x; A) \rangle \text{ for all } P \in \text{Stab}(P(N); x). \quad (7.2.22)$$

Let $\tilde{y} := (\hat{y}^1, \ldots, \hat{y}^K)$ where $\hat{y}^k$ represents a permutation of y^k so that the components are in a decreasing order. There is a vector $v \in \mathbb{R}^N$ with equal components within every block S_k (or in other words $\text{Stab}(P(N); x)$ is a subgroup of $\text{Stab}(P(N); v)$) so that $v + \tilde{y}$ lies in $\mathbb{R}^N_{\geq}$. Lemma 7.2.19 shows

$$\text{diag}(v + \tilde{y}) \in \partial\langle v + \tilde{y}, \lambda\rangle(\text{diag}\, x),$$

or equivalently, for any matrix $B \in S(N)$,

$$\langle \text{diag}(v + \tilde{y}), B \rangle \leq \langle v + \tilde{y}, \lambda'(\text{diag}\, x; B) \rangle. \qquad (7.2.23)$$

On the other hand Lemma 7.2.20 shows

$$\langle \text{diag}\, v, B \rangle = \langle v, \lambda'(\text{diag}\, x; B) \rangle. \qquad (7.2.24)$$

Subtracting equation (7.2.24) from inequality (7.2.23) we have

$$\langle \text{diag}\, \tilde{y}, B \rangle \leq \langle \tilde{y}, \lambda'(\text{diag}\, x; B) \rangle. \qquad (7.2.25)$$

Writing $\text{diag}^* A = a = (a^1, \ldots, a^K)$, there is a matrix $Q \in \text{Stab}(P(N); x)$ satisfying

$$Q\,\text{diag}^*(A) = \text{diag}^*(Q^\top A Q) = (\hat{a}^1, \ldots, \hat{a}^K).$$

Now choosing $B = Q^\top A Q$ in inequality (7.2.25) and using Lemma 7.2.13 repeatedly shows

$$\langle y, a \rangle \leq \langle (\hat{y}^1, \ldots, \hat{y}^K, \hat{a}^1, \ldots, \hat{a}^K \rangle$$
$$= \langle \operatorname{diag} \tilde{y}, Q^\top A Q \rangle$$
$$\leq \langle \operatorname{diag} \tilde{y}, \lambda'(\operatorname{diag} x; Q^\top A Q) \rangle$$
$$= \langle \operatorname{diag} \tilde{y}, \lambda'(Q^\top \operatorname{diag} x Q; Q^\top A Q) \rangle$$
$$= \langle \operatorname{diag} \tilde{y}, \lambda'(\operatorname{diag} x; A) \rangle,$$

where the last equality follows directly from that λ is unitary invariant and the definition of the directional derivative (Exercise 7.2.25). Now choosing the matrix P in inequality (7.2.22) so that $P^\top y = \tilde{y}$ gives a contradiction.   ●

Theorem 7.2.22 *Let* $\phi = f \circ \lambda \colon S(N) \to \mathbb{R} \cup \{+\infty\}$ *be a spectral function and let* $x, y \in \mathbb{R}^N$. *Then*

$$y \in \partial f(x)$$

if and only if

$$\operatorname{diag} y \in \partial \phi(\operatorname{diag} x).$$

Here $\partial = \partial_F, \partial_L$ *or* ∂^∞.

Proof. The "only if" part is easy and is left as an exercise (Exercise 7.2.27). We prove the "if" part. Again we prove the Theorem for $\partial = \partial_F$. The conclusion for the limiting and singular subdifferential follows naturally from a limiting process.

First we discuss the special case when $x \in \mathbb{R}^N_\geq$. By the chain rule for permutation invariant functions of Lemma 7.2.16, $S := \{Py \mid P \in \operatorname{Stab}(P(N); x)\} \subset \partial_F f(x)$. The support function for the convex hull of S, $\sigma_{\operatorname{conv}(S)}$, is sublinear with a global Lipschitz constant $\|y\|$ (Exercise 7.2.26).

Fix $\varepsilon > 0$. It follows from Proposition 3.1.3 that for small $z \in \mathbb{R}^N$,

$$f(x + z) \geq f(x) + \sigma_{\operatorname{conv} S}(z) - \varepsilon \|z\|. \qquad (7.2.26)$$

On the other hand, using Theorem 7.2.12, small matrices $C \in S(N)$ must satisfy

$$\|\lambda(\operatorname{diag} x + C) - x - \lambda'(\operatorname{diag} x; C)\| \leq \varepsilon \|C\|,$$

and hence by inequality (7.2.26),

$$f(\lambda(\operatorname{diag} x + C)) = f(x + (\lambda(\operatorname{diag} x + C) - x))$$
$$\geq f(x) - \varepsilon \|\lambda(\operatorname{diag} x + C) - x\|$$
$$+ \sigma_{\operatorname{conv} S}\Big(\lambda'(\operatorname{diag} x; C) + [\lambda(\operatorname{diag} x + C) - x - \lambda'(\operatorname{diag} x; C)]\Big)$$
$$\geq f(x) + \sigma_{\operatorname{conv} S}(\lambda'(\operatorname{diag} x; C)) - (1 + \|y\|)\varepsilon \|C\|,$$

using the Lipschitz property of $\sigma_{\operatorname{conv} S}$.

By Theorem 7.2.21,

$$\operatorname{diag}^{*} C \in \operatorname{conv}\{P\lambda'(\operatorname{diag} x; C) \mid P \in \operatorname{Stab}(P(N); x)\}. \qquad (7.2.27)$$

Since $\operatorname{conv} S$ is clearly invariant under the group $\operatorname{Stab}(P(N); x)$, so is its support function, whence

$$\sigma_{\operatorname{conv} S}(P\lambda'(\operatorname{diag} x; C)) = \sigma_{\operatorname{conv} S}(\lambda'(\operatorname{diag} x; C)),$$

for any $P \in \operatorname{Stab}(P(N); x)$. This, combined with the convexity of $\sigma_{\operatorname{conv} S}$ and relation (7.2.27), demonstrates

$$\sigma_{\operatorname{conv} S}(\operatorname{diag}^{*} C) \leq \sigma_{\operatorname{conv} S}(\lambda'(\operatorname{diag} x; C)).$$

So the argument above shows

$$\begin{aligned}
f \circ \lambda(\operatorname{diag} x + C) &\geq f(x) + \sigma_{\operatorname{conv} S}(\operatorname{diag}^{*} C) - (1 + \|y\|)\varepsilon\|C\| \\
&\geq f(x) + \langle y, \operatorname{diag}^{*} C \rangle - (1 + \|y\|)\varepsilon\|C\| \\
&\geq f \circ \lambda(\operatorname{diag} x) + \langle \operatorname{diag} y, C \rangle - (1 + \|y\|)\varepsilon\|C\|,
\end{aligned}$$

and since ε was arbitrary

$$\operatorname{diag} y \subset \partial_{F} f \circ \lambda(\operatorname{diag} x).$$

We now turn to the general case when x is not necessarily in $\mathbb{R}^{N}_{\geq}$. Fix a matrix $P \in P(N)$ with $Px \in \mathbb{R}^{N}_{\geq}$. Then $Py \in \partial_{F} f(Px)$ by Lemma 7.2.16. Thus, the special case proved above shows that

$$\operatorname{diag} Py \in \partial_{F}(f \circ \lambda)(\operatorname{diag} Px).$$

By Exercise 7.2.9 we can rewrite the last inclusion as

$$P^{\top}(\operatorname{diag} y)P \in \partial_{F}(f \circ \lambda)(P^{\top} \operatorname{diag} xP).$$

This is

$$\operatorname{diag} y \in \partial_{F}(f \circ \lambda)(\operatorname{diag} x),$$

using chain rule for spectral functions of Lemma 7.2.15.   ●

Now we are ready for our main representation theorem.

Theorem 7.2.23 (Representation of the Subdifferential for Spectral Functions) *Let $\phi = f \circ \lambda\colon S(N) \to \mathbb{R} \cup \{+\infty\}$ be a spectral function and let $A \in S(N)$. Then*

$$\partial\phi(A) = \{U^{\top}(\operatorname{diag} y)U \mid y \in \partial f(\lambda(A)), U \in O(N), A = U^{\top}(\operatorname{diag} \lambda(A))U\}.$$

Here $\partial = \partial_{F}, \partial_{L}$ or ∂^{∞}.

Proof. For any vector $y \in \partial f(\lambda(A))$, Theorem 7.2.22 shows

$$\operatorname{diag} y \in \partial(f \circ \lambda)(\operatorname{diag} \lambda(A)).$$

Thus, for any $U \in O(N)$ with $U^\top \lambda(A) U = A$ we have

$$U^\top(\operatorname{diag} y)U \in \partial(f \circ \lambda)(A),$$

by Lemma 7.2.15.

On the other hand any $B \in \partial(f \circ \lambda)(A)$ commutes with A, by the commutativity Lemma 7.2.17. Therefore, A and B diagonalize simultaneously: there is a matrix $U \in O(N)$ and a vector $y \in \mathbb{R}^N$ such that $U^\top(\operatorname{diag} \lambda(A))U = A$ and $U^\top(\operatorname{diag} y)U = B$. The chain rule for spectral function of Lemma 7.2.15 shows

$$\operatorname{diag} y \in \partial(f \circ \lambda)(\operatorname{diag} \lambda(A)),$$

whence $y \in \partial f(\lambda(A))$, by Theorem 7.2.22.   ●

7.2.5 The kth Largest Eigenvalue Function

We end this section by applying Theorem 7.2.23 to calculate the subdifferential of the kth largest eigenvalue function λ_k of symmetric matrices. Consider the kth order statistic function $\phi_k \colon \mathbb{R}^N \to \mathbb{R}$ defined by $\phi_k(x) =$ the kth largest component of $(x_1, \ldots, x_N)$. Then ϕ_k is permutation invariant. Moreover, $\phi_k(x) = \lambda_k(\operatorname{diag} x)$ and $\lambda_k = \phi_k \circ \lambda$ (Exercise 7.2.28). Thus, calculating the subdifferential for ϕ_k is crucial.

Proposition 7.2.24 (Fréchet Subdifferential of Order Statistics)
Let $\phi_k \colon \mathbb{R}^N \to \mathbb{R}$ be the kth order statistic function. Then, for any $x \in \mathbb{R}^N$,

$$\partial_F \phi_k(x) = \begin{cases} \operatorname{conv}\{e^n \mid x_n = \phi_k(x)\} & \text{if } \phi_{k-1}(x) > \phi_k(x), \\ \emptyset & \text{otherwise.} \end{cases}$$

Here $\{e^n \mid n = 1, \ldots, N\}$ is the standard basis of $\mathbb{R}^N$.

Proof. Define a set of indices $I := \{n \mid x_n = \phi_k(x)\}$. If the inequality $\phi_{k-1}(x) > \phi_k(x)$ holds then clearly, close to the point x, the function ϕ_k is given by $y \in \mathbb{R}^N \to \max_{n \in I} y_n$. The subdifferential at x of this second function (which is convex) is just $\operatorname{conv}\{e^n \mid n \in I\}$.

On the other hand, if $\phi_{k-1}(x) = \phi_k(x)$, suppose $y \in \partial_F \phi_k(x)$ and therefore satisfies

$$\phi_k(x + z) \geq \phi_k(x) + \langle y, z \rangle + o(z), \text{ as } z \to 0.$$

For any index $n \in I$, all small positive h satisfy $\phi_k(x + he^n) = \phi_k(x)$ so that $y_n \leq 0$, but also

$$\phi_k\left(x - h \sum_{n \in I} e^n\right) = \phi_k(x) - h,$$

which implies a contradiction $\sum_{n \in I} y_n \geq 1$.   ●

We now turn to the limiting subdifferentials. We use $\#(S)$ to signify the number of elements in the set S and denote supp $y = \{n \mid y_n \neq 0\}$ for $y \in \mathbb{R}^N$.

Theorem 7.2.25 (Limiting Subdifferentials of Order Statistics)
Let $\phi_k \colon \mathbb{R}^N \to \mathbb{R}$ be the kth order statistic function. Then for any $x \in \mathbb{R}^N$,

$$\partial^\infty \phi_k(x) = \{0\},$$

$$\partial_L \phi_k(x) = \{y \in \text{conv}\{e^n \mid x_n = \phi_k(x)\} \mid \#(\text{supp } y) \leq \alpha\},$$

where $\alpha = 1 - k + \#\{n \mid x_n \geq \phi_k(x)\}$, and $\partial_C \phi_k(x) = \text{conv}\{e^n \mid x_n = \phi_k(x)\}$.

Proof. The singular subdifferential is easy and the Clarke subdifferential can be derived by taking convex hull of the limiting subdifferential (Exercise 7.2.29). So we focus on the limiting subdifferential.

Let x^i be a sequence converging to x. Then for any $x_n \neq \phi_k(x)$, when i is sufficiently large, $x_n^i \neq \phi_k(x^i)$. By Proposition 7.2.24,

$$\partial_F \phi_k(x^i) \subset \text{conv}\{e^n \mid x_n^i = \phi_k(x^i)\} \subset \text{conv}\{e^n \mid x_n = \phi_k(x)\},$$

when $\phi_{k-1}(x^i) > \phi_k(x^i)$, and therefore $\#\{n \mid x_n^i = \phi_k(x^i)\} \leq \alpha$. Thus,

$$\partial_L \phi_k(x) \subset \{y \subset \text{conv}\{e^n \mid x_n = \phi_k(x)\} \mid \#(\text{supp } y) \leq \alpha\}.$$

On the other hand, let $y \in \text{conv}\{e^n \mid x_n = \phi_k(x)\}$ and $\#(\text{supp } y) \leq \alpha$. Denote $J_1 = \{n \mid x_n < \phi_k(x)\}$. Then $y_n = 0$ for any $n \in J_1$. Choose a subset J_2 of $\{n \mid x_n \geq \phi_k(x)\}$ with exactly $k - 1$ indices n for which $y_n = 0$. For $h > 0$, define $x(h) = x + h \sum_{n \in J_2} e^n$. Then when h is small enough we have $\phi_k(x(h)) < \phi_{k-1}(x(h))$ so that by Proposition 7.2.24

$$y \in \text{conv}\{e^n \mid n \notin J_1 \cup J_2\} \subset \text{conv}\{e^n \mid x(h)_n = \phi_k(x(h))\} = \partial_F \phi_k(x(h)).$$

Taking limits as $h \to 0$ yields $y \in \partial_L \phi_k(x)$.   ●

Combining Theorem 7.2.25 and Theorem 7.2.23, we have the following representation of subdifferentials for the kth largest eigenvalue function.

Theorem 7.2.26 *Let $A \in S(N)$. Then for $k \in \{1, \ldots, N\}$,*

$$\partial^\infty \lambda_k(A) = \{0\},$$

$$\partial_C \lambda_k(A) = \text{conv}\{uu^T \mid u \in \mathbb{R}^N, \|u\| = 1, Au = \lambda_k(A)u\},$$

and

$$\partial_L \lambda_k(A) = \{B \in \partial_C \lambda_k(A) \mid \text{rank } B \leq \alpha\},$$

where $\alpha = 1 - k + \#\{n \mid \lambda_n(A) \geq \lambda_k(A)\}$.

Proof. Exercise 7.2.30.   ●

7.2.6 Commentary and Exercises

For additional details on manifolds of matrices we refer to [27, 136]. Ky Fan's minimax theorem is the culmination of earlier research by Courant, Fisher and Weyl. Bhatia [22] provides an excellent exposition of this minimax theorem and extensive commentary on its history. The representation of the subdifferentials for spectral functions of symmetric matrices appeared in Lewis [179]. Generalizations to spectral functions of nonsymmetric matrices can be found in [76, 181]. The exposition here follows [175, 176, 177], viewing the spectral functions as a special class of nonsmooth functions on matrix manifolds.

Exercise 7.2.1 Prove Proposition 7.2.1.

Exercise 7.2.2 Prove the claims in Example 7.2.2.

Exercise 7.2.3 Prove the claims in Example 7.2.3.

Exercise 7.2.4 Show that, for $A, B \in E(N)$,

$$\mathrm{tr}(AB) = \mathrm{tr}(BA).$$

∗**Exercise 7.2.5** Prove the claims in Example 7.2.5. Hint:

(i)   Define $f \colon GL(N, M) \to S(M)$ by $f(A) := A^\top A - I$ and show that, for any $B \in T_A(GL(N, M)) = E_{N \times M}$,

$$(f_*)_A(B) = A^\top B + B^\top A.$$

(ii) Verify that for every $A \in GL(N, M)$, $(f_*)_A$ is a surjection onto $S(M)$.

(iii) Use the fiber theorem to identify $\mathrm{St}(N, M) = f^{-1}(0)$ as a submanifold of $GL(N, M)$ and therefore a submanifold of $E_{N \times M}$.

(iv) Use the fiber theorem again to compute the tangent space $T_A(\mathrm{St}(N, M))$ as the kernel of $(f_*)_A$.

(v)   Derive the normal cone representation of (7.2.6) using a method similar to that in Example 7.2.4.

Exercise 7.2.6 Let $C \in A(N)$ and let $A, B \in S(N)$. Show that

$$\langle [A, C], B \rangle = \langle C, [A, B] \rangle.$$

Exercise 7.2.7 Prove Lemma 7.2.7.

Exercise 7.2.8 Prove formula (7.2.11).

Exercise 7.2.9 Let $x \in \mathbb{R}^N$ and let $P \in P(N)$. Show that

$$\mathrm{diag}\, Py = P^\top \, \mathrm{diag}\, yP.$$

Exercise 7.2.10 Let $A \in S(N)$ and let $P \in P(N)$. Show that

$$P \, \mathrm{diag}^* A = \mathrm{diag}^* P^\top AP.$$

Exercise 7.2.11 Verify the formula (7.2.14).

Exercise 7.2.12 Let $A \in S(N)$, $B \in E_{N \times k}$ and $X \in S(k)$. Suppose that $AB = BX$. Prove that any eigenvalue of X must also be an eigenvalue of A.

Exercise 7.2.13 Prove Corollary 7.2.9.

Exercise 7.2.14 Prove Corollary 7.2.10.

Exercise 7.2.15 Show that the sum of the m largest (resp. smallest) eigenvalues of a symmetric matrix is a continuous convex (resp. concave) function of the data. Deduce that the kth largest eigenvalue, as the difference of two continuous convex functions, is a Lipschitz function of the data.

Exercise 7.2.16 Let λ_2 denote second largest eigenvalue of a symmetric doubly stochastic matrix. Show that λ_2 is a convex function of the data.

Exercise 7.2.17 Let X be a Banach space and let $f \colon X \to \mathbb{R}$ be a Lipschitz function with a Lipschitz constant L. Suppose that for any $h \in X$, $f'(\bar{x}; h)$ exists. Show that $h \to f'(\bar{x}; h)$ is also a Lipschitz function with the same Lipschitz constant L.

Exercise 7.2.18 Show that the iteration $\langle x, P_n y \rangle$ defined in the proof of Lemma 7.2.13 is increasing with n.

Exercise 7.2.19 Prove (ii) of Lemma 7.2.13. Hint: "Only if" is the part that we need to work on. Suppose that $\langle x, Py \rangle = \langle x, y \rangle$.

(i) Partition $\{1, \ldots, N\}$ into sets $S_1, \ldots, S_M$ so that $x_n = s_m$ for all $n \in S_m$ where s_m decreases strictly with m.

(ii) Choose a permutation with matrix $Q \in P(N)$, fixing each index set S_m, and permuting the components $\{(Py)_n \mid n \in S_m\}$ into nonincreasing order.

(iii) Show that Q has the property that we are looking for.

Exercise 7.2.20 Prove that if A and B in $O(N)$ have a simultaneous ordered spectral decomposition then

$$\langle A, B \rangle = \langle \lambda(A), \lambda(B) \rangle.$$

Exercise 7.2.21 Let $U \in O(N)$. Define a mapping $u \colon E(N) \to E(N)$ by $u(B) = U^\top B U$.

(i) Show that u is a diffeomorphism, in fact, a one-to-one linear mapping.

(ii) Show that for any $A \in E(N)$, $T_A(E(N)) = E(N)$ and for any $B \in E(N)$, $(u_*)_A B = U^\top B U$.

(iii) Verify that for any $A \in E(N)$, $T_A^*(E(N)) = E(N)$ and for any $B \in E(N)$, $(u^*)_A B = U B U^\top$.

Exercise 7.2.22 Prove Lemma 7.2.16.

Exercise 7.2.23 Let $w \in \mathbb{R}^N_{\geq}$. Show that the function $A \to \langle w, \lambda \rangle(A) := \langle w, \lambda(A) \rangle$ is sublinear and therefore convex on $S(N)$. Hint: Positive homogeneity is obvious. To show the subadditive property let $A, B \in S(N)$ and let $U \in O(N)$ diagonalize $A + B$. Then

$$
\begin{aligned}
\langle w, \lambda \rangle(A + B) &= \langle \operatorname{diag} w, \operatorname{diag} \lambda(A + B) \rangle \\
&= \langle \operatorname{diag} w, U^\top (A + B) U \rangle \\
&= \langle U(\operatorname{diag} w) U^\top, A + B \rangle \\
&= \langle U(\operatorname{diag} w) U^\top, A \rangle + \langle U(\operatorname{diag} w) U^\top, B \rangle \\
&= \langle \operatorname{diag} w, U^\top A U \rangle + \langle \operatorname{diag} w, U^\top B U \rangle \\
&\leq \langle w, \lambda(A) \rangle + \langle w, \lambda(B) \rangle.
\end{aligned}
$$

The last inequality is due to the von Neumann–Theobald inequality.

Exercise 7.2.24 Prove Lemma 7.2.20. Reference: [181].

Exercise 7.2.25 Let $\phi \colon S(N) \to \mathbb{R}$ be a unitarily invariant function and let $U \in O(N)$. Then the directional derivative $\phi'(A; B)$ exists if and only if $\phi'(U^\top A U; U^\top B U)$ exists. Moreover, when this directional derivative exists

$$
\phi'(A; B) = \phi'(U^\top A U; U^\top B U).
$$

Exercise 7.2.26 Let $y \in \mathbb{R}^N$, let T be a subset of $P(N)$ and let $S := \{Py : P \in T\}$. Show that the support function of $\operatorname{conv} S$ is given by

$$
\sigma_{\operatorname{conv} S}(z) = \max\{\langle z, Py \rangle \mid P \in T\}.
$$

Moreover, this function is sublinear and Lipschitz with global Lipschitz constant $\|y\|$.

Exercise 7.2.27 Prove the "only if" part of Theorem 7.2.22.

Exercise 7.2.28 Show that the kth order statistic function ϕ_k permutation invariant and show that $\phi_k(x) = \lambda_k(\operatorname{diag} x)$ and $\lambda_k = \phi_k \circ \lambda$.

Exercise 7.2.29 Prove the formula for the singular and Clarke subdifferentials in Theorem 7.2.25.

Exercise 7.2.30 Prove Theorem 7.2.26.

7.3 Convex Spectral Functions of Compact Operators

The pattern we observed in the previous section also extends to infinite dimensional spectral functions. We start with the easier case when the spectral functions are convex. We also restrict ourselves to operators from the complex Hilbert space $\ell_2^{\mathbb{C}}$ to itself.

7.3.1 Operator and Spectral Sequence Spaces

Let $\mathbf{C}$ be the space of complex numbers and let $\mathcal{N}$ be the set of natural numbers. Let $\ell_2^{\mathbf{C}}$ be the space of sequences (c_i) such that $c_i \in \mathbf{C}$ for each $i \in \mathcal{N}$ and such that $\sum_{i \in \mathcal{N}} |c_i|^2 < \infty$. We denote by ℓ^2 the standard real square summable sequence space (coefficients in $\mathbb{R}$) with canonical basis $\{e^i\}$. Thus, the jth component of the ith element of this basis is given by the Kronecker delta: $e_i^j = \delta_{ij}$. We also consider the real normed sequence spaces $\ell_p (1 \le p \le \infty)$ and c_0 (the space of real sequences converging to 0 with the norm of ℓ_∞). Note that $\ell_1 \subset \ell_2 \subset c_0 \subset \ell_\infty$.

We also consider the space of bounded self-adjoint operators on $\ell_2^{\mathbf{C}}$, which we denote by B_{sa}. We will need a number of well-known basic properties of B_{sa} and its subspaces, which we state without proof below. Additional details for these preliminaries can be found in [128, 216, 239]. We say an operator $T \in B_{sa}$ is *positive* (denoted $T \ge 0$) if $\langle Tx, x \rangle \ge 0$ for all $x \in \ell_2^{\mathbf{C}}$. To each $T \in B_{sa}$ one can associate a unique positive operator $|T| = (T^*T)^{\frac{1}{2}} \in B_{sa}$. Now to each positive $T \in B_{sa}$ one can associate the (possibly infinite) value

$$\operatorname{tr}(T) := \sum_{i=1}^{\infty} \langle Te^i, e^i \rangle, \tag{7.3.1}$$

which we call the trace of T. The trace is actually independent of the orthonormal basis $\{e^i\}$ chosen.

Within B_{sa} we consider the *trace class operators*, denoted by B_1, which are those self-adjoint operators T for which $\operatorname{tr}(|T|) < \infty$. Since any self-adjoint operator T can be decomposed as $T = T_+ - T_-$ where $T_+ \ge 0$ and $T_- \ge 0$ the trace operator can be extended to any $T \in B_1$ by $\operatorname{tr}(T) = \operatorname{tr}(T_+) - \operatorname{tr}(T_-)$. We let B_2 be the self-adjoint Hilbert–Schmidt operators, which are those $T \in B_{sa}$ such that $T^2 = (T^*T) \in B_1$. This gives $B_1 \subset B_2 \subset B_0 \subset B_{sa}$, where B_0 is the space of compact, self-adjoint operators. Now any compact, self-adjoint operator T is diagonalizable. That is, there exists a unitary operator U and $\lambda \in c_0$ such that $(U^*TU \, x)_i = \lambda_i x_i$ for all $i \in \mathcal{N}$ and all $x \in B_0$. This and the fact that $\operatorname{tr}(ST) = \operatorname{tr}(TS)$ makes proving Lidskii's Theorem (difficult in general) easy for self-adjoint operators. Lidskii's theorem (see [239, p. 45]) states

$$\operatorname{tr}(T) = \sum_{i=1}^{\infty} \lambda_i(T)$$

where $(\lambda_i(T))$ is any *spectral sequence* of T, which is any sequence of eigenvalues of T (counted with multiplicities). Define $B_p \subset B_0$ for $p \in [1, \infty)$ by writing $T \in B_p$ if $\|T\|_p = (\operatorname{tr}(|T|^p))^{1/p} < \infty$. When T is self-adjoint we have (see [128, p. 94])

$$\|T\|_p = (\operatorname{tr}(|T|^p))^{1/p} = \left(\sum_{i=1}^{\infty} |\lambda_i(T)|^p \right)^{1/p}.$$

In this case for $p, q \in (1, \infty)$ and $p^{-1} + q^{-1} = 1$, we get that B_p and B_q are paired, and the sesquilinear form $\langle S, T \rangle := \operatorname{tr}(ST)$ implements the duality on $B_p \times B_q$. These spaces are the Schatten p-spaces. We also consider the space B_0 paired with B_1. Spaces similar to B_p, $p = 0$ or $p \in [1, \infty)$ can also be defined for non self-adjoint operators. Details can be found in [128]. We need these spaces of non self-adjoint operators only in Propositions 7.3.19 and 7.3.20, and in Theorem 7.3.21.

Further, for each $x \in \ell_2^{\mathbf{C}}$ define the operator $x \odot x \in B_1$ by

$$(x \odot x)y = \langle x, y \rangle x.$$

For each $x \in \ell_\infty$ we define the operator $\operatorname{diag} x \in B_{sa}$ pointwise by

$$\operatorname{diag} x := \sum_{i=1}^{\infty} x_i (e^i \odot e^i).$$

For $p \in [1, \infty)$, if we have $x \in \ell_p$, then $\operatorname{diag} x \in B_p$ and $\| \operatorname{diag} x \|_p = \|x\|_p$. If we have $x \in c_0$, then $\operatorname{diag} x \in B_0$ and $\| \operatorname{diag} x \| = \|x\|_\infty$. This motivates:

Definition 7.3.1 (Spectral Sequence Space) *For $p \in [1, \infty)$ we say ℓ_p is the spectral sequence space for B_p and c_0 is the spectral sequence space for B_0.*

Definition 7.3.2 (Paired Banach Spaces) *We say that V and W are paired Banach spaces $V \times W$, if $V = \ell_p$ and $W = \ell_q$ and $p, q \in [1, \infty)$ satisfy $p^{-1} + q^{-1} = 1$ ($p = 1, q = \infty$) or $V = c_0$ (with the supremum norm) and $W = \ell_1$ or vice versa. We denote the norms on V and W by $\| \cdot \|_V$ and $\| \cdot \|_W$ respectively. Similarly, we say $\mathcal{V}$ and $\mathcal{W}$ are paired Banach spaces $\mathcal{V} \times \mathcal{W}$ where $\mathcal{V} = B_p$ and $\mathcal{W} = B_q$ and $p, q \in [1, \infty)$ satisfy $p^{-1} + q^{-1} = 1$ ($p = 1, q = \infty$) or where $\mathcal{V} = B_0$ (with the operator norm) and $\mathcal{W} = B_1$ (or vice versa). We denote the norms on $\mathcal{V}$ and $\mathcal{W}$ by $\| \cdot \|_{\mathcal{V}}$ and $\| \cdot \|_{\mathcal{W}}$ respectively. We always take V to be the spectral sequence space for the operator space $\mathcal{V}$ and W that for $\mathcal{W}$. In this way fixing $V \times W$ fixes $\mathcal{V} \times \mathcal{W}$ and vice versa.*

7.3.2 Unitarily and Rearrangement Invariant Functions

We will always use the setting of paired Banach spaces in Definition 7.3.2. The relationship between V and $\mathcal{V}$ in Definition 7.3.2 is akin to that of $S(N)$ and $\mathbb{R}^N$ in the previous section. Let $\mathcal{B}$ be the set of all bijections from $\mathcal{N}$ to $\mathcal{N}$. If $x = (x_1, x_2, \dots) \in V$ and $\pi \in \mathcal{B}$ then we call $x_\pi = (x_{\pi(1)}, x_{\pi(2)}, \dots)$ a rearrangement of x. We sometimes also call $\pi \in \mathcal{B}$ a rearrangement. We say sequences x and y in V are *rearrangement equivalent* if there is a rearrangement $\pi \in \mathcal{B}$ such that $x_{\pi(i)} = y_i$ for all $i \in \mathcal{N}$. Let $\mathcal{U}$ be the set of all unitary operators on $\ell_2^{\mathbf{C}}$. Similarly, operators S and T in B_{sa} are *unitarily equivalent* if there is a unitary operator $U \in \mathcal{U}$ such that $U^* T U = S$. We say a function $\phi : \mathcal{V} \to \mathbb{R} \cup \{+\infty\}$ is *unitarily invariant* if $\phi(U^* T U) = \phi(T)$ for all $T \in \mathcal{V}$ and all $U \in \mathcal{U}$. We have seen in the previous section that a unitarily invariant function ϕ can be represented as $\phi = f \circ \lambda$ where $f : \mathbb{R}^N \to \mathbb{R} \cup \{+\infty\}$ is a rearrangement invariant function and $\lambda : S(N) \to \mathbb{R}^N$ is the spectral mapping. We now proceed to derive an analogous result for unitarily invariant functions on $\mathcal{V}$. A function $f : \mathcal{V} \to \mathbb{R} \cup \{+\infty\}$ is *rearrangement invariant* if $f(x_\pi) = f(x)$ for any rearrangement π. The definition of the spectral mapping is less straightforward. We need the following construction that gives us a tool similar to arranging the components of a vector in $\mathbb{R}^N$ according to lexicographic order.

For fixed $x \in \mathcal{V}$ let

$$I_>(x) := \{i \mid x_i > 0\}, \quad I_=(x) := \{i \mid x_i = 0\}, \quad I_<(x) := \{i \mid x_i < 0\}.$$

Now, define the mapping $\Phi : \mathcal{V} \to \mathcal{V}$ by means of the infinite algorithm:

(0) Initialize $j = 1$.

(1) If $I_>(x) \neq \emptyset$,   (i)   choose $i \in I_>(x)$ maximizing x_i,
                                   (ii) define $\Phi(x)_j := x_i$,
                                   (iii) update $I_>(x) := I_>(x) \backslash \{i\}$ and $j := j + 1$.

(2) If $I_=(x) \neq \emptyset$,   (i)   choose $i \in I_=(x)$,
                                   (ii) define $\Phi(x)_j := 0$,
                                   (iii) update $I_=(x) := I_=(x) \backslash \{i\}$ and $j := j + 1$.

(3) If $I_<(x) \neq \emptyset$,   (i)   choose $i \in I_<(x)$ minimizing x_i,
                                   (ii) define $\Phi(x)_j := x_i$,
                                   (iii) update $I_<(x) := I_<(x) \backslash \{i\}$ and $j := j + 1$.

(4) Go to (1).

Informally, we start with the largest positive component of x, followed by a 0 component and then the smallest negative component and so on. If any of the sets $I_>(x), I_=(x)$ or $I_<(x)$ is empty or exhausted we skip the corresponding step. We summarize some useful properties of Φ in the following proposition whose easy proof is left as an exercise.

Proposition 7.3.3 *The mapping Φ defined above has the following properties.*

(i)   *For each $x \in \ell_2$ there exists a $\pi \in \mathcal{B}$ with $(\Phi(x))_i = x_{\pi(i)}$ for all $i \in \mathcal{N}$.*
(ii) *For any $x, y \in \ell_2$ we have $\Phi(x) = \Phi(y)$ if and only if there exists a $\pi \in \mathcal{B}$ with $y_i = x_{\pi(i)}$ for all $i \in \mathcal{N}$.*
(iii) $\Phi^2 = \Phi$.
(iv) $f : \ell_2 \to \mathbb{R} \cup \{+\infty\}$ *is rearrangement invariant if and only if $f = f \circ \Phi$.*

Proof. Exercise 7.3.1. •

What is important here is that Φ is constant on the equivalence classes of V that are induced by rearrangements on V and that the function Φ is the identity for some canonical element of each equivalence class.

Define the *eigenvalue mapping* $\lambda: \mathcal{V} \to V$ as follows. For any $T \in \mathcal{V}$ let $\mu(T)$ be any spectral sequence of T. Then $\lambda(T) = \Phi(\mu(T))$, so that λ gives us a canonical spectral sequence for any given compact self-adjoint operator T.

Proposition 7.3.4 *The mapping λ is unitarily invariant, and $\Phi = \lambda \circ \mathrm{diag}$.*

Proof. Exercise 7.3.2 •

Moreover, λ and diag act as inverses in the following sense.

Proposition 7.3.5 (Inverse Relation)

(i) $(\lambda \circ \mathrm{diag})(x)$ *is rearrangement equivalent to x for all $x \in \ell^2$. That is, x and $(\lambda \circ \mathrm{diag})(x)$ are in the same rearrangement invariant equivalence class.*

(ii) $(\mathrm{diag} \circ \lambda)(T)$ *is unitarily equivalent to T for all $T \in \mathcal{V}$.*

Proof. Exercise 7.3.3 •

The proof of the next result can be found in [216, p. 107].

Theorem 7.3.6 *For all $T \in B_0$ there exists $U \in \mathcal{U}$ with $T = U^* \mathrm{diag}(\lambda(T))U$.*

For any rearrangement invariant $f: V \to \mathbb{R} \cup \{+\infty\}$ we have that $f \circ \lambda$ is uniformly invariant, and for any unitarily invariant $\phi: \mathcal{V} \to \mathbb{R} \cup \{+\infty\}$ we have that $f \circ \mathrm{diag}$ is rearrangement invariant. Thus, the maps λ and diag allow us to move between rearrangement invariant functions on V and unitarily invariant functions on $\mathcal{V}$, as the easy results below show.

Theorem 7.3.7 (Unitary Invariance) *Let $\phi: \mathcal{V} \to \mathbb{R} \cup \{+\infty\}$. The following are equivalent:*

(i) ϕ *is unitarily invariant;*
(ii) $\phi = \phi \circ \mathrm{diag} \circ \lambda$;
(iii) $\phi = f \circ \lambda$ *for some rearrangement invariant $f: V \to \mathbb{R} \cup \{+\infty\}$.*

If (iii) *holds then $f = \phi \circ \mathrm{diag}$.*

Proof. Exercise 7.3.4 •

Symmetrically we have:

Theorem 7.3.8 (Rearrangement Invariance) *Let $f: V \to \mathbb{R} \cup \{+\infty\}$. The following are equivalent:*

(i) f *is rearrangement invariant;*

(ii) $f = f \circ \lambda \circ \mathrm{diag}$;

(iii) $f = \phi \circ \mathrm{diag}$ *for some unitarily invariant* $\phi \colon \mathcal{V} \to \mathbb{R} \cup \{+\infty\}$.

If (iii) *holds then* $\phi = f \circ \lambda$.

Proof.   Exercise 7.3.5   ●

Let V and $\mathcal{V}$ be as in Definition 7.3.2. We have the following useful relations (see Exercise 7.3.6): $\|\,\mathrm{diag}(x)\|_{\mathcal{V}} = \|x\|_V$, for all $x \in V$ and $\|\lambda(T)\|_V = \|T\|_{\mathcal{V}}$, for all $T \in \mathcal{V}$.

Having set the stage, we turn to explore the relationship between unitarily invariant functions on $\mathcal{V}$ and its counterpart on V.

7.3.3 The von Neumann Inequality

We now develop an infinite dimensional analogue of the inequality in Theorem 7.2.14. We start with bilinear forms.

Definition 7.3.9 (Bilinear Forms) *Let* $U \in \mathcal{U}$. *We say*

$$B_U(x,y) = \mathrm{tr}[U^*(\mathrm{diag}\,x)U\,(\mathrm{diag}\,y)], (x,y) \in V \times W$$

is the bilinear form generated by the unitary operator U. *Similarly, let* $\pi \in \mathcal{B}$. *We say*

$$P_\pi(x,y) = \sum_{i=1}^{\infty} x_{\pi(i)} y_i, (x,y) \in V \times W$$

is the bilinear form generated by the rearrangement π.

Our next lemma characterizes the bilinear form generated by a unitary operator.

Lemma 7.3.10 *Let* $U \in \mathcal{U}$. *Define* $u^j := U e^j$. *Then* $u^j \in \ell_2^{\mathbf{C}}$, *and the "infinite matrix"* $(|u_i^j|^2)_{i,j \in \mathcal{N}}$ *is doubly stochastic. That is,* $\sum_{i=1}^{\infty} |u_i^j|^2 = 1$ *for each* $j \in \mathcal{N}$ *and* $\sum_{j=1}^{\infty} |u_i^j|^2 = 1$ *for each* $i \in \mathcal{N}$. *Further,*

$$B_U(x,y) = \sum_{i,j=1}^{\infty} x_i \, |u_i^j|^2 \, y_j \leq \|x\|_V \|y\|_W \qquad (7.3.2)$$

for $(x,y) \in V \times W$.

Proof. It is clear that $u^j \in \ell_2^{\mathbf{C}}$. In fact, we know $\{u^j\}_{j=1}^{\infty}$ forms an orthonormal basis for $\ell_2^{\mathbf{C}}$ (Exercise 7.3.7). Thus, $\sum_{i=1}^{\infty} |u_i^j|^2 = 1$ for each $j \in \mathcal{N}$. Further, since $\langle e^j, U^* e^i \rangle = \langle U e^j, e^i \rangle = \langle u^j, e^i \rangle = u_i^j$, we get $U^* e^i = \sum_{j=1}^{\infty} (u_i^j)^* e^j$. Taking the norm of this equality gives $\|U^* e^i\| = \|\sum_{j=1}^{\infty} (u_i^j)^* e^j\| = \sum_{j=1}^{\infty} |u_i^j|^2$. Since $\|U^* e^i\| = 1$ (U is unitary), we have

$(|u_i^j|^2)_{i,j \in \mathcal{N}}$ is doubly stochastic. We derive equation (7.3.2) by considering the following equalities

$$\operatorname{tr}[U^*(\operatorname{diag} x)U\,(\operatorname{diag} y)] = \sum_{j=1}^{\infty}\Big\langle U^*(\operatorname{diag} x)U\,(\operatorname{diag} y)e^j, e^j\Big\rangle$$

$$= \sum_{j=1}^{\infty}\Big\langle (\operatorname{diag} x)U\,(y_j e^j), U e^j\Big\rangle$$

$$= \sum_{j=1}^{\infty} y_j \langle (\operatorname{diag} x)u^j, u^j\rangle$$

$$= \sum_{j=1}^{\infty} y_j \Big\langle \sum_i x_i u_i^j e^i, u^j\Big\rangle$$

$$= \sum_{i,j=1}^{\infty} x_i\,|u_i^j|^2\,y_j.$$

Now the sesquilinear form $\langle S, T\rangle := \operatorname{tr}(ST)$ implements the duality on $V \times W$. Thus, since $\|U^*(\operatorname{diag} x)U\|_V = \|\operatorname{diag} x\|_V = \|x\|_V$ for any $U \in \mathcal{U}$ and any $x \in V$, we get that

$$B_U(x,y) = \langle\, U^*(\operatorname{diag} x)U\,,\, \operatorname{diag} y\,\rangle$$
$$\leq \|U^*(\operatorname{diag} x)U\|_V\|\operatorname{diag} y\|_W$$
$$= \|x\|_V\|y\|_W,$$

and we are done.   •

Now we can prove the von Neumann type inequality.

Theorem 7.3.11 *Let $(x,y) \in V \times W$. Then*

$$\sup_{\pi \in \mathcal{B}} P_\pi(x,y) = \sup_{U \in \mathcal{U}} B_U(x,y).$$

Proof. Given any $\pi \in \mathcal{B}$ we can define $U \in \mathcal{U}$ by $U e^j = e^{\pi(j)}$ for all $j \in \mathcal{N}$. Then $U^*(\operatorname{diag} x)U = \operatorname{diag}(x_{\pi(j)})$, so that we get the inequality

$$\sup_{\pi \in \mathcal{B}} P_\pi(x,y) \leq \sup_{U \in \mathcal{U}} B_U(x,y) \qquad (7.3.3)$$

for $(x,y) \in V \times W$. Now let us define two functions on $V \times W$. These are

$$b(x,y) := \sup_{U \in \mathcal{U}} B_U(x,y) \quad \text{and} \quad p(x,y) := \sup_{\pi \in \mathcal{B}} P_\pi(x,y).$$

For fixed $x \in V$ we know that as a supremum of a family of linear functions both p and b are convex in y. Lemma 7.3.10 and (7.3.3) together give the inequality $p(x, y) \le b(x, y) \le \|x\|_V \|y\|_W$, so for fixed $x \in V$ both p and b are everywhere finite, lsc and hence continuous (Theorem 4.1.3), convex functions of y. The same is true if we hold y fixed and consider b and p as functions of x.

Consider $F := \{x \in \ell_\infty \mid x_j = 0 \text{ eventually}\}$, the set of real finitely non-zero sequences. We know F is norm dense in both V and W. If we show for fixed $x \in F$ that $b(x, y) \le p(x, y)$ for all $y \in F$, then since F is norm dense in W and b and p are continuous functions (in y for fixed x), we get $b(x, y) \le p(x, y)$ for all $y \in W$. This holds for arbitrary $x \in F \subset V$, so $b(x, y) \le p(x, y)$ for all $(x, y) \in F \times W$. Now fix $y \in W$. Since b and p are continuous functions in x the same density arguments give $b(x, y) \le p(x, y)$ for all $x \in V$. As this y is arbitrary, we get $b(x, y) \le p(x, y)$ for all $(x, y) \in V \times W$, which together with (7.3.3) gives the results. Thus, it suffices to show $b(x, y) \le p(x, y)$ for any $(x, y) \in F \times F$.

Fix $(x, y) \in F \times F$ and choose $N \in \mathcal{N}$ such that $x_n = y_n = 0$ for all $n \ge N$. For $U \in \mathcal{U}$ define the doubly stochastic "infinite matrix" as in Lemma 7.3.10. Let $\mathcal{A}(N)$ be the set of doubly stochastic $N \times N$ matrices. Define $A = (a_{mn})_{m,n=1}^N$ by

$$a_{mn} = |u_m^n|^2 \qquad m, n = 1, \ldots, N-1,$$

$$a_{Nn} = 1 - \sum_{m=1}^{N-1} a_{mn} \quad n = 1, \ldots, N-1,$$

$$a_{mN} = 1 - \sum_{n=1}^{N-1} a_{mn} \quad n = 1, \ldots, N-1,$$

$$a_{NN} = 1 - \sum_{n=1}^{N-1} a_{Nn} \quad \left(= 1 - \sum_{m=1}^{N-1} a_{mN}\right).$$

Clearly, $A \in \mathcal{A}(N)$ and for our $(x, y) \in F \times F$ we have $B_U(x, y) = x^\top A y$, where we abuse notation and interpret x and y in $\mathbb{R}^N$ in the natural way. Let $P(N)$ be the set of all $N \times N$ permutation matrices. Birkhoff's Theorem 2.4.10 says that the convex hull of $P(N)$ is exactly $\mathcal{A}(N)$. Thus, we can write $A = \sum_{n=1}^k \lambda_n P_n$ where $P_n \in P(N)$ and $\lambda_n \ge 0$ for $n = 1, \ldots, k$ with $\sum_{n=1}^k \lambda_n = 1$. By Lemma 7.2.13 we have $x^\top P y \le \bar{x}^\top \bar{y}$ for any $P \in P(N)$, so that

$$x^\top A y = \sum_{n=1}^k \lambda_n \left(x^\top P_n y\right) \le \sum_{n=1}^k \lambda_n \left(\bar{x}^\top \bar{y}\right) = \bar{x}^\top \bar{y}.$$

If we choose $\pi \in \mathcal{B}$ (independent of U) such that $P_\pi(x, y) = \bar{x}^\top \bar{y}$, we obtain the inequality $B_U(x, y) \le P_\pi(x, y)$. We take the supremum over $U \in \mathcal{U}$ to get

$$\sup_{U \in \mathcal{U}} B_U(x, y) \le P_\pi(x, y),$$

which means $b(x, y) \le p(x, y)$ for all $(x, y) \in F \times F$. ●

Unlike in the finite dimensional space, the supremum in Theorem 7.3.11 may not be attained (Exercise 7.3.8).

7.3.4 Conjugacy of Unitarily Invariant Functions

Next we turn to conjugacy of unitarily invariant convex functions. Let X be a Banach space. Recall that the *Fenchel conjugate* of an arbitrary function $f: X \to \mathbb{R} \cup \{+\infty\}$, which we denote by $f^*: X^* \to \mathbb{R} \cup \{+\infty\}$, is

$$f^*(y) = \sup_{x \in X}\{\langle x, y \rangle - f(x)\}.$$

Note that the Fenchel conjugate of any function is a convex function. We have seen that when f is a proper lsc convex function, so is f^*. In this section when considering the second conjugate $f^{**}: X \to \mathbb{R} \cup \{+\infty\}$ of f, we restrict the domain of f^{**} to the original space X as opposed to considering the space X^{**}. Again we fix two paired spectral sequences spaces V and W, as in Definition 7.3.2, with their corresponding paired operator spaces $\mathcal{V}$ and $\mathcal{W}$.

Theorem 7.3.12 (Conjugacy and Diagonals) *Let* $\phi: \mathcal{V} \to \mathbb{R} \cup \{+\infty\}$ *be a unitarily invariant function. Then we have the identity*

$$\phi^* \circ \operatorname{diag} = (\phi \circ \operatorname{diag})^*.$$

Proof. Choose $y \in W$. It follows from the definition that

$$(\phi^* \circ \operatorname{diag})(y) = \phi^*(\operatorname{diag} y) = \sup\{\operatorname{tr}[X(\operatorname{diag} y)] - f(X) \mid X \in \mathcal{V}\}.$$

Since we can write any X as $U^* \operatorname{diag}(x)U$ for some appropriate $x \in V$ and $U \in \mathcal{U}$, we can write the right hand side of the above equality as $\sup\{\operatorname{tr}[U^*(\operatorname{diag} x)\ U(\operatorname{diag} y)] - \phi(U^*(\operatorname{diag} x)U) \mid U \in \mathcal{U},\ x \in V\}$. By the definition of B_U and the fact that ϕ is unitarily invariant we have

$$\sup\{\operatorname{tr}[U^*(\operatorname{diag} x)\ U(\operatorname{diag} y)] - \phi(U^*(\operatorname{diag} x)U) \mid U \in \mathcal{U},\ x \in V\}$$
$$= \sup\{\sup_{U \in \mathcal{U}} B_U(x, y) - f(\operatorname{diag} x) \mid x \in V\}.$$

By virtue of Theorem 7.3.11, we have

$$(\phi^* \circ \operatorname{diag})(y) = \sup\{\sup_{\pi \in \mathcal{B}} P_\pi(x, y) - \phi(\operatorname{diag} x) \mid x \in V\}$$
$$= \sup\left\{ \sum_{j=1}^{\infty} x_{\pi(j)} y_j - f(\operatorname{diag}(x_{\pi(j)})) \,\Big|\, x \in V,\ \pi \in \mathcal{B}\right\}$$
$$= \sup\{\langle z, y \rangle - f(\operatorname{diag} z) \mid z \in V\}, \ (\text{set } z = (x_{\pi(j)})_{j=1}^{\infty})$$
$$= (f \circ \operatorname{diag})^*(y).$$

Corollary 7.3.13 (Convexity) *Let* $\phi\colon V \to \mathbb{R}\cup\{+\infty\}$ *be a unitarily invariant function. Then* ϕ *is proper, convex, and weakly lsc if and only if* $\phi \circ \mathrm{diag}$ *is.*

Proof. Unitary invariance gives $\phi = \phi^{**}$ if and only if $\phi \circ \mathrm{diag} = \phi^{**} \circ \mathrm{diag}$. The results follow since $\phi^{**} \circ \mathrm{diag} = (\phi \circ \mathrm{diag})^{**}$. $\quad\bullet$

Lemma 7.3.14 (Invariance and Conjugacy) *Let* $\phi\colon V \to \mathbb{R}\cup\{+\infty\}$ ($f\colon V \to \mathbb{R}\cup\{+\infty\}$) *be unitarily invariant. Then* $\phi^*(f^*)$ *is unitarily (rearrangement) invariant.*

Proof. This lemma follows directly from the definitions. We leave the proof as an exercise (Exercise 7.3.9). $\quad\bullet$

Theorem 7.3.15 *Let* $f\colon V \to \mathbb{R}\cup\{+\infty\}$ *be a rearrangement invariant function. Then we have*

$$(f \circ \lambda)^* = f^* \circ \lambda.$$

Proof. Theorems 7.3.8 and 7.3.12 allow us to write

$$f^* = (f \circ \lambda \circ \mathrm{diag})^* = (f \circ \lambda)^* \circ \mathrm{diag}.$$

If we compose this expression with λ and observe that $(\mathrm{diag} \circ \lambda)(T)$ is unitarily equivalent to T for all $T \in B_0$, then Lemma 7.3.14 allows us to write

$$f^* \circ \lambda = (f \circ \lambda)^* \circ \mathrm{diag} \circ \lambda = (f \circ \lambda)^*,$$

and we are done. $\quad\bullet$

Putting together Theorems 7.3.7 and 7.3.15 as well as Theorems 7.3.8 and 7.3.12 gives

Theorem 7.3.16 *Let* $\phi\colon V \to \mathbb{R}\cup\{+\infty\}$ (*resp.,* $f\colon V \to \mathbb{R}\cup\{+\infty\}$) *be a unitarily invariant function (resp., rearrangement invariant). Then we have* $\phi = f \circ \lambda$ (*resp.,* $f = \phi \circ \mathrm{diag}$) *for some rearrangement invariant (resp., unitarily invariant) function* $f\colon V \to \mathbb{R}\cup\{+\infty\}$ (*resp.,* $\phi\colon V \to \mathbb{R}\cup\{+\infty\}$). *Further we get the formula*

$$\phi^* = f^* \circ \lambda \;(f^* = \phi^* \circ \mathrm{diag}).$$

Proof. Exercise 7.3.10. $\quad\bullet$

7.3.5 Subdifferentials of Unitarily Invariant Functions

Now we proceed to examine the subdifferential of a unitarily invariant convex function.

Proposition 7.3.17 *Suppose $R \in B_{sa}$. Then for $t \in \mathbb{R}$ we have*

$$U_t := \exp(itR) \in \mathcal{U}.$$

Further, we have $\|U_t - I\| \to 0$ and $\|t^{-1}(U_t - I) - iR\| \to 0$ as $t \to 0$ (where $\| \cdot \|$ denotes the uniform operator norm and $i = \sqrt{-1}$).

Proof. It is not hard to check that $U_t \in \mathcal{U}$ (Exercise 7.3.11). We also have

$$\|U_t - I\| = \left\| \sum_{j=1}^{\infty} \left(\frac{(it)^j}{j!} \right) R^j \right\|$$

$$\leq \sum_{j=1}^{\infty} \left(\frac{|t|^j}{j!} \right) \|R^j\|$$

$$\leq \exp(\|tR\|) - 1.$$

Thus, we have $U_t \to I$ uniformly as $t \to 0$. Further,

$$\|t^{-1}(U_t - I) - iR\| = \left\| t^{-1} \sum_{j=2}^{\infty} \left(\frac{(it)^j}{j!} \right) R^j \right\|$$

$$\leq t^{-1} \sum_{j=2}^{\infty} \left(\frac{|t|^j}{j!} \right) \|R^j\|$$

$$\leq t^{-1} \left[\exp(\|tR\|) - (1 + \|tR\|) \right],$$

which gives $t^{-1}(U_t - I) \to iR$ uniformly as $t \to 0$.   ●

Next we state a technical lemma on approximating bounded operators with unitary operators, whose proof can be found in [216, p. 98].

Lemma 7.3.18 *If $\tilde{A}$ is a bounded (not necessarily self-adjoint) operator for which $\|\tilde{A}\| < 1$, then there is some $N > 2$ and $\{U_n\}_{n=1}^N \subset \mathcal{U}$ such that*

$$\tilde{A} = \frac{1}{N} \sum_{n=1}^{N} U_n.$$

We use this in the proof of the following estimate.

Proposition 7.3.19 *Let X be any (non-self-adjoint) Schatten p-space ($1 \leq p < \infty$) or the space of compact or bounded operators. Let A be a bounded operator and let $T \in X$. Then both $\|AT\|$ and $\|TA\|$ are bounded by $\|T\|_X \|A\|$.*

Proof. If X is the space of compact or bounded operators, then this is immediate. We consider the case when X is a (non-self-adjoint) Schatten p-space ($1 \leq p < \infty$). If $A = 0$, the results are also immediate, so assume $A \neq 0$. Fix $\delta \in (0,1)$, let $\tilde{A} = A/(1 + \delta)\|A\|$ and apply Lemma 7.3.18 to $\tilde{A}$, giving

$$\|AT\|_p = (1+\delta)\|A\| \cdot \|\tilde{A}T\|_p$$

$$= (1+\delta)\|A\| \cdot \left\|\frac{1}{N}\sum_{i=1}^{N} U_i T\right\|_p$$

$$\leq (1+\delta)\|A\|\frac{1}{N}\sum_{i=1}^{N} \|U_i T\|_p$$

$$= (1+\delta)\|A\| \cdot \|T\|_p,$$

where we use the fact that $\|UT\|_p = \|T\|_p$ for all $U \in \mathcal{U}$. Since this holds for each $\delta \in (0,1)$, we obtain the results we want. $\qquad\bullet$

Proposition 7.3.20 *Let X be any (non self-adjoint) Schatten p-space ($1 \leq p < \infty$) or the space of compact or bounded operators. Let A_t be a family of bounded operators which converge uniformly to A as $t \to 0$. Let $T_t \subset X$ be a family of operators which converge in the norm on X to T as $t \to 0$. Then we have*

$$\lim_{t\to 0} \|A_t T_t - AT\|_X = 0 = \lim_{t\to 0}\|T_t A_t - TA\|_X.$$

Proof. If we consider the inequalities

$$\|A_t T_t - AT\|_X = \|A_t(T_t - T) + (A_t - A)T\|_X$$
$$\leq \|A_t(T_t - T)\|_X + \|(A_t - A)T\|_X$$
$$\leq \|A_t\| \cdot \|T_t - T\|_X + \|A_t - A\| \cdot \|T\|_X,$$

we get the first equality by Proposition 7.3.19. The second can be proven similarly and is left as Exercise 7.3.12. $\qquad\bullet$

Again we return to the paired spectral sequence spaces V and W, as in Definition 7.3.2, with their corresponding paired operator spaces $\mathcal{V}$ and $\mathcal{W}$.

Theorem 7.3.21 *Let $f \colon \mathcal{V} \to \mathbb{R} \cup \{+\infty\}$ be a unitarily invariant lsc convex function. Suppose that $(S,T) \in \mathcal{V} \times \mathcal{W}$ satisfy $T \in \partial f(S)$. Then $TS = ST$.*

Proof. Let $\tilde{\mathcal{V}}$ be the non self-adjoint extension of $\mathcal{V}$ and let $U_t = \exp[-t(ST - TS)] \in \mathcal{U}$. That is, in Proposition 7.3.17 we have $R = -i(TS - ST) \in B_{sa}$. Thus, $U_t \to I$ uniformly and $t^{-1}(U_t - I) \to (ST - TS)$ uniformly as $t \to 0$. By Proposition 7.3.20 we obtain

$$t^{-1}(U_t - I)S \to -(ST - TS)S \qquad (7.3.4)$$

in the norm on $\tilde{\mathcal{V}}$. Upon taking the adjoint of this we get

$$\|t^{-1}S(U_t^* - I) \to -S(TS - ST)\|_{\tilde{\mathcal{V}}} \to 0. \qquad (7.3.5)$$

Now if we apply Proposition 7.3.20 with the left hand side of (7.3.4) playing the role of T_t and U_t^* playing the role of A_t, we obtain

$$\left\| [t^{-1}(U_t - I)S]U_t^* + [(ST - TS)S]I \right\|_{\tilde{\mathcal{V}}} \to 0,$$

which when we add to (7.3.5) gives

$$\left\| t^{-1}(U_t S U_t^* - S) + (2STS - S^2T - TS^2) \right\|_{\mathcal{V}} \to 0. \qquad (7.3.6)$$

Hence

$$\mathrm{tr}[t^{-1}(U_t S U_t^* - S)T] + \mathrm{tr}[2STST - S^2T^2 - TS^2T]. \qquad (7.3.7)$$

If we examine the right hand side of (7.3.7), we see that since $(S, T) \in \mathcal{V} \times \mathcal{W}$, we have $STST, S^2T^2$ and $TS^2T \in B_1$. Further, we have

$$
\begin{aligned}
\|TS - ST\|^2 &= \mathrm{tr}[(TS - ST)^*(TS - ST)] \\
&= \mathrm{tr}[(ST - TS)(TS - ST)] \\
&= -\mathrm{tr}[TSTS + STST - ST^2S - TS^2T] \\
&= -\mathrm{tr}[TSTS] - \mathrm{tr}[STST] + \mathrm{tr}[ST^2S] + \mathrm{tr}[TS^2T] \\
&= -\mathrm{tr}[2STST] + \mathrm{tr}[S^2T^2] + \mathrm{tr}[TS^2T] \\
&= -\mathrm{tr}[2STST - S^2T^2 - TS^2T]. \qquad (7.3.8)
\end{aligned}
$$

If we use the unitary invariance of f, then for any $t \in \mathbb{R}$ we get the inequality

$$
\begin{aligned}
t \times \mathrm{tr}[t^{-1}(U_t S U_t^* - S)T] &= \langle T, U_t S U_t^* - S \rangle \\
&\leq f(U_t S U_t^*) - f(S) = 0,
\end{aligned}
$$

which together with (7.3.7) and (7.3.8) implies $\|TS - ST\|^2 = 0$, or $TS = ST$. ●

Combining this result with the fact that commuting elements of B_0 are simultaneously diagonalizable [216] we have the following result.

Lemma 7.3.22 *Let* $f \colon \mathcal{V} \to \mathbb{R} \cup \{+\infty\}$ *be a unitarily invariant lsc convex function. For* $x \in \mathcal{V}$ *we have* $y \in \partial(f \circ \mathrm{diag})(x)$ *if and only if* $\mathrm{diag}\, y \in \partial f(\mathrm{diag}\, x)$.

Proof. Assume first that $\mathrm{diag}\, y \in \partial f(\mathrm{diag}\, x)$. Then for each $z \in \mathcal{V}$ we have

$$
\begin{aligned}
\langle y, z - x \rangle &= \mathrm{tr}[(\mathrm{diag}\, y)(\mathrm{diag}\, z - \mathrm{diag}\, x)] \\
&\leq f(\mathrm{diag}\, z) - f(\mathrm{diag}\, x)
\end{aligned}
$$

so that $y \in \partial(f \circ \mathrm{diag})(x)$.

To see the other implication given $Q \in \mathcal{V}$, choose $z \in V$ and $U \in \mathcal{U}$ so that $Q = U^*(\mathrm{diag}\, z)U$. Now we use Theorem 7.3.11, the assumption that $y \in \partial(f \circ \mathrm{diag})(x)$ and the unitary invariance of f (twice) to obtain

$$\operatorname{tr}[(\operatorname{diag} y)(Q - \operatorname{diag} x)] = \operatorname{tr}[(\operatorname{diag} y)(U^* \operatorname{diag} z \, U)] - \langle y, x \rangle$$
$$= B_U(z, y) - \langle y, x \rangle$$
$$\leq \sup_{\pi \in \mathcal{B}} P_\pi(z, y) - \langle y, x \rangle$$
$$= \sup\{\langle y, (z_{\pi(j)}) - x \rangle \mid \pi \in \mathcal{B}\}$$
$$\leq \sup\{f(\operatorname{diag}(z_{\pi(j)})) - f(\operatorname{diag} x) \mid \pi \in \mathcal{B}\}$$
$$= f(\operatorname{diag} z) - f(\operatorname{diag} x)$$
$$= f(Q) - f(\operatorname{diag} x),$$

which gives us $\operatorname{diag}(y) \in \partial f(\operatorname{diag} x)$. •

The next result can be viewed as a generalization of the chain rule in Lemma 7.2.15. We give a direct proof.

Lemma 7.3.23 *Let $f : \mathcal{V} \to \mathbb{R} \cup \{+\infty\}$ be a unitarily invariant lsc convex function. Then, for $(S, T) \in \mathcal{V} \times \mathcal{W}$ we have $T \in \partial f(S)$ if and only if $U^* T U \in \partial f(U^* S U)$.*

Proof. By definition, $T \in \partial f(S)$ if and only if

$$f(S + UHU^*) - f(S) \geq \langle T, UHU^* \rangle \quad \text{for all } H \in \mathcal{V}. \tag{7.3.9}$$

Since f is unitarily invariant $f(S+UHU^*)-f(S) = f(U^*SU+H)-f(U^*SU)$. Moreover, one can directly check that $\langle T, UHU^* \rangle = \langle U^*TU, H \rangle$. Thus, (7.3.9) is equivalent to

$$f(U^*SU + H) - f(U^*SU) \geq \langle U^*TU, H \rangle \quad \text{for all } H \in \mathcal{V}.$$

That is $U^*TU \in \partial f(U^*SU)$. •

We now prove the central result of this subsection.

Theorem 7.3.24 (Convex Subgradient) *Let $f : \mathcal{V} \to \mathbb{R} \cup \{+\infty\}$ be a unitarily invariant lsc convex function. For $(S, T) \in \mathcal{V} \times \mathcal{W}$ we have $T \in \partial f(S)$ if and only if there exist $U \in \mathcal{U}$ and $(x, y) \in V \times W$ with $S = U^*(\operatorname{diag} x)U$, $T = U^*(\operatorname{diag} y)U$ and $y \in \partial(f \circ \operatorname{diag})(x)$.*

Proof. Assume first the existence of U, x and y with the stated properties. Since $y \in \partial(f \circ \operatorname{diag})(x)$, Lemma 7.3.22 tells us that $\operatorname{diag} y \in \partial f(\operatorname{diag} x)$. The unitary invariance of f gives $T = U^*(\operatorname{diag} y)U \in \partial f(U^*(\operatorname{diag} x)U) = \partial f(S)$.

To see the reverse implication note that $T \in \partial f(S)$ implies $TS = ST$ by Theorem 7.3.21, which means S and T are simultaneously diagonalizable. Thus, there exists $U \in \mathcal{U}$, $(x, y) \in V \times W$ with $S = U^*(\operatorname{diag} x)U$, $T = U^*(\operatorname{diag} y)U$, so that $U^*(\operatorname{diag} y)U \in \partial f(U^*(\operatorname{diag} x)U)$. Now the unitary invariance allows us to rewrite this as $\operatorname{diag} y \in \partial f(\operatorname{diag} x)$, which by Lemma 7.3.22 implies $y \in \partial(f \circ \operatorname{diag})(x)$ as required. •

The next theorem shows that the differentiability of a convex unitarily invariant function f is characterized by that of $f \circ \mathrm{diag}$.

Theorem 7.3.25 *Let $f \colon \mathcal{V} \to \mathbb{R} \cup \{+\infty\}$ be a unitarily invariant lower semicontinuous convex function. Then f is Gâteaux differentiable at $A \in \mathcal{V}$ if and only if $f \circ \mathrm{diag}$ is Gâteaux differentiable at $\lambda(A) \in V$.*

Proof. Note that a lsc convex function is Gâteaux differentiable at a point if and only if its subdifferential is a singleton at that point. Assume that f is Gâteaux differentiable at A. Let $y, z \in \partial(f \circ \mathrm{diag})(\lambda(A))$ and let $u \in \mathcal{U}$ be such that $U^* \lambda(A) U = A$. By Theorem 7.3.24, $U^*(\mathrm{diag}\, y)U$ and $U^*(\mathrm{diag}\, z)U \in \partial f(A)$. Since f is Gâteaux differentiable at A, $U^*(\mathrm{diag}\, y)U = U^*(\mathrm{diag}\, z)U$ which implies that $y = z$. Thus, $\partial(f \circ \mathrm{diag})(\lambda(A))$ is a singleton and $f \circ \mathrm{diag}$ is Gâteaux differentiable at $\lambda(A)$.

To prove the converse we assume that F is Gâteaux differentiable at $\lambda := \lambda(A)$. Then $\partial(f \circ \mathrm{diag})(\lambda) = \{\mu\}$ is a singleton.

First we observe that if $\lambda_i = \lambda_j$ then $\mu_i = \mu_j$. In fact, when $\lambda_i = \lambda_j$ the permutation invariance and the Gâteaux differentiability of $f \circ \mathrm{diag}$ implies that

$$\mu_i t + o(t) = f \circ \mathrm{diag}(\lambda + te_i) - f \circ \mathrm{diag}(\lambda)$$
$$= f \circ \mathrm{diag}(\lambda + te_j) - f \circ \mathrm{diag}(\lambda) = \mu_j t + o(t).$$

Therefore, $\mu_i = \mu_j$.

Next we show that if $B \in \mathcal{U}$ commutes with $\mathrm{diag}\, \lambda$, i.e.,

$$B \,\mathrm{diag}\, \lambda = \mathrm{diag}\, \lambda B, \qquad (7.3.10)$$

then B also commute with $\mathrm{diag}\, \mu$. Obviously it suffices to show that for arbitrary natural numbers i and j,

$$\langle e_i, B(\mathrm{diag}\, \mu)e_j\rangle = \langle e_i, (\mathrm{diag}\, \mu)Be_j\rangle,$$

or equivalently

$$\mu_j \langle e_i, Be_j\rangle = \mu_i \langle e_i, Be_j\rangle. \qquad (7.3.11)$$

When $\lambda_i = \lambda_j$ we have $\mu_i = \mu_j$ which immediately leads to (7.3.11). It remains to consider the case when $\lambda_i \neq \lambda_j$. Note that it follows from (7.3.10) that

$$\lambda_j \langle e_i, Be_j\rangle = \langle e_i, B(\mathrm{diag}\, \lambda)e_j\rangle = \langle e_i, (\mathrm{diag}\, \lambda)Be_j\rangle = \lambda_i \langle e_i, Be_j\rangle,$$

which implies that $\langle e_j, Be_i\rangle = 0$. Thus, both sides of (7.3.11) are 0.

Now consider any two elements S and T of $\partial f(A)$. By Theorem 7.3.24 there exist unitary operators $U, V \in \mathcal{U}$ satisfying

$$U^*(\mathrm{diag}\, \lambda)U = V^*(\mathrm{diag}\, \lambda)V = A \qquad (7.3.12)$$

such that $S = U^*(\text{diag}\,\mu)U$ and $T = V^*(\text{diag}\,\mu)V$. It follows from (7.3.12) that $B := VU^*$ commutes with $\text{diag}\,\lambda$. Therefore, B also commutes with $\text{diag}\,\mu$. Then we have $S = U^*(\text{diag}\,\mu)U = V^*B(\text{diag}\,\mu)U = V^*(\text{diag}\,\mu)BU = V^*(\text{diag}\,\mu)V = T$. That is to say $\partial f(A)$ is a singleton, or f is Gâteaux differentiable at A.   ●

7.3.6 Generalizations to Nonconvex Functions

In this subsection we show that the method we used in the previous sections is also applicable to nonconvex functions. More concretely, we show Theorem 7.3.24 extends to nonconvex Lipschitz functions with the convex subdifferential replaced by the Gâteaux subdifferential. Let X be a Banach space and let $f\colon X \to R$ be a Lipschitz function. Recall that the Gâteaux subdifferential $\partial_G f(x)$ of f at $x \in X$ is defined by

$$\partial_G f(x) := \{x^* \in X^* : \liminf_{t \to 0} \frac{f(x+th) - f(x)}{t} \geq \langle x^*, h \rangle, \ \text{ for all } h \in X\}.$$

We continue to use the paired Banach spaces $V \times W$ and their corresponding paired operator spaces $\mathcal{V} \times \mathcal{W}$ as described in Definition 7.3.2.

The proof of the extension of Theorem 7.3.24 follows the same strategy as Section 5. First we extend Theorem 7.3.21.

Theorem 7.3.26 (Commutativity) *Let $f\colon \mathcal{V} \to \mathbb{R} \cup \{+\infty\}$ be a unitarily invariant locally Lipschitz function. Let $S \in \mathcal{V}$ and $T \subset \mathcal{W}$ satisfy $T \in \partial_G f(S)$. Then $TS = ST$.*

Proof. Define, as in the proof of Theorem 7.3.21, $U_t = \exp[-t(ST-TS)] \in \mathcal{U}$. Then it follows from (7.3.6) that

$$U_t S U_t^* = S - t(2STS - S^2T - TS^2) + o(t).$$

Since f is unitarily invariant, locally Lipschitz and $T \in \partial_G f(S)$ we have

$$\begin{aligned}
0 &= t^{-1}[f(U_t S U_t^*) - f(S)] \\
&= t^{-1}[f(S - t(2STS - S^2T - TS^2) + o(t)) - f(S)] \\
&= t^{-1}[f(S - t(2STS - S^2T - TS^2)) - f(S)] \\
&\quad + t^{-1}[f(S - t(2STS - S^2T - TS^2) + o(t)) \\
&\quad\quad - f(S - t(2STS - S^2T - TS^2))] \\
&= t^{-1}[f(S - t(2STS - S^2T - TS^2)) - f(S)] + o(1) \\
&\geq -\langle T, 2STS - S^2T - TS^2 \rangle + o(1) \\
&= -\text{tr}(2STST - S^2T^2 - TS^2T) + o(1) \\
&= \|TS - ST\|^2 + o(1).
\end{aligned}$$

Taking limits as $t \to 0$ yields $\|TS - ST\| = 0$, i.e., $TS = ST$.   ●

Lemma 7.3.27 *Let $f\colon \mathcal{V} \to \mathbb{R}$ be a unitarily invariant locally Lipschitz function. For $x \in \mathcal{V}$ we have $y \in \partial_G(f \circ \mathrm{diag})(x)$ if and only if $\mathrm{diag}\, y \in \partial_G f(\mathrm{diag}\, x)$.*

Proof. The idea is to approximate operators and functions on $\mathcal{V}$ by their finite dimensional restrictions. It is not hard to check that results similar to Lemma 7.2.17 apply to these restrictions for the Gâteaux subdifferential. Then the general result is derived by taking limits.

Observe that the "if" part is easy and left as Exercise 7.3.14. We concentrate on the "only if" part. We will need the following notations and simple facts.

Let $\{e^j\}$ be the standard basis in ℓ_2. Define subspace E_n of $\ell_2^{\mathbb{C}}$ by $E_n := \mathrm{span}\{e^1, e^2, \ldots, e^n\}$. Then $E_n^\perp := \mathrm{cl}\,\mathrm{span}\{e^{n+1}, e^{n+2}, \ldots\}$. Let H_n denote the Hermitian operators on the space E_n. Define the eigenvalue mapping $\lambda_n\colon H_n \to \mathbb{R}^n$ taking operators to their eigenvalue sequence in decreasing order and define the diagonal mapping $\mathrm{diag}_n\colon \mathbb{R}^n \to H_n$ in the natural way: for $y \in \mathbb{R}^n$ and $\mu \in C^n$,

$$(\mathrm{diag}_n y) \sum_{k=1}^{n} \mu_k e^k = \sum_{k=1}^{n} \mu_k y_k e^k.$$

Let $P_n\colon \ell_2^{\mathbb{C}} \to E_n$ denote the orthogonal projection. It is easy to check that the adjoint $P_n^*\colon E_n \to \ell_2^{\mathbb{C}}$ is just inclusion (Exercise 7.3.15). Given any $Z \in \mathcal{V}$ we have, relative to the decomposition $\ell_2^{\mathbb{C}} = E_n \oplus E_n^\perp$, the block decomposition

$$P_n^* P_n Z P_n^* P_n = \begin{pmatrix} P_n Z P_n^* & 0 \\ 0 & 0 \end{pmatrix}.$$

Furthermore, we can check the following as an exercise: (Exercise 7.3.16)

$$\lim_{n \to \infty} \|Z - P_n^* P_n Z P_n^* P_n\|_{\mathcal{V}} = 0.$$

Finally, let $P_n^V\colon \mathcal{V} \to \mathbb{R}^n$ and $P_n^W\colon W \to \mathbb{R}^n$ denote the natural truncation maps, and note the identity

$$\mathrm{diag}_n(P_n^V z) = P_n(\mathrm{diag}\, z)P_n^*, \quad \text{for } z \in \mathcal{V}.$$

The analogous identity also holds in W.

Suppose that elements $x \in \mathcal{V}$ and $y \in W$ satisfy $y \in \partial_G(f \circ \mathrm{diag})(x)$. Define a permutation-invariant locally Lipschitz function $\phi\colon \mathbb{R}^n \to \mathbb{R} \cup \{+\infty\}$ by $\phi(z) := (f \circ \mathrm{diag})(z_1, \ldots, z_n, x_{n+1}, x_{n+2}, \ldots)$. It is straightforward to check $P_n^W y \in \partial_G \phi(P_n^V x)$ (Exercise 7.3.17). Then, it follows from Theorem 7.2.22 (note that the proof of that theorem applies to the Gâteaux subdifferential) that $\mathrm{diag}_n P_n^W y \in \partial_G(\phi \circ \lambda_n)(\mathrm{diag}_n P_n^V x)$. Denote the Lipschitz constant of f by L. Then for any operator $Z \in \mathcal{V}$ and small real $t > 0$ we have

$$f(\text{diag}\,x + tZ) - f(\text{diag}\,x) \geq -Lt\|Z - P_n^* P_n Z P_n^* P_n\|_V$$
$$+ f(\text{diag}\,x + t P_n^* P_n Z P_n^* P_n) - f(\text{diag}\,x).$$

By using the above block decomposition we know

$$f(\text{diag}\,x + t P_n^* P_n Z P_n^* P_n) - f(\text{diag}\,x)$$
$$= \phi(\lambda_n(\text{diag}_n(P_n^V x) + t P_n Z P_n^*)) - \phi(\lambda_n \text{diag}(P_n^V x))$$
$$\geq t\langle \text{diag}_n P_n^W y, P_n Z P_n^* \rangle + \|Z\|_V o(t)$$
$$\geq t\langle P_n \text{diag}\,y P_n^*, P_n Z P_n^* \rangle + \|Z\|_V o(t)$$
$$\geq t\langle \text{diag}\,y, Z \rangle - t\|y\|_W \|Z - P_n^* P_n Z P_n^* P_n\|_V + \|Z\|_V o(t).$$

Hence we deduce

$$f(\text{diag}\,x + tZ) - f(\text{diag}\,x)$$
$$\geq t\langle \text{diag}\,y, Z \rangle - t(\|y\|_W + L)\|Z - P_n^* P_n Z P_n^* P_n\|_V + \|Z\|_V o(t).$$

Taking the limit as $n \to \infty$ shows

$$\liminf_{t \to 0} \frac{f(\text{diag}\,x + tZ) - f(\text{diag}\,x)}{t} \geq \langle \text{diag}\,y, Z \rangle.$$

Therefore, $\text{diag}\,y \in \partial_G f(\text{diag}\,x)$, as required   ●

The next lemma is elementary.

Lemma 7.3.28 *Let* $f\colon \mathcal{V} \to \mathbb{R}$ *be a unitarily invariant locally Lipschitz function. For* $S \in \mathcal{V}$ *and* $T \in \mathcal{W}$ *we have* $T \in \partial_G f(S)$ *if and only if* $U^* T U \in \partial_G f(U^* S U)$.

Proof. The proof of this lemma is similar to that of the proof of Lemma 7.3.23 with appropriate notational changes. We leave the details to Exercise 7.3.15.   ●

Theorem 7.3.29 *Let* $f\colon \mathcal{V} \to \mathbb{R}$ *be a unitarily invariant locally Lipschitz function. For* $S \in \mathcal{V}$ *and* $T \in \mathcal{W}$ *we have* $T \in \partial_G f(S)$ *if and only if there exists* $U \in \mathcal{U}$, $x \in V$ *and* $y \in W$ *with* $S = U^*(\text{diag}(x))U$, $T = U^*(\text{diag}\,y)U$ *and* $y \in \partial_G(f \circ \text{diag})(x)$.

Proof. Assume first the existence of U, x and y with the stated properties. Since $y \in \partial_G(f \circ \text{diag})(x)$, Lemma 7.3.27 tells us that $\text{diag}\,y \in \partial_G f(\text{diag}\,x)$. The unitary invariance of f gives $T = U^*(\text{diag}\,y)U \in \partial_G f(U^*(\text{diag}\,x)U) = \partial_G f(S)$.

To see the reverse implication note that $T \in \partial_G f(S)$ implies $TS = ST$ by Theorem 7.3.26, which means T and S are simultaneously diagonalizable. Thus, there exist $U \in \mathcal{U}$, $x \in V$ and $y \in W$ with $S = U^*(\text{diag}\,x)U$,

$T = U^*(\text{diag } y)U$, so that $U^*(\text{diag } y)U \in \partial_G f(U^*(\text{diag } x)U)$. Now the unitary invariance allows us to rewrite this as $\text{diag } y \in \partial_G f(\text{diag } x)$, which by Lemma 7.3.27 implies $y \in \partial_G(f \circ \text{diag})(x)$ as required.   ●

7.3.7 Applications and Examples

We illustrate some of the results with a number of examples.

Example 7.3.30 We know that the norm of ℓ_p for $p \in [1, \infty)$ is Gâteaux differentiable away from the origin. As an immediate consequence of Theorem 7.3.25, so is the norm of B_p.

Next we take up the case of the Calderón norm.

Example 7.3.31 For $p \in (1, \infty)$ and q satisfying $p^{-1} + q^{-1} = 1$ the symmetric norm defined by

$$|||x|||_p = \sup_n \left(n^{-\frac{1}{q}} \sum_{i=1}^{n} \lambda_i(\text{diag } x) \right)$$

is the Calderón norm on ℓ_p (see [239]). Clearly, this is a rearrangement invariant continuous function on this sequence space, so that $||| \cdot |||_{\mathcal{V},p} = ||| \cdot |||_p \circ \lambda(\cdot)$ is a unitarily invariant function on B_p. Now apply Theorem 7.3.16 to get (as in the case of the Schatten p-norms) the very pleasant formula for the dual norm

$$(||| \cdot |||_{\mathcal{V},p})^* = (||| \cdot |||_p \circ \lambda)^*$$
$$= (||| \cdot |||_p)^* \circ \lambda$$
$$= ||| \cdot |||_{\mathcal{V},q}.$$

A generalization can be found in Exercise 7.3.18.

Now let us consider an example that has been the subject of much interest in recent years.

Example 7.3.32 In [116] the author looks at self-concordant barriers, which play a central role in many types of optimization problems. A number of the details here can be put into quite an appealing setting by the theory developed thus far. We start with the Taylor expansion

$$\ln(1+t) = t - \frac{1}{2}t^2 + \frac{1}{3}t^3 - \frac{1}{4}t^4 + \cdots$$
$$= -\sum_{n=1}^{\infty} \frac{(-1)^n}{n} t^n.$$

Thus, the function $g \colon \mathbb{R} \to \mathbb{R}$ given by

$$g(t) = \begin{cases} t - \ln(1 + t) & t > -1, \\ +\infty & t \leq -1. \end{cases}$$

is convex. Further, g has the conjugate $g^*(s) = g(-s)$ (Exercise 7.3.19). Now we define the function f on ℓ_2 by writing

$$f(x) = \sum_{i=1}^{\infty} g(x_i).$$

Clearly, f is finite and continuous at 0 since $|g(t)| \leq kt^2$ if $|t| \leq 1/2$. Using our technique we can derive a function ϕ on B_2 in the obvious way:

$$\phi = f \circ \lambda,$$

and Theorem 7.3.15 shows $\phi^*(T) = \phi(-T)$. Notice that if $T \in B_2$ and $I + T \geq 0$, then

$$\phi(T) = \operatorname{tr}(T) - \ln(\det(I + T)).$$

Example 7.3.33 The previous example can be extended to B_p for appropriate p by subtracting more terms of the series for log: $\ln(1+t) - (t - t^2/2), \ldots$. This will be convex or concave alternatively. Consider for example $g(t) := \ln(1 + t) - (t - t^2/2)$. Direct calculation yields that $g'(t) = t^2/(1 + t)$ and $g''(t) = 1 - 1/(1 + t)^2$. Therefore, for $t > 0$, g is increasing and convex. The dual of g is

$$g^*(s) = st(s) - \ln(1 + t(s)) + t(s) - t^2(s)/2,$$

where $t(s) = \frac{s + \sqrt{s^2 + 4s}}{2}$. Since $|g(t)| \leq kt^3$ for t small, the (permutation invariant) function

$$f(x) := \sum_{i=1}^{\infty} g_i(x_i)$$

is well defined, continuous and convex on ℓ_3. Then $\phi := f \circ \lambda$ is a continuous unitarily invariant convex function on B_3. Consequently $\phi^* = f^* \circ \lambda$ is a continuous unitarily invariant convex function on $B_{3/2}$, where $f^*(x) = \sum_{i=1}^{\infty} g^*(x_i)$.

Our last example concerns the kth largest eigenvalues of a positive self-adjoint operator.

Example 7.3.34 The kth largest eigenvalue of a positive self-adjoint operator arises naturally in many applications. Consider $0 < S \in B_p$. Let us denote the kth largest eigenvalue of S by $\mu_k(S)$. Then $\mu_k = \phi_k \circ \lambda$ where $\phi_k \colon \ell_2 \to \mathbb{R}$ is defined by $\phi_k(x) :=$ kth largest component of x. Observe that ϕ_k can be expressed as the difference of two convex continuous permutation invariant functions. In fact, define $\sigma_k(x) := \max\{\sum_{i=1}^{k} x_{n_i} : n_i \in \mathcal{N}\}$. Then σ_k is continuous, permutation invariant and convex and $\phi_k = \sigma_k - \sigma_{k-1}$. Thus, ϕ_k is a locally Lipschitz permutation invariant function on ℓ_p, and therefore μ_k is a unitarily invariant locally Lipschitz function on B_p. Using Theorem 7.3.29, the characterization of the subdifferentials for the kth largest eigenvalue of symmetric matrices in Theorem 7.2.26 can be generalized to that of μ_k. In fact, replacing the Fréchet subdifferential ∂_F in Theorem 7.2.25 by the Gâteaux subdifferential ∂_G, the results and the proofs of Theorem 7.2.25 remain valid. So we have

Theorem 7.3.35 *Let* $x \in \ell_p, p > 1$ *with positive components. Then*

$$\partial_G \phi_k(x) = \begin{cases} \mathrm{conv}\{e^i \mid x_i = \phi_k(x)\} & \text{if } \phi_{k-1}(x) > \phi_k(x), \\ \emptyset & \text{otherwise}, \end{cases}$$

$$\partial_C \phi_k(x) = \mathrm{conv}\{e^i \mid x_i = \phi_k(x)\},$$

and

$$\partial_L \phi_k(x) = \{y \in \mathrm{conv}\{e^i \mid x_i = \phi_k(x)\} : \#(\mathrm{supp}\, y) \le \alpha\},$$

where $\alpha = 1 - k + \#\{i \mid x_i \ge \phi_k(x)\}$.

Proof. Exercise 7.3.20.   ●

Combining Theorems 7.3.29 and 7.3.35 we have

Theorem 7.3.36 *Let* $S \in B_p, p > 1$ *be a positive operator. Then*

$$\partial_C \mu_k(S) = \mathrm{conv}\{u \odot u \mid u \in \ell_p, \|u\| = 1, Su = \mu_k(S)u\},$$

and

$$\partial_L \mu_k(S) = \{Y \in \partial_C \mu_k(S) \mid \mathrm{rank}\, Y \le 1 - k + \#\{i \mid \mu_i(S) \ge \mu_k(S)\}\}.$$

Proof. Exercise 7.3.21.   ●

7.3.8 Commentary and Exercises

Our main reference for this section is [59]. We recommend [128, 216, 239] for additional information on various operator spaces and corresponding sequence spaces discussed here. The Gâteaux differentiability characterization in Theorem 7.3.25 cannot be generalized to Fréchet differentiability. In fact, the eigenvalue mapping λ defined in this section may not even be continuous (Exercise 7.3.13). One way to overcome this difficulty is to eliminate all the zeros in the definition of the eigenvalue mapping. Of course, an eigenvalue mapping defined this way is not "faithful". We refer to [55] for additional details.

Exercise 7.3.1 Prove Proposition 7.3.3.

Exercise 7.3.2 Prove Proposition 7.3.4.

Exercise 7.3.3 Prove Proposition 7.3.5.

Exercise 7.3.4 Prove Theorem 7.3.7.

Exercise 7.3.5 Prove Theorem 7.3.8.

Exercise 7.3.6 Let V and $\mathcal{V}$ be as in Definition 7.3.2. Prove that

$$\| \operatorname{diag}(x) \|_{\mathcal{V}} = \|x\|_V, \quad \text{for all } x \in V$$

and

$$\| \lambda(T) \|_V = \|T\|_{\mathcal{V}}, \quad \text{for all } T \in \mathcal{V}.$$

Exercise 7.3.7 Let $U \in \mathcal{U}$ and let $u^j := U e^j$. Show that $\{u^j\}$ forms an orthonormal basis for ℓ_2^C. Reference: [216, p. 95].

Exercise 7.3.8 Consider the two sequences

$$x = \left(0, \frac{1}{\sqrt{2}}, \frac{1}{\sqrt{2^2}}, \frac{1}{\sqrt{2^3}}, \dots \right)$$
$$y = \left(\frac{1}{\sqrt{2}}, \frac{1}{\sqrt{2^2}}, \frac{1}{\sqrt{2^3}}, \dots \right)$$

Show that $\sup_{\pi \in \mathcal{B}} P_\pi(x, y) = 1$, but this cannot be attained by applying any bijection.

Exercise 7.3.9 Prove Lemma 7.3.14.

Exercise 7.3.10 Prove Theorem 7.3.16.

Exercise 7.3.11 Let $R \in B_{sa}$. Prove that, for $t \in \mathbb{R}$,

$$U_t := \exp(itR) \in \mathcal{U}.$$

Reference: [216, p. 214].

Exercise 7.3.12 Prove that $T_t A_t$ converges to TA in the norm on X as $t \to 0$ in Proposition 7.3.20.

Exercise 7.3.13 Consider "diagonal" operators A_n with diagonal elements

$$(-1, -1/2, -1/4, \dots, -2^{1-n}, 0, -2^{-n}, \dots)$$

and A with diagonal elements

$$(-1, -1/2, -1/4, \dots).$$

Show that A_n converges to A in B_2 but $\lambda(A_n)$ does not converge to $\lambda(A)$ in ℓ^2.

Exercise 7.3.14 Prove the "if" part of Lemma 7.3.27.

Exercise 7.3.15 Let $P_n \colon \ell_2^C \to E_n$ denote the orthogonal projection. Prove that $P_n^* \colon E_n \to \ell_2^C$ is the inclusion mapping.

Exercise 7.3.16 Prove that for any $Z \in \mathcal{V}$,

$$\lim_{n \to \infty} \|Z - P_n^* P_n Z P_n^* P_n\|_{\mathcal{V}} = 0.$$

Reference: [128, Theorem 6.3, p. 90].

Exercise 7.3.17 Suppose that $x \in V$ and $y \in W$ satisfy $y \in \partial_G(f \circ \mathrm{diag})(x)$. Define $\phi \colon \mathbb{R}^n \to \mathbb{R} \cup \{+\infty\}$ by

$$\phi(z) := (f \circ \mathrm{diag})(z_1, \ldots, z_n, x_{n+1}, x_{n+2}, \ldots).$$

Show that ϕ is a permutation invariant locally Lipschitz function and $P_n^W y \in \partial_G \phi(P_n^V x)$.

*__Exercise 7.3.18__ For an element x of c_0, we rearrange the components $|x_i|$ into decreasing order to obtain a new element $\bar{x}$ of c_0. Show that

(i)   for a fixed decreasing sequence $t_i \to 0+$ (not identically zero),

$$x \in V \to \sup_{n \in \mathcal{N}} \left\{ \frac{\sum_{i=1}^n \bar{x}_i}{\sum_{i=1}^n t_i} \right\},$$

and

$$y \in W \to \sum_{i=1}^\infty t_i \bar{y}_i$$

are a pair of dual norms;

(ii)   their compositions with λ are dual norms on the spaces $\mathcal{V}$ and $\mathcal{W}$;

(iii) when $t_i = i^{1/q} - (1-i)^{1/q} (1 < q < \infty)$ they are the Calderón norms.

Exercise 7.3.19 Let

$$g(t) = \begin{cases} t - \ln(1+t) & t > -1, \\ +\infty & t \le -1. \end{cases}$$

Show that g is convex and has the conjugate $g^*(s) = g(-s)$.

Exercise 7.3.20 Prove Theorem 7.3.35.

Exercise 7.3.21 Prove Theorem 7.3.36.

References

1. Antonio Ambrosetti. Variational methods and nonlinear problems: classical results and recent advances. In M. Matzeu and A. Vignoli, editors, *Topological nonlinear analysis*, volume 15 of *Progr. Nonlinear Differential Equations Appl.*, pages 1–36. Birkhäuser Boston, Boston, MA, 1995.
2. Antonio Ambrosetti and Paul H. Rabinowitz. Dual variational methods in critical point theory and applications. *J. Functional Analysis*, 14:349–381, 1973.
3. T. Ando. Majorization, doubly stochastic matrices, and comparison of eigenvalues. *Linear Algebra Appl.*, 118:163–248, 1989.
4. Edgar Asplund. Fréchet differentiability of convex functions. *Acta Math.*, 121:31–47, 1968.
5. Jean-Pierre Aubin. Lipschitz behavior of solutions to convex minimization problems. *Math. Oper. Res.*, 9:87–111, 1984.
6. Jean-Pierre Aubin. *Viability theory*. Systems & Control: Foundations & Applications. Birkhäuser Boston Inc., Boston, MA, 1991.
7. Jean-Pierre Aubin and Ivar Ekeland. *Applied nonlinear analysis*. Pure and Applied Mathematics (New York). John Wiley & Sons Inc., New York, 1984.
8. Jean-Pierre Aubin and Hélène Frankowska. *Set-valued analysis*, volume 2 of *Systems & Control: Foundations & Applications*. Birkhäuser Boston Inc., Boston, MA, 1990.
9. A. Auslender. Differentiable stability in nonconvex and nondifferentiable programming. *Math. Programming Stud.*, pages 29–41, 1979.
10. Didier Aussel, Jean-Noël Corvellec, and Marc Lassonde. Mean value property and subdifferential criteria for lower semicontinuous functions. *Trans. Amer. Math. Soc.*, 347:4147–4161, 1995.
11. E. N. Barron and R. Jensen. Semicontinuous viscosity solutions for Hamilton-Jacobi equations with convex Hamiltonians. *Comm. Partial Differential Equations*, 15:1713–1742, 1990.
12. H. H. Bauschke. *Projection Algorithms and Monotone Operators*. Ph. D. Thesis, Simon Fraser University, 1996.
13. H. H. Bauschke and J. M. Borwein. On the convergence of von Neumann's alternating projection algorithm for two sets. *Set-Valued Anal.*, 1:185–212, 1993.

14. Heinz H. Bauschke and J. M. Borwein. On projection algorithms for solving convex feasibility problems. *SIAM Rev.*, 38:367–426, 1996.

15. Heinz H. Bauschke and J. M. Borwein. Maximal monotonicity of dense type, local maximal monotonicity, and monotonicity of the conjugate are all the same for continuous linear operators. *Pacific J. Math.*, 189:1–20, 1999.

16. Heinz H. Bauschke and J. M. Borwein. Joint and separate convexity of the Bregman distance. In S. Reich D. Butnariu, Y. Censor, editor, *Inherently parallel algorithms in feasibility and optimization and their applications (Haifa, 2000)*, volume 8 of *Stud. Comput. Math.*, pages 23–36. North-Holland, Amsterdam, 2001.

17. Heinz H. Bauschke, J. M. Borwein, and Patrick L. Combettes. Bregman monotone optimization algorithms. *SIAM J. Control Optim.*, 42:596–636, 2003.

18. Heinz H. Bauschke and Patrick L. Combettes. Iterating Bregman retractions. *SIAM J. Optim.*, 13:1159–1173, 2003.

19. Heinz H. Bauschke, Eva Matoušková, and Simeon Reich. Projection and proximal point methods: convergence results and counterexamples. *Nonlinear Anal.*, 56:715–738, 2004.

20. M. S. Bazaraa, J. J. Goode, and M. Z. Nashed. On the cones of tangents with applications to mathematical programming. *J. Optimization Theory Appl.*, 13:389–426, 1974.

21. Joël Benoist. Intégration du sous-différentiel proximal: un contre exemple. *Canad. J. Math.*, 50:242–265, 1998.

22. Rajendra Bhatia. *Matrix analysis*, volume 169 of *Graduate Texts in Mathematics*. Springer-Verlag, New York, 1997.

23. Garrett Birkhoff. Three observations on linear algebra. *Univ. Nac. Tucumán. Revista A.*, 5:147–151, 1946.

24. Errett Bishop and R. R. Phelps. A proof that every Banach space is subreflexive. *Bull. Amer. Math. Soc.*, 67:97–98, 1961.

25. Errett Bishop and R. R. Phelps. The support functionals of a convex set. In V. L. Klee, editor, *Proc. Sympos. Pure Math., Vol. VII*, pages 27–35. Amer. Math. Soc., Providence, R.I., 1963.

26. Jean-Marc Bonnisseau and Bernard Cornet. Valuation equilibrium and Pareto optimum in nonconvex economies. *J. Math. Econom.*, 17:293–308, 1988.

27. William M. Boothby. *An introduction to differentiable manifolds and Riemannian geometry*, volume 120 of *Pure and Applied Mathematics*. Academic Press Inc., Orlando, FL, 1986.

28. D. Borwein, J. M. Borwein, and Xianfu Wang. Approximate subgradients and coderivatives in $\mathbf{R}^n$. *Set-Valued Anal.*, 4:375–398, 1996.

29. J. M. Borwein. A note on the existence of subgradients. *Math. Programming*, 24:225–228, 1982.

30. J. M. Borwein. Stability and regular points of inequality systems. *J. Optim. Theory Appl.*, 48:9–52, 1986.

31. J. M. Borwein. Minimal CUSCOS and subgradients of Lipschitz functions. In *Fixed point theory and applications (Marseille, 1989)*, volume 252 of *Pitman Res. Notes Math. Ser.*, pages 57–81. Longman Sci. Tech., Harlow, 1991.

32. J. M. Borwein. Maximal monotonicity via convex analysis. *Proc. Amer. Math. Soc.*, to appear.

33. J. M. Borwein. Maximal monotonicity via convex analysis. *J. Conv. Anal.*, to appear.

34. J. M. Borwein and David Bailey. *Mathematics by experiment.* A K Peters Ltd., Natick, MA, 2004.
35. J. M. Borwein, David Bailey, and Roland Girgensohn. *Experimentation in mathematics.* A K Peters Ltd., Natick, MA, 2004.
36. J. M. Borwein, R. Choksi, and P. Maréchal. Probability distributions of assets inferred from option prices via the principle of maximum entropy. *SIAM J. Optim.*, 14:464–478, 2003.
37. J. M. Borwein, M. Fabián, I. Kortezov, and P. D. Loewen. The range of the gradient of a continuously differentiable bump. *J. Nonlinear Convex Anal.*, 2:1–19, 2001.
38. J. M. Borwein, M. Fabián, and J. P. Revalski. A one perturbation variational principle and applications. *Set-valued Analysis*, Accepted April, 2003.
39. J. M. Borwein and Marián Fabián. A note on regularity of sets and of distance functions in Banach space. *J. Math. Anal. Appl.*, 182:566–570, 1994.
40. J. M. Borwein and S. Fitzpatrick. Local boundedness of monotone operators under minimal hypotheses. *Bull. Australian Math. Soc.*, 39:439–441, 1988.
41. J. M. Borwein and S. Fitzpatrick. Characterization of clarke subgradients among one-dimensional multifunctions. In *Proceedings of the Optimization Miniconference II*, pages 61–73. University of Balarat Press, 1995.
42. J. M. Borwein and S. Fitzpatrick. Duality inequalities and sandwiched functions. *Nonlinear Anal.*, 46:365–380, 2001.
43. J. M. Borwein, S. Fitzpatrick, and J. R. Giles. The differentiability of real functions on normed linear space using generalized subgradients. *J. Math. Anal. Appl.*, 128:512–534, 1987.
44. J. M. Borwein, Simon Fitzpatrick, and Roland Girgensohn. Subdifferentials whose graphs are not norm × weak* closed. *Canad. Math. Bull.*, 46:538–545, 2003.
45. J. M. Borwein and J. R. Giles. The proximal normal formula in Banach space. *Trans. Amer. Math. Soc.*, 302:371–381, 1987.
46. J. M. Borwein, R. Girgensohn, and Xianfu Wang. On the construction of Hölder and proximal subderivatives. *Canad. Math. Bull.*, 41:497–507, 1998.
47. J. M. Borwein and Roland Girgensohn. A class of exponential inequalities. *Math. Inequal. Appl.*, 6:397–411, 2003.
48. J. M. Borwein and Alexander D. Ioffe. Proximal analysis in smooth spaces. *Set-Valued Anal.*, 4:1–24, 1996.
49. J. M. Borwein and Alejandro Jofré. A nonconvex separation property in Banach spaces. *Math. Methods Oper. Res.*, 48:169–179, 1998.
50. J. M. Borwein, I. Kortezov, and H. Wiersma. A C^1-function that is even on a sphere and has no critical points in the ball. *J. Nonlinear Convex Anal.*, 3:1–16, 2002.
51. J. M. Borwein and A. S. Lewis. Partially finite convex programming. I. Quasi relative interiors and duality theory. *Math. Programming*, 57:15–48, 1992.
52. J. M. Borwein and A. S. Lewis. Partially-finite programming in L_1 and the existence of maximum entropy estimates. *SIAM J. Optim.*, 3:248–267, 1993.
53. J. M. Borwein and A. S. Lewis. A survey of convergence results for maximum entropy methods. In A. Mohammad-Djafari and G. Demoments, editors, *Maximum entropy and Bayesian methods (Paris, 1992)*, volume 53 of *Fund. Theories Phys.*, pages 39–48. Kluwer Acad. Publ., Dordrecht, 1993.
54. J. M. Borwein, A. S. Lewis, and R. D. Nussbaum. Entropy minimization, *DAD* problems, and doubly stochastic kernels. *J. Funct. Anal.*, 123:264–307, 1994.

55. J. M. Borwein, A. S. Lewis, and Q. J. Zhu. Convex spectral functions of compact operators, Part II: lower semicontinuity and rearrangement invariance. In *Proceedings of the Optimization Miniconference VI*, pages 1–18. Kluwer Academic Publ., 2000.

56. J. M. Borwein and Adrian S. Lewis. *Convex analysis and nonlinear optimization*. CMS Books in Mathematics/Ouvrages de Mathématiques de la SMC, 3. Springer-Verlag, New York, 2000.

57. J. M. Borwein and W. B. Moors. Lipschitz functions with minimal clarke subdifferential mappings. In *Proc. Optimization Miniconference III (1996)*, pages 5–12. Univ. of Ballarat Press, 1997.

58. J. M. Borwein and D. Preiss. A smooth variational principle with applications to subdifferentiability and to differentiability of convex functions. *Trans. Amer. Math. Soc.*, 303:517–527, 1987.

59. J. M. Borwein, J. Read, A. S. Lewis, and Q. J. Zhu. Convex spectral functions of compact operators. *J. Nonlinear Convex Anal.*, 1:17–35, 2000.

60. J. M. Borwein and H. M. Strójwas. Directionally Lipschitzian mappings on Baire spaces. *Canad. J. Math.*, 36:95–130, 1984.

61. J. M. Borwein and H. M. Strójwas. Proximal analysis and boundaries of closed sets in Banach space. I. Theory. *Canad. J. Math.*, 38:431–452, 1986.

62. J. M. Borwein and H. M. Strójwas. Proximal analysis and boundaries of closed sets in Banach space. II. Applications. *Canad. J. Math.*, 39:428–472, 1987.

63. J. M. Borwein, J. S. Treiman, and Q. J. Zhu. Necessary conditions for constrained optimization problems with semicontinuous and continuous data. *Trans. Amer. Math. Soc.*, 350:2409–2429, 1998.

64. J. M. Borwein, J. S. Treiman, and Q. J. Zhu. Partially smooth variational principles and applications. *Nonlinear Anal.*, 35:1031–1059, 1999.

65. J. M. Borwein and Xianfu Wang. Lipschitz functions with maximal Clarke subdifferentials are generic. *Proc. Amer. Math. Soc.*, 128:3221–3229, 2000.

66. J. M. Borwein and Xianfu Wang. Lipschitz functions on the line with prescribed Hölder subdifferentials. *Adv. Stud. Contemp. Math. (Kyungshang)*, 7:93–117, 2003.

67. J. M. Borwein and Q. J. Zhu. Variational methods in convex analysis. *JOGO, to appear*.

68. J. M. Borwein and Q. J. Zhu. Viscosity solutions and viscosity subderivatives in smooth Banach spaces with applications to metric regularity. *SIAM J. Control Optim.*, 34:1568–1591, 1996.

69. J. M. Borwein and Q. J. Zhu. Limiting convex examples for nonconvex subdifferential calculus. *J. Convex Anal.*, 5:221–235, 1998.

70. J. M. Borwein and Q. J. Zhu. Multifunctional and functional analytic techniques in nonsmooth analysis. In F. H. Clarke and R. J. Stern, editors, *Nonlinear analysis, differential equations and control (Montreal, QC, 1998)*, volume 528 of *NATO Sci. Ser. C Math. Phys. Sci.*, pages 61–157. Kluwer Acad. Publ., Dordrecht, 1999.

71. J. M. Borwein and Q. J. Zhu. A survey of subdifferential calculus with applications. *Nonlinear Anal.*, 38:687–773, 1999.

72. J. M. Borwein and D. M. Zhuang. On Fan's minimax theorem. *Math. Programming*, 34:232–234, 1986.

73. J. M. Borwein and D. M. Zhuang. Verifiable necessary and sufficient conditions for openness and regularity of set-valued and single-valued maps. *J. Math. Anal. Appl.*, 134:441–459, 1988.

74. L. M. Brègman. Finding the common point of convex sets by the method of successive projection. *Dokl. Akad. Nauk SSSR*, 162:487–490, 1965.

75. James V. Burke, Michael C. Ferris, and Maijian Qian. On the Clarke sub-differential of the distance function of a closed set. *J. Math. Anal. Appl.*, 166:199–213, 1992.

76. James V. Burke and Michael L. Overton. Variational analysis of the abscissa mapping for polynomials. *SIAM J. Control Optim.*, 39:1651–1676, 2001.

77. C. Carathéodory. Uber den Variabilitäts-beraich der Fourierschen Konstanten von positiven harmonischen Funktionen. *Rendiconti del Circolo Matematico de Palermo*, 32:218–239, 1911.

78. Lixin Cheng and Marián Fabián. The product of a Gâteaux differentiability space and a separable space is a Gâteaux differentiability space. *Proc. Amer. Math. Soc.*, 129:3539–3541, 2001.

79. I. Chryssochoos and R. B. Vinter. Optimal control problems on manifolds: a dynamic programming approach. *J. Math. Anal. Appl.*, 287:118–140, 2003.

80. F. H. Clarke. *Necessary Conditions for Nonsmooth Problems in Optimal Control and the Calculus of Variations*. Ph.D. Thesis, University of Washington, 1973.

81. F. H. Clarke. Generalized gradients and applications. *Trans. Amer. Math. Soc.*, 205:247–262, 1975.

82. F. H. Clarke. A new approach to Lagrange multipliers. *Math. Oper. Res.*, 1:165–174, 1976.

83. F. H. Clarke. On the inverse function theorem. *Pacific J. Math.*, 64:97–102, 1976.

84. F. H. Clarke. *Optimization and nonsmooth analysis*. Canadian Mathematical Society Series of Monographs and Advanced Texts. John Wiley & Sons Inc., New York, 1983.

85. F. H. Clarke. *Methods of dynamic and nonsmooth optimization*, volume 57 of *CBMS-NSF Regional Conference Series in Applied Mathematics*. Society for Industrial and Applied Mathematics (SIAM), Philadelphia, PA, 1989.

86. F. H. Clarke. Lecture notes in nonsmooth analysis. 1996.

87. F. H. Clarke and Yu. S. Ledyaev. Mean value inequalities. *Proc. Amer. Math. Soc.*, 122:1075–1083, 1994.

88. F. H. Clarke and Yu. S. Ledyaev. Mean value inequalities in Hilbert space. *Trans. Amer. Math. Soc.*, 344:307–324, 1994.

89. F. H. Clarke, Yu. S. Ledyaev, and R. J. Stern. Complements, approximations, smoothings and invariance properties. *J. Convex Anal.*, 4:189–219, 1997.

90. F. H. Clarke, Yu. S. Ledyaev, R. J. Stern, and P. R. Wolenski. Qualitative properties of trajectories of control systems: a survey. *J. Dynam. Control Systems*, 1:1–48, 1995.

91. F. H. Clarke, Yu. S. Ledyaev, R. J. Stern, and P. R. Wolenski. *Nonsmooth analysis and control theory*, volume 178 of *Graduate Texts in Mathematics*. Springer-Verlag, New York, 1998.

92. F. H. Clarke, Yu. S. Ledyaev, and P. R. Wolenski. Proximal analysis and minimization principles. *J. Math. Anal. Appl.*, 196:722–735, 1995.

93. M. G. Crandall, L. C. Evans, and P.-L. Lions. Some properties of viscosity solutions of Hamilton-Jacobi equations. *Trans. Amer. Math. Soc.*, 282:487–502, 1984.

94. Michael G. Crandall and Pierre-Louis Lions. Viscosity solutions of Hamilton-Jacobi equations. *Trans. Amer. Math. Soc.*, 277:1–42, 1983.

95. Josef Daneš. A geometric theorem useful in nonlinear functional analysis. *Boll. Un. Mat. Ital. (4)*, 6:369–375, 1972.

96. Klaus Deimling. *Nonlinear functional analysis.* Springer-Verlag, Berlin, 1985.

97. Frank Deutsch. *Best approximation in inner product spaces.* CMS Books in Mathematics/Ouvrages de Mathématiques de la SMC, 7. Springer-Verlag, New York, 2001.

98. Robert Deville, Gilles Godefroy, and Václav Zizler. A smooth variational principle with applications to Hamilton-Jacobi equations in infinite dimensions. *J. Funct. Anal.*, 111:197–212, 1993.

99. Robert Deville, Gilles Godefroy, and Václav Zizler. *Smoothness and renormings in Banach spaces*, volume 64 of *Pitman Monographs and Surveys in Pure and Applied Mathematics.* Longman Scientific & Technical, Harlow, 1993.

100. Robert Deville and Milen Ivanov. Smooth variational principles with constraints. *Arch. Math. (Basel)*, 69:418–426, 1997.

101. Joseph Diestel. *Geometry of Banach spaces—selected topics.* Springer-Verlag, Berlin, 1975.

102. Joseph Diestel. *Sequences and series in Banach spaces*, volume 92 of *Graduate Texts in Mathematics.* Springer-Verlag, New York, 1984.

103. A. V. Dmitruk, A. A. Miljutin, and N. P. Osmolovskiĭ. Ljusternik's theorem and the theory of the extremum. *Uspekhi Mat. Nauk*, 35:11–46, 215, 1980.

104. Szymon Dolecki. A general theory of necessary optimality conditions. *J. Math. Anal. Appl.*, 78:267–308, 1980.

105. David Downing and W. A. Kirk. A generalization of Caristi's theorem with applications to nonlinear mapping theory. *Pacific J. Math.*, 69:339–346, 1977.

106. I. Ekeland. On the variational principle. *J. Math. Anal. Appl.*, 47:324–353, 1974.

107. Ivar Ekeland. Nonconvex minimization problems. *Bull. Amer. Math. Soc. (N.S.)*, 1:443–474, 1979.

108. E. El Haddad and R. Deville. The viscosity subdifferential of the sum of two functions in Banach spaces. I. First order case. *J. Convex Anal.*, 3:295–308, 1996.

109. M. Fabián. On minimum principles. *Acta Polytechnica*, 20:109–118, 1983.

110. M. Fabián. Subdifferentials, local ϵ-supports and Asplund spaces. *J. London Math. Soc. (2)*, 34:568–576, 1986.

111. M. Fabián and N. V. Zhivkov. A characterization of Asplund spaces with the help of local ϵ-supports of Ekeland and Lebourg. *C. R. Acad. Bulgare Sci.*, 38:671–674, 1985.

112. M. Fabián and V. Zizler. A "nonlinear" proof of Pitt's compactness theorem. *Proc. Amer. Math. Soc.*, 131:3693–3694, 2003.

113. Marián Fabián. Subdifferentiability and trustworthiness in the light of a new variational principle of Borwein and Preiss. *Acta Univ. Carolin. Math. Phys.*, 30:51–56, 1989.

114. Marián Fabián. *Gâteaux differentiability of convex functions and topology.* Canadian Mathematical Society Series of Monographs and Advanced Texts. John Wiley & Sons Inc., New York, 1997.

115. Marián Fabián, Petr Habala, Petr Hájek, Vicente Montesinos Santalucía, Jan Pelant, and Václav Zizler. *Functional analysis and infinite-dimensional geometry.* CMS Books in Mathematics/Ouvrages de Mathématiques de la SMC, 8. Springer-Verlag, New York, 2001.

116. Leonid Faybusovich. Infinite-dimensional semidefinite programming: regularized determinants and self-concordant barriers. In *Topics in semidefinite and interior-point methods (Toronto, ON, 1996)*, volume 18 of *Fields Inst. Commun.*, pages 39–49. Amer. Math. Soc., Providence, RI, 1998.

117. W. Fenchel. On conjugate convex functions. *Canad. J. Math.*, 1:73–77, 1949.

118. Luis A. Fernández. On the limits of the Lagrange multiplier rule. *SIAM Rev.*, 39:292–297, 1997.

119. Jesús Ferrer. Rolle's theorem fails in l_2. *Amer. Math. Monthly*, 103:161–165, 1996.

120. Simon Fitzpatrick. Representing monotone operators by convex functions. In *Workshop/Miniconference on Functional Analysis and Optimization*, Proc. Centre Math. Anal. Austral. Nat. Univ., 20, pages 59–65, Canberra, 1988. Austral. Nat. Univ.

121. Simon Fitzpatrick and Bruce Calvert. In a nonreflexive space the subdifferential is not onto. *Math. Z.*, 189:555–560, 1985.

122. Wendell H. Fleming and Raymond W. Rishel. *Deterministic and stochastic optimal control.* Springer-Verlag, Berlin, 1975.

123. Jacques Gauvin. The generalized gradient of a marginal function in mathematical programming. *Math. Oper. Res.*, 4:458–463, 1979.

124. Jacques Gauvin and Jon W. Tolle. Differential stability in nonlinear programming. *SIAM J. Control Optimization*, 15:294–311, 1977.

125. N. Ghoussoub. A general mountain pass principle for locating and classifying critical points. *Ann. Inst. H. Poincaré Anal. Non Linéaire*, 6:321–330, 1989.

126. John R. Giles. *Convex analysis with application in the differentiation of convex functions*, volume 58 of *Research Notes in Mathematics*. Pitman (Advanced Publishing Program), Boston, Mass., 1982.

127. Kazimierz Goebel and W. A. Kirk. *Topics in metric fixed point theory*, volume 28 of *Cambridge Studies in Advanced Mathematics*. Cambridge University Press, Cambridge, 1990.

128. I. C. Gohberg and M. G. Kreĭn. *Introduction to the theory of linear nonselfadjoint operators.* Translated from the Russian by A. Feinstein. Translations of Mathematical Monographs, Vol. 18. American Mathematical Society, Providence, R.I., 1969.

129. P. Gordan. Uber die Auflösung linearer gleichungen mit reelen Coefficienten. *Mathematische Annalen*, 6:23–28, 1873.

130. S. Gossez. On a convexity property of the range of a maximal monotone operators. *Proc. Amer. Math. Soc.*, 55:358–360, 1976.

131. Monique Guignard. Generalized Kuhn-Tucker conditions for mathematical programming problems in a Banach space. *SIAM J. Control*, 7:232–241, 1969.

132. J. Hadamard. Résolution d'une question relative aux déterminants. *Bull. Sci. Math.*, 2:240–248, 1893.

133. Hubert Halkin. Mathematical programming without differentiability. In D. L. Russell, editor, *Calculus of variations and control theory (Proc. Sympos., Math. Res. Center, Univ. Wisconsin, Madison, Wis., 1975)*, pages 279–287. Math. Res. Center, Univ. Wisconsin, Publ. No. 36. Academic Press, New York, 1976.

134. P. Hall. On representation of subsets. *Journal of London Mathematical Society*, 10:26–30, 1935.

135. Andreas H. Hamel. Phelps' lemma, Daneš' drop theorem and Ekeland's principle in locally convex spaces. *Proc. Amer. Math. Soc.*, 131:3025–3038, 2003.

136. Uwe Helmke and John B. Moore. *Optimization and dynamical systems.* Communications and Control Engineering Series. Springer-Verlag London Ltd., London, 1994.

137. J.-B. Hiriart-Urruty. What conditions are satisfied at points minimizing the maximum of a finite number of differentiable functions? In *Nonsmooth optimization: methods and applications (Erice, 1991)*, pages 166–174. Gordon and Breach, Montreux, 1992.

138. Roger A. Horn and Charles R. Johnson. *Matrix analysis.* Cambridge University Press, Cambridge, 1985.

139. Roger A. Horn and Charles R. Johnson. *Topics in matrix analysis.* Cambridge University Press, Cambridge, 1991.

140. Hein S. Hundal. An alternating projection that does not converge in norm. *Nonlinear Anal.*, 57:35–61, 2004.

141. Donald H. Hyers, George Isac, and Themistocles M. Rassias. *Topics in nonlinear analysis & applications.* World Scientific Publishing Co. Inc., River Edge, NJ, 1997.

142. A. D. Ioffe. Nonsmooth analysis: differential calculus of nondifferentiable mappings. *Trans. Amer. Math. Soc.*, 266:1–56, 1981.

143. A. D. Ioffe. Approximate subdifferentials and applications. I. The finite-dimensional theory. *Trans. Amer. Math. Soc.*, 281:389–416, 1984.

144. A. D. Ioffe. Necessary conditions in nonsmooth optimization. *Math. Oper. Res.*, 9:159–189, 1984.

145. A. D. Ioffe. Approximate subdifferentials and applications. III. The metric theory. *Mathematika*, 36:1–38, 1989.

146. A. D. Ioffe. Proximal analysis and approximate subdifferentials. *J. London Math. Soc. (2)*, 41:175–192, 1990.

147. A. D. Ioffe and V. L. Levin. Subdifferentials of convex functions. *Trudy Moskov. Mat. Obšč.*, 26:3–73, 1972.

148. A. D. Ioffe and R. T. Rockafellar. The Euler and Weierstrass conditions for nonsmooth variational problems. *Calc. Var. Partial Differential Equations*, 4:59–87, 1996.

149. Alexander D. Ioffe. Regular points of Lipschitz functions. *Trans. Amer. Math. Soc.*, 251:61–69, 1979.

150. Alexander D. Ioffe. Variational methods in local and global non-smooth analysis. In F. H. Clarke and R. J. Stern, editors, *Nonlinear analysis, differential equations and control (Montreal, QC, 1998)*, volume 528 of *NATO Sci. Ser. C Math. Phys. Sci.*, pages 447–502. Kluwer Acad. Publ., Dordrecht, 1999.

151. Alexander D. Ioffe and Jean-Paul Penot. Subdifferentials of performance functions and calculus of coderivatives of set-valued mappings. *Serdica Math. J.*, 22:257–282, 1996.

152. G. J. O. Jameson. *Topology and normed spaces.* Chapman and Hall, London, 1974.

153. A. Jofré and J. Rivera. A nonconvex separation property and some applications. *to appear in Math. Programming.*

154. Alejandro Jofré. A second-welfare theorem in nonconvex economies. In M. Théra, editor, *Constructive, experimental, and nonlinear analysis (Limoges, 1999)*, volume 27 of *CMS Conf. Proc.*, pages 175–184. Amer. Math. Soc., Providence, RI, 2000.

155. Fritz John. Extremum problems with inequalities as subsidiary conditions. In *Studies and Essays Presented to R. Courant on his 60th Birthday, January 8, 1948*, pages 187–204. Interscience Publishers, Inc., New York, N. Y., 1948.

156. C. Jordan. Mémoire sur les formes bilinéaires. *J. Math. Pures et Appl.*, 2:35–54, 1874.

157. Elyes Jouini. Functions with constant generalized gradients. *J. Math. Anal. Appl.*, 148:121–130, 1990.

158. W. Karush. Minima of functions of several variables with inequalities as side conditions. Master's thesis, University of Chicago, 1939.

159. M. Ali Khan. Ioffe's normal cone and the foundations of welfare economics: the infinite-dimensional theory. *J. Math. Anal. Appl.*, 161:284–298, 1991.

160. W. A. Kirk and J. Caristi. Mappings theorems in metric and Banach spaces. *Bull. Acad. Polon. Sci. Sér. Sci. Math. Astronom. Phys.*, 23:891–894, 1975.

161. M. D. Kirszbraun. Über die zusammenziehende und Lipschitzsche Tansformationen. *Fund. Math.*, 22:77–108, 1934.

162. E. Klein and A. C. Thompson. *Theory of Correspondences, Including Applications to Mathematical Economics*. Wiley, New York, 1984.

163. D. König. *Theorie der Endlichen und Unendlichen Graphen*. Akademische Verlagsgesellschaft, Leipzig, 1936.

164. A. Y. Kruger. Subdifferentials of nonconvex functions and generalized directional derivatives. *VINITI, Minsk*, 2661-2677, 1977.

165. A. Y. Kruger. Generalized differentials of nonsmooth functions. *VINITI, Minsk*, 1332-81, 1981.

166. A. Y. Kruger. On Fréchet subdifferential. *J. Math. Sci. (N. Y.)*, 116:3325–3358, 2003.

167. A. Y. Kruger and B. S. Mordukhovich. Extremal points and the Euler equation in nonsmooth optimization problems. *Dokl. Akad. Nauk BSSR*, 24:684–687, 763, 1980.

168. H. W. Kuhn and A. W. Tucker. Nonlinear programming. In *Proceedings of the Second Berkeley Symposium on Mathematical Statistics and Probability, 1950*, pages 481–492, Berkeley and Los Angeles, 1951. University of California Press.

169. V. Lakshmikantham and Yu. S. Ledyaev. Proximal (sub-)differential inequalities. *Nonlinear Anal.*, 39:111–133, 2000.

170. Marc Lassonde. First-order rules for nonsmooth constrained optimization. *Nonlinear Anal.*, 44:1031–1056, 2001.

171. Gérard Lebourg. Valeur moyenne pour gradient généralisé. *C. R. Acad. Sci. Paris Sér. A-B*, 281:Ai, A795–A797, 1975.

172. Yu. S. Ledyaev. On generic existence and uniqueness in nonconvex optimal control problems. *Set-Valued Analysis*, 12:147–162, 2004.

173. Yu. S. Ledyaev and J. S. Treiman. Sub and supergradients of envelops, semicontinuous closures and limits of functions. *preprint*, 2002.

174. Yu. S. Ledyaev and Q. J. Zhu. Implicit multifunction theorems. *Set-Valued Anal.*, 7:209–238, 1999.

175. Yu. S. Ledyaev and Q. J. Zhu. Nonsmooth analysis on smooth manifolds. *submitted to Transactions of AMS*, 2003.

176. Yu. S. Ledyaev and Q. J. Zhu. Techniques for nonsmooth analysis on smooth manifolds: Part I: The local theory. In *Optimal control, stabilization and nonsmooth analysis*, volume 301 of *Lecture Notes in Control and Inform. Sci.*, pages 283–297. Springer–Verlag, Berlin, 2004.

177. Yu. S. Ledyaev and Q. J. Zhu. Techniques for nonsmooth analysis on smooth manifolds: Part II: Using deformation and flows. In *Optimal control, stabilization and nonsmooth analysis*, volume 301 of *Lecture Notes in Control and Inform. Sci.*, pages 299–311. Springer–Verlag, Berlin, 2004.

178. Yu. S. Ledyaev and Q. J. Zhu. Multidirectional mean value inequalities and weak monotonicity. *Journal of London Mathematical Society*, 71:187–202, 2005.

179. A. S. Lewis. Nonsmooth analysis of eigenvalues. *Math. Program.*, 84:1–24, 1999.

180. A. S. Lewis and D. Ralph. A nonlinear duality result equivalent to the Clarke-Ledyaev mean value inequality. *Nonlinear Anal.*, 26:343–350, 1996.

181. Adrian S. Lewis and Michael L. Overton. Eigenvalue optimization. In *Acta numerica, 1996*, volume 5 of *Acta Numer.*, pages 149–190. Cambridge Univ. Press, Cambridge, 1996.

182. Y. X. Li and S. Z. Shi. A generalization of Ekeland's ε-variational principle and of its Borwein-Preiss smooth version. *Preprint*, 1991.

183. Joram Lindenstrauss and Lior Tzafriri. *Classical Banach spaces. I Sequence spaces*. Ergebnisse der Mathematik und ihrer Grenzgebiete, Vol. 92. Springer-Verlag, Berlin, 1977.

184. L. A. Liusternik and L. Schnirelman. *Méthode topologique dans les problémes variationelles*. Hermann, Paris, 1934.

185. Philip D. Loewen. *Optimal control via nonsmooth analysis*, volume 2 of *CRM Proceedings & Lecture Notes*. American Mathematical Society, Providence, RI, 1993.

186. Philip D. Loewen. A mean value theorem for Fréchet subgradients. *Nonlinear Anal.*, 23:1365–1381, 1994.

187. Dinh The Luc. Characterisations of quasiconvex functions. *Bull. Austral. Math. Soc.*, 48:393–406, 1993.

188. Dinh The Luc. A strong mean value theorem and applications. *Nonlinear Anal.*, 26:915–923, 1996.

189. Zhi-Quan Luo, Jong-Shi Pang, and D. Ralph. *Mathematical programs with equilibrium constraints*. Cambridge University Press, Cambridge, 1996.

190. O. L. Mangasarian and S. Fromovitz. The Fritz John necessary optimality conditions in the presence of equality and inequality constraints. *J. Math. Anal. Appl.*, 17:37–47, 1967.

191. Eva Matoušková and Simeon Reich. The Hundal example revisited. *J. Nonlinear Convex Anal.*, 4:411–427, 2003.

192. Yozo Matsushima. *Differentiable manifolds*, volume 9 of *Pure and Applied Mathematics*. Marcel Dekker Inc., New York, 1972.

193. Philippe Michel and Jean-Paul Penot. Calcul sous-différentiel pour des fonctions lipschitziennes et non lipschitziennes. *C. R. Acad. Sci. Paris Sér. I Math.*, 298:269–272, 1984.

194. George J. Minty. On some aspects of the theory of monotone operators. In *Theory and Applications of Monotone Operators (Proc. NATO Advanced Study Inst., Venice, 1968)*, pages 67–82. Edizioni "Oderisi", Gubbio, 1969.

195. B. S. Mordukhovich. Maximum principle in the problem of time optimal response with nonsmooth constraints. *Prikl. Mat. Meh.*, 40:1014–1023, 1976.

196. B. S. Mordukhovich. Metric approximations and necessary conditions for optimality for general classes of nonsmooth extremal problems. *Dokl. Akad. Nauk SSSR*, 254:1072–1076, 1980.

197. B. S. Mordukhovich. Nonsmooth analysis with nonconvex generalized differentials and conjugate mappings. *Dokl. Akad. Nauk BSSR*, 28:976–979, 1984.

198. B. S. Mordukhovich. *Metody approksimatsii v zadachakh optimizatsii i upravleniya*. "Nauka", Moscow, 1988.

199. B. S. Mordukhovich. Complete characterization of openness, metric regularity, and Lipschitzian properties of multifunctions. *Trans. Amer. Math. Soc.*, 340:1–35, 1993.

200. B. S. Mordukhovich. Generalized differential calculus for nonsmooth and set-valued mappings. *J. Math. Anal. Appl.*, 183:250–288, 1994.

201. B. S. Mordukhovich. Sensitivity analysis for constraint and variational systems by means of set-valued differentiation. *Optimization*, 31:13–46, 1994.

202. B. S. Mordukhovich. Coderivatives of set-valued mappings: calculus and applications. *Nonlinear Anal.*, 30:3059–3070, 1997.

203. B. S. Mordukhovich. An abstract extremal principle with applications to welfare economics. *J. Math. Anal. Appl.*, 251:187–216, 2000.

204. B. S. Mordukhovich. *Variational Analysis and Generalized Differentiation*. Springer-Verlag, to appear.

205. B. S. Mordukhovich and Yong Heng Shao. Differential characterizations of covering, metric regularity, and Lipschitzian properties of multifunctions between Banach spaces. *Nonlinear Anal.*, 25:1401–1424, 1995.

206. B. S. Mordukhovich and Yong Heng Shao. Extremal characterizations of Asplund spaces. *Proc. Amer. Math. Soc.*, 124:197–205, 1996.

207. B. S. Mordukhovich and Yong Heng Shao. Nonconvex differential calculus for infinite-dimensional multifunctions. *Set-Valued Anal.*, 4:205–236, 1996.

208. B. S. Mordukhovich and Yong Heng Shao. Nonsmooth sequential analysis in Asplund spaces. *Trans. Amer. Math. Soc.*, 348:1235–1280, 1996.

209. B. S. Mordukhovich and Yong Heng Shao. Fuzzy calculus for coderivatives of multifunctions. *Nonlinear Anal.*, 29:605–626, 1997.

210. B. S. Mordukhovich, J. S. Treiman, and Q. J. Zhu. An extended extremal principle with applications to multiobjective optimization. *SIAM J. Optim.*, 14:359–379, 2003.

211. B. S. Mordukhovich and Bingwu Wang. Necessary suboptimality and optimality conditions via variational principles. *SIAM J. Control Optim.*, 41:623–640, 2002.

212. Marston Morse. Relations between the critical points of a real function of n independent variables. *Trans. Amer. Math. Soc.*, 27:345–396, 1925.

213. Huynh Van Ngai and Michel Théra. A fuzzy necessary optimality condition for non-Lipschitz optimization in Asplund spaces. *SIAM J. Optim.*, 12:656–668, 2002.

214. J. V. Outrata. A generalized mathematical program with equilibrium constraints. *SIAM J. Control Optim.*, 38:1623–1638, 2000.

215. R. S. Palais and S. Smale. A generalized Morse theory. *Bull. Amer. Math. Soc.*, 70:165–172, 1964.

216. Gert K. Pedersen. *Analysis now*, volume 118 of *Graduate Texts in Mathematics*. Springer-Verlag, New York, 1989.

217. Jean-Paul Penot. The drop theorem, the petal theorem and Ekeland's variational principle. *Nonlinear Anal.*, 10:813–822, 1986.

218. Jean-Paul Penot. On the mean value theorem. *Optimization*, 19:147–156, 1988.

219. Jean-Paul Penot. Metric regularity, openness and lipschitzean behavior of multifunctions. *Nonlinear Anal.*, 13:629–643, 1989.

220. Jean-Paul Penot. A mean value theorem with small subdifferential. *J. Optimization Theory Appl.*, 94:209–221, 1997.

221. Robert R. Phelps. *Convex functions, monotone operators and differentiability*, volume 1364 of *Lecture Notes in Mathematics*. Springer-Verlag, Berlin, 1989.

222. David Preiss. Gâteaux differentiable functions are somewhere Fréchet differentiable. *Rend. Circ. Mat. Palermo (2)*, 33:122–133, 1984.

223. B. N. Pshenichnyi. *Necessary conditions for an extremum*, volume 4 of *Translated from the Russian by Karol Makowski. Translation edited by Lucien W. Neustadt. Pure and Applied Mathematics*. Marcel Dekker Inc., New York, 1971.

224. Paul H. Rabinowitz. Critical point theory and applications to differential equations: a survey. In M. Matzeu and A. Vignoli, editors, *Topological nonlinear analysis*, volume 15 of *Progr. Nonlinear Differential Equations Appl.*, pages 464–513. Birkhäuser Boston, Boston, MA, 1995.

225. M. L. Radulescu and F. H. Clarke. The multidirectional mean value theorem in Banach spaces. *Canad. Math. Bull.*, 40:88–102, 1997.

226. S. Reich and S. Simons. Fenchel duality, fitzpatrick functions and the Kirszbraun-Valentine extension theorem. *preprint*, 2004.

227. Stephen M. Robinson. Regularity and stability for convex multivalued functions. *Math. Oper. Res.*, 1:130–143, 1976.

228. Stephen M. Robinson. Stability theory for systems of inequalities. II. Differentiable nonlinear systems. *SIAM J. Numer. Anal.*, 13:497–513, 1976.

229. R. T. Rockafellar. *Convex analysis*. Princeton Mathematical Series, No. 28. Princeton University Press, Princeton, N.J., 1970.

230. R. T. Rockafellar. On the maximality of sums of nonlinear monotone operators. *Trans. Amer. Math. Soc.*, 149:75–88, 1970.

231. R. T. Rockafellar. Clarke's tangent cones and the boundaries of closed sets in $\mathbf{R}^n$. *Nonlinear Anal.*, 3:145–154 (1978), 1979.

232. R. T. Rockafellar. Directionally Lipschitzian functions and subdifferential calculus. *Proc. London Math. Soc. (3)*, 39:331–355, 1979.

233. R. T. Rockafellar. Generalized directional derivatives and subgradients of nonconvex functions. *Canad. J. Math.*, 32:257–280, 1980.

234. R. T. Rockafellar. Proximal subgradients, marginal values, and augmented Lagrangians in nonconvex optimization. *Math. Oper. Res.*, 6:424–436, 1981.

235. R. T. Rockafellar. *The theory of subgradients and its applications to problems of optimization*, volume 1 of *R & E*. Heldermann Verlag, Berlin, 1981.

236. R. T. Rockafellar. Lagrange multipliers and subderivatives of optimal value functions in nonlinear programming. *Math. Programming Stud.*, pages 28–66, 1982.

237. R. T. Rockafellar and Roger J.-B. Wets. *Variational analysis*, volume 317 of *Grundlehren der Mathematischen Wissenschaften [Fundamental Principles of Mathematical Sciences]*. Springer-Verlag, Berlin, 1998.

238. S. Rolewicz. On drop property. *Studia Math.*, 85:27–35, 1986.

239. Barry Simon. *Trace ideals and their applications*, volume 35 of *London Mathematical Society Lecture Note Series*. Cambridge University Press, Cambridge, 1979.

240. S. Simons. *Minimax and monotonicity*, volume 1693 of *Lecture Notes in Mathematics*. Springer-Verlag, Berlin, 1998.

241. S. Simons and C. Zălinescu. A new proof for Rockafellar's characterization of maximal monotone operators. *Proc. Amer. Math. Soc.*, 132:2969–2972, 2004.

242. H. Stark and Y. Yang. *Vector Space Projections: A Numerical Approach to Signal and Image Processing, Neural Nets, and Optics.* John Wiley and Sons, New York, 1998.

243. Charles Stegall. Optimization of functions on certain subsets of Banach spaces. *Math. Ann.*, 236:171–176, 1978.

244. Karl R. Stromberg. *Introduction to classical real analysis.* Wadsworth International, Belmont, Calif., 1981.

245. H. J. Sussmann. A strong version of the maximum principle under weak hypotheses. In *Proceedings of the 33rd IEEE conference on decision and control,* volume 2, pages 1950–1956. IEEE, 1994.

246. H. J. Sussmann. Transversality conditions and a strong maximum principle for systems of differential inclusions. In *Proceedings of the 37th IEEE conference on decision and control,* volume 1, pages 1–6. IEEE, 1998.

247. Lionel Thibault. On subdifferentials of optimal value functions. *SIAM J. Control Optim.*, 29:1019–1036, 1991.

248. Lionel Thibault and Dariusz Zagrodny. Integration of subdifferentials of lower semicontinuous functions on Banach spaces. *J. Math. Anal. Appl.*, 189:33–58, 1995.

249. J. S. Treiman. Generalized gradients, Lipschitz behavior and directional derivatives. *Canad. J. Math.*, 37:1074–1084, 1985.

250. J. S. Treiman. Finite-dimensional optimality conditions: B-gradients. *J. Optim. Theory Appl.*, 62:139–150, 1989.

251. J. S. Treiman. Optimal control with small generalized gradients. *SIAM J. Control Optim.*, 28:720–732, 1990.

252. A. N. Tykhonov. Solution of incorrectly formulated problems and the regularization method. *Soviet Math. Dokl.*, 4:1035–1038, 1963.

253. Corneliu Ursescu. Multifunctions with convex closed graph. *Czechoslovak Math. J.*, 25(100):438–441, 1975.

254. F. A. Valentine. A Lipschitz condition preserving extension for a vector function. *Amer. J. Math.*, 67:83–93, 1945.

255. Jon Vanderwerff and Q. J. Zhu. A limiting example for the local "fuzzy" sum rule in nonsmooth analysis. *Proc. Amer. Math. Soc.*, 126:2691–2697, 1998.

256. V. M. Veliov. Sufficient conditions for viability under imperfect measurement. *Set-Valued Anal.*, 1:305–317, 1993.

257. V. M. Veliov. Lipschitz continuity of the value function in optimal control. *J. Optim. Theory Appl.*, 94:335–363, 1997.

258. A. Verona and M. E. Verona. Regular maximal monotone operators and the sum rule. *J. Convex Analysis*, 7:115–128, 2000.

259. L. P. Vlasov. Approximative properties of sets in normed linear spaces. *Uspehi Mat. Nauk*, 28:3–66, 1973.

260. John von Neumann. *Functional Operators. II. The Geometry of Orthogonal Spaces.* Annals of Mathematics Studies, no. 22. Princeton University Press, Princeton, N. J., 1950. Reprint of mimeographed lecture notes first distributed in 1933.

261. John von Neumann and Oskar Morgenstern. *Theory of Games and Economic Behavior.* Princeton University Press, Princeton, N. J., 1947. 2d ed.

262. D. E. Ward. Dini derivatives of the marginal function of a non-Lipschitzian program. *SIAM J. Optim.*, 6:198–211, 1996.

352   References

263. J. Warga. Derivative containers, inverse functions, and controllability. In D. L. Russell, editor, *Calculus of variations and control theory (Proc. Sympos., Math. Res. Center, Univ. Wisconsin, Madison, Wis., 1975; dedicated to Laurence Chisholm Young on the occasion of his 70th birthday)*, pages 13–45; errata, p. 46. Math. Res. Center, Univ. Wisconsin, Publ. No. 36. Academic Press, New York, 1976.

264. J. Warga. Fat homeomorphisms and unbounded derivate containers. *J. Math. Anal. Appl.*, 81:545–560, 1981.

265. J. Warga. Controllability, extremality, and abnormality in nonsmooth optimal control. *J. Optim. Theory Appl.*, 41:239–260, 1983.

266. J. Warga. Optimization and controllability without differentiability assumptions. *SIAM J. Control Optim.*, 21:837–855, 1983.

267. H. Wiersma. Duality inequalities in nonsmooth optimization. Master's thesis, Simon Fraser University, 2003.

268. J. J. Ye. Constraint qualifications and necessary optimality conditions for optimization problems with variational inequality constraints. *SIAM J. Optim.*, 10:943–962, 2000.

269. J. J. Ye and Q. J. Zhu. Multiobjective optimization problem with variational inequality constraints. *Math. Program.*, 96:139–160, 2003.

270. Dariusz Zagrodny. Approximate mean value theorem for upper subderivatives. *Nonlinear Anal.*, 12:1413–1428, 1988.

271. Eberhard Zeidler. *Nonlinear functional analysis and its applications. I.* Springer-Verlag, New York, 1986.

272. Q. J. Zhu. Clarke-Ledyaev mean value inequalities in smooth Banach spaces. *Nonlinear Anal.*, 32:315–324, 1998.

273. Q. J. Zhu. The equivalence of several basic theorems for subdifferentials. *Set-Valued Anal.*, 6:171–185, 1998.

274. Q. J. Zhu. Subderivatives and their applications. In W. Chen and S. Hu, editors, *Proceedings of International Conference on Dynamical Systems and Differential Equations (Springfield, MO, 1996)*, pages 379–394. Southwest Missouri State University, Springfield, Missouri, 1998.

275. Q. J. Zhu. Hamiltonian necessary conditions for a multiobjective optimal control problem with endpoint constraints. *SIAM J. Control Optim.*, 39:97–112, 2000.

276. Q. J. Zhu. Necessary conditions for constrained optimization problems in smooth Banach spaces and applications. *SIAM J. Optim.*, 12:1032–1047, 2002.

277. Q. J. Zhu. Lower semicontinuous Lyapunov functions and stability. *J. Nonlinear Convex Anal.*, 4:325–332, 2003.

278. Q. J. Zhu. Nonconvex separation theorem for multifunctions, subdifferential calculus and applications. *Set-Valued Analysis*, 12:275–290, 2004.

279. Q. J. Zhu. A variational proof of birkhoff's theorem on doubly stochastic matrices. *Mathematical Inequalities and Applications*, 7:309–313, 2004.

Index

$(X, \|\cdot\|)$, 5
(X, d), 2
2^Y, 3
$A(N)$, 299
$A^\top$, 78
B, 2
B_0, 317
$B_U(x, y)$, 321
B_X, 2
B_p, 317
$B_r(S)$, 3
$B_r(x)$, 2
B_{sa}, 317
D_F^*, 220
D_L^*, 220
$E(N)$, 23
E_f, 166
$E_{N \times M}$, 299
$F \circ G$, 222
F^{-1}, 4
F_T, 177
$GL(N)$, 299
$GL(N, M)$, 299
G_δ, 243
$K(x^*, \varepsilon)$, 10
K°, 83
$K^{\circ\circ}$, 136
N_C, 190
$N_F(S; x)$, 41
N_L, 196
$O(N)$, 299
$P(N)$, 22, 302
P_C, 141

$P_\gamma(a, b)$, 11
$P_\pi(x, y)$, 321
$S(N)$, 299
$S(f, S, \alpha)$, 267
$T(Y; y)$, 292
$T^*(Y; y)$, 292
$T_y^*(Y)$, 292
$T_D(S; x)$, 200
T_C, 190
$T_P(S; x)$, 209
$T_y(Y)$, 291
X^*, 2
X^N, 48
$[A, B]$, 299
$[a, C]$, 12
$[x, y]$, 15
$\#(S)$, 23
$\Gamma(a, b)$, 275
$\mathbf{C}$, 317
$\bigwedge[f_1, \ldots, f_N](S)$, 47
$\mathcal{N}$, 317
χ_S, 45
δ_{ij}, 317
γ_C, 111
$\iota(S; \cdot)$, 3
ι_S, 3
$\lambda(A)$, 302
$\langle \cdot, \cdot \rangle$, 2
$\liminf_{i \to \infty} F_i$, 169
$\limsup_{i \to \infty} F_i$, 169
$\overline{\mathrm{conv}}\, S$, 3
$\text{e-}\liminf_{i \to \infty} f_i$, 170
$\text{e-}\limsup_{i \to \infty} f_i$, 171

e-$\lim_{i\to\infty} f_i$, 171
s-$\liminf_{x\to\bar{x}} F(x)$, 173
s-$\limsup_{x\to\bar{x}} F(x)$, 173
w*-$\lim_{i\to\infty} x_i^*$, 193
Ker, 292
St(N, M), 301
argmin, 166
bd S, 3
conv S, 3
core(S), 114
diag, 303
diag*, 303
diam S, 3
dom f, 3
epi f, 3
graph, 3
graph F, 3
int S, 3
range F, 165
range f, 3
tr(A), 302
$\overline{S}$, 3
∂, 117
∂^F, 44
∂^∞, 196
∂_C^∞, 191
∂_C, 189
∂_F, 39
∂_L, 196
∂_P, 46
∂_{VF}, 40
ϕ^*, 292
ϕ_*, 292
π, 292
π^*, 292
$\lim_{i\to\infty} F_i$, 169
$\rho(\cdot,\cdot)$, 30
$\sigma(A)$, 302
σ_C, 111
τ_n, 62
$\underline{f}$, 52
$\underline{d}(S;\cdot)$, 3
d_S, 3
d_f, 149
e_n, 66
f', 39
$f\Box g$, 3
f^*, 134
f°, 188

f^{**}, 134
f^{-1}, 3
f_+, 98
$l(y)$, 21
s_p, 47
$star(S)$, 213
$x\odot x$, 318
$x\prec y$, 21
x^+, 69
x^-, 69
$x^\downarrow$, 21
$\mathcal{A}(N)$, 25
$\mathcal{B}$, 318
$\mathcal{K}$, 223
$\mathbb{R}$, 3
$\mathbb{R}\cup\{+\infty\}$, 3
$\mathbb{R}^N$, 19
$\mathbb{R}_+^N$, 21
$\mathbb{R}_\geq^N$, 302
dom$F$, 4
Stab, 293
det(A), 302
ext, 29
pr$_\eta(S;v)$, 232
sgn, 76
supp ϕ, 33

analytic center, 139
approximate
  chain rule, *87*, 90, 93, 107
    strong, 87
    weak, 88
  extremal principle, *104*, 107, 254
  Fermat principle, *19*, 20, 27
  local sum rule, *54*, 71, 74, 81, 87, 102,
    107, 197, 225, 245, 253, 254
    Lipschitz, 245
    strong, *54*, 57, 60, 74
    strong, failure of, 74
    weak, *57*, 62
  mean value theorem, *81*, 82, 83, 167,
    193
  multidirectional mean value inequal-
    ity, *95*, 253, 254
  multiplier rule, 62
    weak, 64
  nonlocal sum rule, *48*, 51, 55, 95, 105,
    128, 253, 254, 260
  projection, 232

approximate critical point, 10
argmin, 166
Asplund space, 243
  and Fréchet smooth space, 245
  separable, 245
  subdifferential characterization, 254
  sum rule characterization, 252
asymptotically regular, 145
atlas, 291
attainment
  in Fenchel problems, 136
attractive set, 89

Baire category theorem, *33*, 203
Banach fixed point theorem, 15
biconjugate, 134
bilinear form, 321
bipolar, 136
Birkhoff theorem, *25*, 323
Bishop–Phelps cone, *10*, 59, 134
Bishop–Phelps theorem, 10
Borwein–Preiss variational principle,
    30, 34
  general form, 32
  in finite dimensional spaces, 19
  subdifferential form, 42
boundary, 3
Bregman distance, 149
bump function, 33
  range, 36

canonical projection, 292
Caristi–Kirk fixed point theorem, *17*
chain rule, *87*, 91, 93
  approximate, *87*, 88, 90, 94, 107
  convex subdifferential, 129
  for permutation invariant function,
      307, 310
  for spectral function, 307, 311
  limiting, 198
  on manifolds, 296
  smooth, *94*, 297
Chebyshev set, 217
Clarke
  directional derivative, 188
  normal cone, 190
    representation, 195
  singular subdifferential, 191
    representation, 195

subdifferential, 189
  maximal, 206
  of distance function, 215
  representation, 193, *195*
  tangent cone, *190*, 209
closed convex hull, 3
closest point, 215, 239
  density, 217
closure, 3
coderivative, 220
  chain rule
    limiting, 227
    strong, 226
  Fréchet, 220
  limiting, 220
  sum rule
    limiting, 226
    strong, 224
    weak, 221
comparison theorem, 50
concave
  conjugate, 139
  function, 111
cone
  bipolar, 136
  polar, 83, 136
cone monotonicity, 83
conjugation, 134
constrained optimization problem, 41,
    62, 65, 108, 109, 190, 211, 212, 298
  and inequality, 69
  equilibrium constraint, 77
  multifunction constraint, 76
  variational inequality constraint, 76
constraint
  linear, 136, 139
constraint qualification, 64, *66*, 120, 161
  calmness, *68*, 72
  Mangasarian–Fromovitz, 64, *67*, 72
contingent cone, 209
contraction, 15
  directional, 16
convex
  feasibility problem, 140
  function, 111
    difference of, 150
    differentiability of, 121
    recognizing, 124
    regularity, 192

hull, 3
multidirectional mean value theorem,
    130
normal cone, 117
program, 135
quasi, 83
separation, 209
set, 111
subdifferential, 117
    calculus, 129, 135, 136
    failure of calculus, 134
    sum rule, 129
    subgradient, *117*, 134
convex series(cs)
    closed, 113
    compact, 113
convexity, 80, 167
core, 114
    versus interior, 116
cotangent
    bundle, 292
    space, 292
    vectors, 292
        limit of, 293
critical sets, 89
cusco, *174*, 190
    minimal, *174*, 215

DAD problem, 159
DC function, *124*, 214
decoupled infimum, 47
decoupling lemma, *127*, 178
decrease principle, *97*, 98
demi-closed, *181*, 184
determinant, 302
Deville–Godefroy–Zizler variational
    principle, 33
diameter, *3*, 267
differentiability
    Fréchet, *39*, 243
    Gâteaux, *121*, 258, 330
    of convex functions, 121
    of distance function, 150
directional contraction, 16
directional derivative, 117
    and subgradients, 119
    Clarke, 188
    of convex function, 118
    sublinear, 118

distance function, *3*, 38, 214, 230
domain, 3
    of subdifferential, 117
    not convex, 123
doubly stochastic
    matrix, *23*, 159
    and majorization, 26
    pattern, 159
drop, *12*, 95
    and flower petal, 13
    theorem, 13
dual space, 2
duality
    entropy maximization, 157
    inequalities, 150
    map, 178
    pairing, 2
    weak, 135, 136

eigenvalue
    function, 38, *312*
    largest, 77
    mapping, 302, 320
Ekeland variational principle, *6*, 14
    alternative forms, 8
    and Bishop–Phelps theorem, 14
    and completeness, 14
    geometric picture, 5
enlargement, 3
entropy
    Boltzmann–Shannon, 149, *158*
    maximization, *157*, 159
        duality, 157, *161*
        in infinite dimensional space, 161
    maximum, 137, 139
epi-convergence, 170
epi-limit, 171
    characterization, 171
    lower, 170
    upper, 171
epigraph of function, 3
epigraphical profile, 166
exact penalization, 38
extremal
    point, 104
    principle, 104
    system, 104
        fixed sets, 108
extremal principle, *104*, 107, 254, 262

and convex separation, 209
approximate, *104*, 107
geometry, 106
limiting, 199
extreme points, 29

Farkas lemma, 29
feasible allocation, 263
Fejér Monotone sequence, 144
Fenchel
biconjugate, 137
conjugate, *134*, 152, 324
examples, 138
of exponential, 137, 139
transformations, 138
duality, *136*, 157
linear constraints, 139
symmetric, 139
problem, 135
Fenchel–Rockafellar Theorem, 121
Fenchel–Young inequality, *135*, 137
Fitzpatrick function, 177
fixed point, 15
fixed point theorem
Banach, 15
Caristi–Kirk, 17
and Banach, 18
Clarke's refinement, 16
error estimate, 18
iteration method, 18
flower petal, 11
theorem, 11
Fréchet
coderivative, 220
derivative, 39
differentiable, *39*, 243
normal cone, 41
smooth space, 39
subderivative, 39
subdifferentiable, 39
subdifferential, *39*, 245
superderivative, 98
superdifferential, *44*, *50*, 98
Fritz John condition, 64
function
bump, 33
Cantor, 45
characteristic, 45
convex

Lipschitz property, 112
regularity, 192
distance, *3*, 38, 150, 214, 230
eigenvalue, 38, *312*
entropy, 158
epigraph of, 3
Fitzpatrick, 177
gauge, 111
gauge-type, 30
graph of, 3
indicator, 3
conjugate of, 139
subdifferential of, 117
Lipschitz, 38, 79
lower semicontinuous (lsc), 2, *3*
Lyapunov, 88, 89
max, 37
subdifferential of, 133
nonexpansive, 217
number of elements, 23
optimal value, 37, 53
order statistic, 312
penalization, 38
set of continuity, 120
sign, 76
spectral, 38
support, *111*, 162

Gâteaux differentiable, 10, *121*
game, 109
gauge-type function, 30
general metric regularity qualification
condition, 223
generic, 33
generic Gâteaux differentiability, 287
Gordan alternative, 20
graph of function, 3
Graves–Lyusternik thoerem, 103

Hadamard's inequality, 78
Hahn–Banach extension, 132
Hamilton–Jacobi equation, 50, 53
Hessian
and convexity, 124
hypermonotone, 182

implicit function theorem, 233
implicit multifunction, 233
indicator function, 3

induced map, 292
inf-convolution, 3, *149*
interior, 3

Kirszbraun–Valentinc theorem, 179
kronecker delta, 317
Ky Fan minimax theorem, 303

Lagrange multiplier rule, 212, 298
Lambert W-function, 150
level sets, 3
  normal cone
    representation, 60
  of majorization, 21
  of preference, 107
  singular normal cone
    representation, 61
lexicographic order, 319
Lidskii theorem, 317
Lie
  bracket, 299
  group, 293
    action, 293
limiting
  chain rule, 198
  coderivative, 220
  extremal principle, 199
  multiplier rule, 198
  normal cone, 196
  subdifferential, *196*, 294
  sum rule, 197
    failure, 200
linearity space, 118
Lipschitz
  and cone monotonicity, 86
  criterion, 82
  property, 79
local coordinate system, 291
local sum rule, 253
  approximate, *54*, 57, 62, 87, 102, 107,
    252
log, 139
lower semicompact, 226
lower semicontinuous (lsc), 3
lsc closure, 52
Lyapunov
  function, 89
  stability, 89

majorization, 21
  and doubly stochastic matrix, 26
  level sets of, 21, *27*
    representation, 22
Mangasarian–Fromovitz condition, 67
manifold, 291
  full rank matrices, 299
  invertible matrices, 299
  orthogonal matrices, 299
  Stiefel, 301
mapping
  attracting, 143
  nonexpansive, 143
marginal price rule, 206
mathematical economics, 206
mathematical program with equilibrium
    constraint, 77
matrix, 299
  doubly stochastic, *23*, 159
  permutation, 22
  skew symmetric, 299
  symmetric, 299
max formula, 120
maximal monotone, *167*, 177
  on a set, 176
maximum eigenvalue, 302
meager, 33
mean value inequality, 102
mean value theorem
  approximate, 81
  Cauchy, 84
  Lagrange, 79
  Rolle, 79
metric regularity, *234*, 236, 239
  tangential conditions, 240
metric space, 2
minimal cusco, 177, 202, 215
minimal usco, 175
  existence, 175
minimax theorem, 99, 139
minimizer
  subdifferential zeroes, 117
monotone
  maximal, 167
  multifunction, 166
  quasi, 83
monotonicity, 79
  cone, 83
  of gradients, 125

mountain pass theorem, 274, *278*
  approximate, 276
multidirectional mean value inequality,
    95, 231, 253
  approximate, 95
  convex, 130
  refined, 102
  two sets, 153
multifunction, 3, *165*, 233
  argmin, 166
  close valued, 166
  closed, 166
  compact valued, 166
  composition, 222
  continuity, *173*
  convex valued, 166
  domain of, *4*, 165
  epigraphical profile, 166
  graph of, *3*, 105
  inverse of, *4*, 165
  maximal monotone, 167, 176
  monotone, 166
    boundedness, 168
  open valued, 166
  profile mapping, 174
  range of, 165
  semicontinuity, 173
  sequential lower limit, 173
  sequential upper limit, 173
  subdifferential, 165
  sublevel set mapping, 174
  sublevel sets, 166
multiobjective optimization, 106
multiplier set, 65
multipliers
  nonexistence, 76

nearest point, *140*, 150
  existence and uniqueness, 140
  in polyhedron, 139
  normal cone characterization, 141
necessary optimality condition, 68, 198
  approximate multiplier rule, 62, 64
  Clarke subdifferential, 190
  comparison, 211
  Fréchet normal cone condition, 41
  Frizt John condition, 64
  Guignard, 241
  Karush–Kuhn–Tucker, 67, *68*, 70, 77

Lagrange multiplier rule, 211, 212,
    298
  multi-objective optimization, 107
  Pshenichnii–Rockafellar, 131, 141
nonconvex separation theorem, 259
  and extremal principle, 262
  for multifunctions, 259
  for sets, 261
nonexpansive
  mapping, 179
    extension, 179
nonexpansive function, 217
nonlocal approximate sum rule, *48*, 51,
    55, 95, 105, 128, 253, 260
nonsmooth analysis, 2
nonsymmetrical minimax theorem, *99*,
    103
norm
  subgradients of, 123
normal cone
  and subgradients, 130
  Clarke, 190
  convex, 117
  epigraph, 42
  Fréchet, 41
  level sets, 60
  limiting, 196
  nonclosed, 201
  of a submanifold, 295
  on manifolds, 294
  sublevel sets, 58
  to intersection, 130
normal upper semicontinuity, 265

one-perturbation variational principle,
    281
open covering with a linear rate, 234
open mapping theorem, 114
optimal principle, 53
optimal value
  dual, 135
  function, 54, 65
  primal, 135
optimality condition, 189
optimization
  constrained, *62*, 190
  multiobjective, 106
    necessary optimality conditions, 107
  subdifferential in, 117

orbit, 293, 306
order statistic function, 312
order-reversing, 134

Pacman set, 210
Painlevé–Kuratowski limit, 169
paired Banach spaces, 318
Palais–Smale condition, 278
Pareto optimal allocation, 264
pass, 275
permutation
    invariant, 302
    matrix, *22*, 302
Pitt's theorem, 272
polar, 190
polar cone, 136
polyhedron, 133
    nearest point in, 139
positive operator, 317
positively homogeneous, 118
preference, 106
preferred neighborhood, 292
preimage, 3
projection, 141
    algorithm, 143
    approximate, 232
    attractive property, 143
    potential function of, 141
    properties, 141
projection algorithm, 143
    asymptotically regular, 145
    convergence, 145
    strong convergence, 148
    weak convergence, 147
proximal
    normal cone, 218
    normal formula, 218
    normal vector, 218
    subderivative, 45
    subdifferential, 45
pseudo-Lipschitzian, 234
pseudoconvex set, 211
pseudotangent cone, 209
Pshenichnii–Rockafellar conditions, 131

quasi
    convexity, *83*, 167
    monotone, 83

Radon–Nikodym property, *267*, 271
real normed sequence space, 317
rearrangement, 319
    equivalent, 318
    invariant, 319
regularity
    function, 192
    set, 210
residual, 33
resolvent, 183

sandwich theorem, 127, *128*
    Lewis–Ralph, 153
sandwiched functions, 151
Schatten p-spaces, 326
segment, 15
self-adjoint operator, 317
sensitivity, 65
separable, 139
separable points, 259
separable reduction, 245, *247*, 249
separation theorem, 132, 209
sequence of sets
    lower limit, 169
    Painlevé–Kuratowski limit, 169
    upper limit, 169
sequential uniform lower semicontinuity,
    73
set-valued function, 3
shadow price, *264*, 265
sign function, 76
singular
    subdifferential, *196*, 294
singular subdifferential
    Clarke, 191
singular value
    largest, 78
slice, *266*, 267
    weak-star, 266
smooth chain rule, 296
solvability, 97, *229*
spectral
    decomposition, 306
    function, 38, 299, *302*
    sequence, 317
    sequence space, 318
spectral radius, 302
stability, 90
stabilizer, 293

stable set, 89
star of a set, 213
Stegall variational principel, 238, *267*
Stiemke's theorem, 28
strictly convex, 124
   and Hessian, 124
strong minimum, 33, 267
strongly exposed, 271
strongly exposing functional, 271
subadditive, 118
subderivative, 39
   proximal, 45
   superderivative representation, 101
subdifferentiable, 39
subdifferential, 39
   of max function, 94
   at optimality, 117
   Clarke, 189
   convex, 117
   density, 297
   Fréchet, 294
   Gâteaux, 331
   limiting, 196, 294
   limiting and Clarke, 196
   monotonicity, 167
   nonempty, 121
   of convex functions, 117
   of infimum, 230
   on manifolds, 294
   proximal, 45
   singular, 196, 294
   spectral function, 307, 311
subgradient
   and normal cone, 130
   construction of, 120
   existence of, 120
   of maximum eigenvalue, 123
   of norm, 123
   unique, 121
sublevel sets, *3*, 166
   normal cone
      representation, 58
   singular normal cone
      representation, 61
sublinear, *118*, 120, 132
submanifold, 292
sum rule
   convex subdifferential, *129*, 132
   limiting, *189*, 197

local, 54, 57
   nonlocal, 48
sun, 219
superderivative
   subderivative representation, 100
superdifferential, 294
supergradient, 294
support, 33

tangent
   bundle, 292
   cone, 209
      Clarke, 190
   space, 292
      bases of, 292
   vectors, 291
trace, 302
trace class operators, 317
transversality, 185, 240

ubiquitous set, 191
unit ball, 2
unitary
   equivalent, 318
   invariant, 302, 318
   mapping, 302
   operator, 318
upper semicontinuous, 44
usco, 174
   minimal, 174

value function, 127
variational principle, 2, 35
   Borwein–Preiss, *30*, 40, 42, 48, 51
   Deville–Godefroy–Zizler, 33
   Ekeland, *6*, 8, 10, 12, 14, 16, 17, 59,
      131, 194, 255, 277
   in finite dimensional spaces, 19
   one-perturbation, 281
   smooth, 19
   Stegall, 238, *267*
   subdifferential form, 42
variational techniques, 1
vector field, 292
Ville's theorem, 28
viscosity
   Fréchet subderivative, 39
   Fréchet subdifferential, 39
   solution, 50

uniqueness, 51
subsolution, 50
supersolution, 50
von Neumann–Theobald inequality,
    306, 308

welfare economy, 263

Yosida approximate, 183